MOLECULAR BIOLOGY
INTELLIGENCE
UNIT

Molecular Mechanisms of Exocytosis

Romano Regazzi, Ph.D.
Department of Cell Biology and Morphology
University of Lausanne,
Lausanne, Switzerland

LANDES BIOSCIENCE / EUREKAH.COM
AUSTIN, TEXAS
U.S.A.

SPRINGER SCIENCE+BUSINESS MEDIA
NEW YORK, NEW YORK
U.S.A.

MOLECULAR MECHANISMS OF EXOCYTOSIS

Molecular Biology Intelligence Unit

Landes Bioscience / Eurekah.com
Springer Science+Business Media, LLC

ISBN: 0-387-39960-7   Printed on acid-free paper.

Springer Science+Business Media, LLC, 233 Spring Street, New York, New York 10013, U.S.A.
http://www.springer.com

Please address all inquiries to the Publishers:
Landes Bioscience / Eurekah.com, 1002 West Avenue, 2nd floor, Austin, Texas, 78701, U.S.A.
Phone: 512/ 863 7762; FAX: 512/ 863 0081
http://www.eurekah.com
http://www.landesbioscience.com

Printed in the United States of America.

9 8 7 6 5 4 3 2 1

Library of Congress Cataloging-in-Publication Data

Library of Congress Cataloging-in-Publication Data

Molecular mechanisms of exocytosis / [edited by] Romano Regazzi.
    p. ; cm. -- (Molecular biology intelligence unit)
Includes bibliographical references and index.
ISBN 0-387-39960-7 (alk. paper)
    1. Exocytosis. 2. Cellular signal transduction.   I. Regazzi, Romano.
II. Series: Molecular biology intelligence unit (Unnumbered)
    [DNLM: 1. Exocytosis--physiology. 2. Cell Membrane--physiology.
3. SNARE Proteins. 4. Signal Transduction.   QU 375 M718 2006]
QH634.2.M65 2006
571.6--dc22
                                                    2006025297

CONTENTS

EDITOR

Romano Regazzi
Department of Cell Biology and Morphology
University of Lausanne
Lausanne, Switzerland
Email: Romano.Regazzi@unil.che
Chapter 3

CONTRIBUTORS

Abderrahmani Amar
Department of Cell Biology
   and Morphology
Service of Internal Medicine
Lausanne, Switzerland
Email: amar.abderrahmani@unil.ch
Chapter 11

Marie-France Bader
Department of Neurotransmission
   and Neuroendocrine Secretion
CNRS UMR-7168/LC2
Strasbourg, France
Email: bader@neurochem.u-strasbg.fr
Chapter 6

Pietro Baldelli
Center of Neuroscience
   and Neuroengineering
Department of Experimental Medicine
Section of Human Physiology
University of Genova
Genova, Italy
Chapter 5

Fabio Benfenati
Center of Neuroscience
   and Neuroengineering
Department of Experimental Medicine
Section of Human Physiology
University of Genova
Genova, Italy
Email: Fabio.Benfenati@unige.it
Chapter 5

Ulrich Blank
INSERM U699
Faculté de Médecine X
Paris, France
Email: ublank@bichat.inserm.fr
Chapter 8

Cristiana Brochetta
INSERM U699
Faculté de Médecine X
Paris, France
Chapter 8

Luke H. Chamberlain
Henry Wellcome Laboratory
   of Cell Biology
Division of Biochemistry
   and Molecular Biology
University of Glasgow
Glasgow, U.K.
Email: l.chamberlain@bio.gla.ac.uk
Chapter 7

Emanuele Cocucci
Department of Neuroscience
Vita-Salute San Raffaele University
   and Scientific Institute San Raffaele
Milano, Italy
Chapter 10

Anna Corradi
Center of Neuroscience
   and Neuroengineering
Department of Experimental Medicine
Section of Human Physiology
University of Genova
Genova, Italy
Chapter 5

Anna Fassio
Center of Neuroscience
    and Neuroengineering
Department of Experimental Medicine
Section of Human Physiology
University of Genova
Genova, Italy
Chapter 5

Mitsunori Fukuda
Fukuda Initiative Research Unit
The Institute of Physical
    and Chemical Research
Wako, Saitama, Japan
Email: mnfukuda@brain.riken.go.jp.
Chapter 4

Thierry Galli
Team "Avenir" INSERM
Membrane Traffic in Neuronal
    and Epithelial Morphogenesis
Institut Jacques Monod
UMR7592, CNRS, Universités
Paris, France
Email: thierry@tgalli.net
Chapter 1

Alexander J.A. Groffen
Department of Functional Genomics
Center for Neurogenomics
    and Cognition Research
Vrije Universiteit (VU) and VU
    Medical Centre
Amsterdam, The Netherlands
Email: sander.groffen@falw.vu.nl
Chapter 2

Jacopo Meldolesi
Department of Neuroscience
Vita-Salute San Raffaele University
    and Scientific Institute San Raffaele
Milano Italy
Email: meldolesi.jacopo@hsr.it
Chapter 10

Christine Salaün
Henry Wellcome Laboratory
    of Cell Biology
Division of Biochemistry
    and Molecular Biology
University of Glasgow
Glasgow, U.K.
Chapter 7

Emmanuel Sotirakis
INSERM Avenir Team
Institut Jacques Monod
Paris, France
Chapter 1

Claudia Nora Tomes
Laboratorio de Biología Celular
    y Molecular
Instituto de Histología y Embriología
    (IHEM-CONICET)
Facultad de Ciencias Médicas
Universidad Nacional de Cuyo
Mendoza, Argentina
Email: ctomes@fcm.uncu.edu.ar
Chapter 9

Flavia Valtorta
Department of Neuroscience
Vita-Salute San Raffaele University
    and Scientific Institute San Raffaele
Milano Italy
Chapter 5

Matthijs Verhage
Department of Functional Genomics
Center for Neurogenomics
    and Cognition Research
Vrije Universiteit (VU)
    and VU Medical Centre
Amsterdam, The Netherlands
Chapter 2

Nicolas Vitale
Department of Neurotransmission
    and Neuroendocrine secretion
CNRS UMR-7168/LC2
Strasbourg, France
Email: bader@neurochem.u-strasbg.fr
Chapter 6

PREFACE

Exocytosis is a fundamental cellular process that is used by eukaryotic cells to release a variety of biological compounds such as peptide hormones and neurotransmitters or to insert specific lipids and proteins in the plasma membrane. In recent years, a multidisciplinary approach, including genetics, in vitro reconstitution of vesicular transport and studies devoted to the definition of the mechanism of action of bacterial neurotoxins, promoted an extraordinary progress in the understanding of the molecular mechanisms of exocytosis. The genetic approach was pioneered in the beginning of the eighties by Peter Novick and Randy Schekman with the identification of a group of genes (SEC genes) required for post-translational events in the yeast secretory pathway. Genes fulfilling analogous functions were then found in other genetic models such as *C. elegans* and *Drosophila*. James Rothman and collaborators designed a strategy aimed at tackling the problem from a different angle and developed a method to reconstitute vesicular transport in a cell-free system. Interestingly, the SNARE and SNARE-associated proteins required for vesicular transport identified using this assay turned out to be closely related to the products of the genes isolated in the genetic screening of Novick and Schekman. The fundamental role of SNAREs in exocytosis was then beautifully demonstrated by the seminal discovery of the groups of Cesare Montecucco and Heiner Niemann that clostridial neurotoxins, the most powerful natural agents that block neurotransmitter release, exert their action by specifically cleaving neuronal SNAREs. The convergence of these findings has revolutionized our knowledge of the molecular mechanisms governing exocytosis and has emphasized the conservation of the protein apparatus driving membrane fusion from yeast to man.

Despite the impressive improvement in the understanding of the process of exocytosis, a number of important questions remain unsolved and are presently the focus of intense investigations. Thus, the cascade of events leading to the docking and fusion of secretory vesicles is still not defined in molecular terms and the mechanisms coupling second messenger generation to the activation of the secretory apparatus is only beginning to emerge. Moreover, the precise roles of lipids in the exocytotic process remain to be clarified. After all, exocytosis depends on the merging of two phospholipid leaflets and their lipid composition directly affects the physico-chemical properties of the two membranes and influences the activity of the protein machinery driving fusion.

The picture emerging after more than two decades of investigations illustrates exocytosis as a complicated and finely tuned process involving a large number of components that transiently associate and dissociate. For these reasons, the understanding of the molecular basis of exocytosis has so far remained the privilege of a relatively small group of specialists. This unique collection of up-to-date reviews intends to introduce researchers and students to the forefront of this rapidly moving and fascinating field. Written by recognized experts in the field, the book aims at clarifying for a general audience the role of the key players in the exocytotic process not only in neuronal and endocrine cells but also in a variety of other cells that use exocytosis to accomplish their specialized tasks.

Romano Regazzi

CHAPTER 1

Exocytosis:
Lessons from SNARE Mutants and Friends

Emmanuel Sotirakis and Thierry Galli*

Abstract

Our understanding of the mechanism of membrane fusion and particularly of exocytosis saw a revolution in the early 90s with the elucidation of the targets of clostridial neurotoxins (the most potent blockers of neurotransmitter release), the identification of secretory mutants in yeast and the proposal of the SNARE hypothesis by Rothman and his coworkers. Since then, the field of membrane fusion has seen hundreds of papers further characterizing the central role of SNARE proteins in membrane fusion, particularly during exocytosis. The purpose of this chapter is to present and discuss the function of SNARE proteins in exocytosis with a particular focus on their mutants and partners.

SNARE Proteins: An Historical Perspective

Söllner and coworkers proposed the SNARE hypothesis in 1993[1] after a series of important observations using an assay reconstituting the transport of newly synthesized proteins in the Golgi apparatus. This assay was based on a cell line defective in N-acetylglucosamine transferase, an enzyme normally found in the medial cisternae of the Golgi apparatus, in which secreted proteins remain sensitive to englucosidase H. Rothman and his coworkers had showed that a viral glycoprotein expressed in these defective cells could be N-acetylglucosaminated when the defective Golgi was incubated in vitro in the presence of a normal Golgi in a temperature dependent manner, provided that cytosol and ATP were added.[2] This assay reconstituting transport between the cis and medial Golgi cisternae in the test tube turned out to be very powerful to identify proteins involved in the budding of vesicles from a donor membrane and their fusion with acceptor membranes. Using this assay, Rothman and coworkers first demonstrated that N-ethylmaleimide (NEM), an alkylating oxidant used to inhibit ATPases, strongly impaired intra-Golgi transport.[3] The target of NEM was purified using biochemical procedures and turned out to be an ATPase, that the authors called NEM-Sensitive Fusion protein (NSF). Realizing that mammalian NSF was the ortholog of the product of the gene responsible for the secretion defect of the yeast strain Sec18,[4] Rothman and coworkers combined yeast genetics and their in vitro assay to show that the binding of NSF to Golgi membranes and the intra-Golgi transport depended upon the product of another secretory mutant gene, Sec17.[5] They identified this new factor as Soluble NSF attachment protein (SNAP) which is now known to exist in three different forms (α, β, γ) in mammals. Rothman and coworkers then searched for membranous receptors of the NSF-SNAP complex and identified it as a complex of three proteins in a bovine brain

*Corresponding Author: Thierry Galli—Team "Avenir" INSERM Membrane Traffic in Neuronal and Epithelial Morphogenesis, Institut Jacques Monod, UMR7592, CNRS, Universités Paris 6 and 7. 2, Place Jussieu, F-75251 Paris Cedex 05, France. Email: thierry@tgalli.net

Molecular Mechanisms of Exocytosis, edited by Romano Regazzi. ©2007 Landes Bioscience and Springer Science+Business Media.

extract.[6] This data constituted the basis for the SNARE model of membrane fusion, as the synaptic complex identified was composed of a synaptic vesicle protein, called synaptobrevin 2 or Vesicle Associated Membrane Protein 2 thus referred to as vesicle or v-SNARE, and two plasma membrane proteins, Synaptosomal Associated Membrane Protein of 25kD (SNAP25) and syntaxin 1, referred to as target or t-SNARE. Strikingly, another line of evidence converged on the same proteins at about the same time with the identification of synaptobrevin 2, SNAP25 and syntaxin 1 as the substrates of the proteolytic activity of clostridial neurotoxins (tetanus and botulinum neurotoxins), the most potent blockers of neurotransmitter release[7-12] (also for review see ref. 13). Yeast genetics also confirmed the SNARE model by showing that yeast orthologs of synaptobrevin 2 (SNC1/2), SNAP25 (SEC9) and syntaxin 1 (SSO1/2) were mutated in strains bearing secretory defects.[14,15] Furthermore, homologs of the synaptic SNAREs were found to operate in all the membrane trafficking pathways from yeast to human,[16] thus generalizing the principle of SNAREmediated fusion. It was also proposed that specific v-t SNARE pairs would account for the specificity of intracellular membrane fusions.[17] Altogether, these data were instrumental to put the SNARE model at the center stage of membrane fusion: as proposed by the authors, the SNARE hypothesis assumes that the specific pairing of a v-SNARE with its cognate t-SNARE, leading to the formation of the four-helical bundle called SNAREpin, is necessary and sufficient to promote fusion of a vesicle with a target membrane.[18] Some variations of this model will be discussed briefly below.

Exocytic SNARE Mutants

SNARE mutants have been obtained in numerous Bilateria species, leading to important observations. The synaptobrevin 2 null mice[19] and *n-syb* null flies[20] have morphologically normal neurons, suggesting that synaptobrevin 2 is neither required for neurite outgrowth, nor for synaptogenesis. In contrast with their normal neuronal morphology, *n-syb* null flies show a block of stimulation-evoked neurotransmitter release and synaptobrevin 2 -/- mice show a defect in neurotransmitter release upon stimulation, both in central and peripheral synapses, leading to the post-natal death. In both cases, spontaneous release still occurs at least in part, in agreement with the recent finding that spontaneous and stimulated secretion may use significantly different mechanisms.[21] The lack of major defect in neuronal differentiation in synaptobrevin mutants may be explained by the prominent role of tetanus neurotoxin insensitive VAMP (TI-VAMP)[22] in neurite outgrowth as supported by the block of neuronal differentiation in TI-VAMP silenced neurons or in neurons expressing TI-VAMP mutant forms.[23,24] This hypothesis is consistent with the absence of effect of tetanus neurotoxin on neuronal differentiation in vitro[25] and further supported by the presence of remaining exocytosis in synaptobrevin 2/cellubrevin (a non neuronal homologue of synaptobrevin) double null chromaffin cells.[26] Furthermore, the silencing of TI-VAMP reduces surface expression of Glut4 triggered by hypertonicity in muscle cells[27] thus suggesting that TI-VAMP mediates the tetanus neurotoxin resistant exocytosis pathway in these cells. Compensatory mechanisms may also account for the lack of phenotype of the cellubrevin null mice, where none of the cellubrevin-mediated fusion events (insulin sensitivity,[28] phagocytosis,[29] platelet exocytosis[30]) seem to be affected, likely due to the presence of synaptobrevin 2.

The SNAP-25 null mice show a phenotype comparable to the synaptobrevin 2[31] null mice that may be explained by the complementation with SNAP-23 at early developmental stages as this was suggested in drosophila.[32]

As for syntaxin 1 mutants, only fly and nematode syntaxin 1 mutants are available. They show an almost complete block of neurotransmission with some residual asynchronous release,[33,34] suggesting that the latter may depend on different SNARE proteins.

Apart from SNARE mutants it is also interesting to note that a mutation in α-SNAP causes hydrocephalus with gait (*hyh*) in mouse, due to a targeting defect of TI-VAMP and of apical proteins involved in regulation of neural fate and of TI-VAMP, causing post-natal death.[35,36]

Altogether, these data suggest that, although some exocytic pathways are highly conserved, certain SNARE can substitute for one another in vivo.

Exocytic SNARE Partners: Pathways for Regulation

The central role of SNARE proteins in membrane fusion and exocytosis in particular suggests that their partners may be important regulators. Thus the identification and characterization of SNARE interactors has been an important focus of recent studies. In addition to SNAPs, several types of partners have been found. Figure 1 presents a comprehensive diagram of SNARE partners. We will discuss here the role of partners during exocytosis, i.e., along the exocytic pathway of a synaptic vesicle, thus excluding partners involved either in tethering or in SNARE complex disassembly (SNAP and NSF) or in endocytosis.

From a more general point of view, two main exocytosis pathways have been defined in neuronal and nonneuronal cells based on their sensitivity to tetanus neurotoxin, the sensitive one utilizing synaptobrevins 1 or 2 or cellubrevin, the second involving TI-VAMP.[37] The brevins are involved in exocytosis of early endosomal secretory vesicles whereas TI-VAMP mediates the exocytosis of late endosomal and lysosomal secretory vesicles. Synaptobrevins are expressed in neurons, neuroendocrine cells, adipocytes and muscle cells whereas cellubrevin has a large cell type expression but is not expressed in neurons.[38] TI-VAMP has so far been found in all cell types and in a vast majority of tissues, except the heart.[39] Synaptobrevins, cellubrevin and TI-VAMP interact with plasma membrane t-SNARE proteins to mediate exocytosis. Synaptobrevin 2 mediates exocytosis by pairing with syntaxin 1[40] and SNAP25[41] in neurons, syntaxin 4 and SNAP23 (an homolog of SNAP-25) in adipocytes.[42] Similarly, TI-VAMP pairs with syntaxin 1 and SNAP25 in neurons,[43] syntaxin 3 and SNAP23 in epithelial cells,[22] and syntaxin 4 and SNAP23 in fibroblasts.[44] SNAP29, another protein closely related to SNAP25 but lacking the palmitoylation anchors, also interacts with both synaptobrevin 2[45] and TI-VAMP[46] but its role in exocytosis is not yet fully understood. The other partners of brevins and TI-VAMP are more divergent. Figure 1 depicts the interactome of both. Here we will discuss the regulatory interactions along the synaptic exocytic pathway (Fig. 2), after the arrival and tethering of the synaptic vesicle at the active zone of the plasma membrane (also for review see ref. 47).

The last few microns covered by the vesicle along actin cables are facilitated by the interaction of synaptobrevin 2 with MyoVa.[48] In addition, interaction between synaptobrevin 2 and cdc42[49] may participate to the polarity of exocytosis because cdc42 is a key regulator of cell polarity and actin dynamics from yeast to mammals.[50] Interestingly, cdc42 also interacts with syntaxin 1,[51] suggesting that the regulation of exocytosis as a whole may be downstream of cdc42 activation as well as actin dynamics. Until the formation of the SNARE complex, synaptobrevin 2 interacts with synaptophysin,[52] thus keeping synaptobrevin 2 unavailable for membrane fusion. Furthermore, its interaction with the membrane sector of the V-ATPase[53] may reflect a mechanism ensuring the coupled sorting of synaptobrevin, synaptophysin[54] and the V-ATPase. A putative role of the V-ATPase in membrane fusion has also been proposed (see below).

On the plasma membrane, Munc18 keeps syntaxin 1 in a closed conformation and avoids its interaction with SNAP25 (for review see ref. 55). Interestingly in the absence of Munc18a expression in nonneuronal cells, ectopically expressed syntaxin 1 does not exit the endoplasmic reticulum and diverts secretory vesicles thus suggesting that Munc18a may chaperone syntaxin 1 to prevent illegitimate fusion reactions.[56-58] Munc13 takes over syntaxin 1 after Munc18 to prime the vesicles for fusion, probably in concert with calmodulin.[59] It is noteworthy to note that calmodulin also interacts with synaptobrevin 2 and may also prime synaptobrevin 2 for fusion by a conformational change, which removes the cis-inhition and allows for

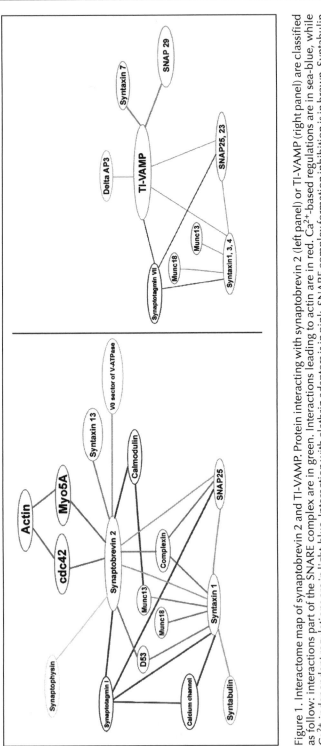

Figure 1. Interactome map of synaptobrevin 2 and TI-VAMP. Protein interacting with synaptobrevin 2 (left panel) or TI-VAMP (right panel) are classified as follow: interactions part of the SNARE complex are in green. Interactions leading to actin are in red. Ca^{2+}-based regulations are in sea-blue, while Ca^{2+}-independent regulations are in light-blue. Interaction with clathrin adaptors is in pink. SNARE complex formation inhibition is in brawn. Syntabulin is involved in correct trafficking of syntaxin. Synaptophysin is known to interact with synaptobrevin in an activity-dependent manner. Interaction with a putative fusion pore is in orange. A color version of this figure is available online at www.Eurekah.com.

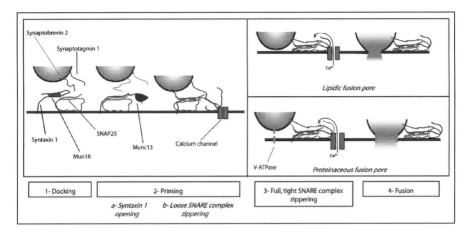

Figure 2. A vesicle undergoing evoked fusion: the machinery. This scheme represents the different steps of calcium-triggered synaptic vesicle fusion. After the vesicle is docked at the plasma membrane, Munc13 releases Munc18 from syntaxin 1 by interacting with its amino terminal domain, thus allowing the interaction of the different SNARE proteins. When this loose SNARE complex is formed, the vesicle is called "primed" and ready for fusion. At the opening of the calcium channel, Synaptotagmin binds two Ca^{2+} ions, changes its conformation and brings the vesicle even closer to the plasma membrane. The SNARE complex thus adopts a "tight" conformation by full zippering, and the vesicle, fusing with the plasma membrane, releases its content. Two hypotheses are currently proposed about the nature of the pore connecting the inside of the vesicle with the outside of the cell: the fusion pore may be either lipidic, or proteinaceous. The involvement of proteins such as the V0 sector of the V-ATPase is thought to help membrane fusion without being sufficient per se to drive fusion: the SNARE machinery brings the driving force necessary for fusion lipid bilayers. A color version of this figure is available online at www.Eurekah.com.

trans-interactions.[60] By releasing Munc18a from syntaxin 1, Munc13 allows the SNARE complex to partially zipper in a loose conformation.[59] At this stage, the vesicle is primed and ready to release its content upon stimulation.

The calcium sensitivity of synaptobrevin 2-mediated exocytosis is most likely not encoded in SNARE proteins because in vitro fusion assays using synaptobrevin 2, syntaxin 1 and SNAP25 did not demonstrate calcium dependency[61-63] in the exception of one study.[64] More likely, synaptotagmin 1, a protein that interacts with the synaptic SNARE proteins during exocytosis, may be the key calcium-dependent regulator[65] (see chapter on synaptotagmin). Moreover, both syntaxin 1 and synaptotagmin 1 interact with calcium channels,[66] possibly to concentrate exocytic zones at close vicinity of the trigger of exocytosis.

When the action potential arrives in the nerve terminal, calcium channels open and calcium concentration rises to mM levels. Synaptotagmin, through its C2A and C2B domains, binds two calcium ions and drives the full zippering of the SNARE complex in a tight conformation (also for review see ref. 67). The energy freed during this zippering process is thought to be sufficient to by-pass the energy barrier of two opposite lipid bilayers intermixing (for a focus on the lipidic side of membrane fusion;[68] for a focus about the critical hemifusion state[69]). At this stage the content of the vesicle is released. Complexin interaction with the assembled SNARE complex[70] is thought to stabilize an intermediate of the SNARE complex. This stabilization may retard the disassembly of SNARE complex by SNAPs and NSF to allow for fast track of exocytosis.[71] D53 is a coiled-coil containing protein specifically expressed in the nervous system, that enhances synaptobrevin 2 and syntaxin 1 interactions in vitro,[72] which suggests a possible regulation of SNARE complex assembly and/or disassembly.

The regulators of TI-VAMP are largely unknown. One key partner is the δ subunit of the molecular coat AP-3 because this interaction is required to target TI-VAMP to late endosomal vesicles in non neuronal cells[46] and to the synapses in the brain (our unpublished observations). TI-VAMP was shown to interact with synaptotagmin 7 to mediate the exocytosis of secretory lysosomes.[44]

SNARE and Friends: Is That All about Fusion?

Several observations have questioned the SNARE model in the last years. First, whereas in vitro fusion of liposomes was shown to strictly depend on the presence of v- and t-SNARE proteins on separate populations, the reaction was found to be too slow by 40 to 10000 times to account for the fastest sub millisecond synaptic response.[73] These data have been clarified by removing the amino-terminal regulatory domain of syntaxin 1, which results in fast fusion in vitro.[64] Moreover, real time wide-field fluorescence microscopy observation of the docking and fusion of single proteoliposomes reconstituted with full-length synaptobrevin into planar lipid bilayers containing anchored syntaxin and SNAP-25 revealed kinetics only 40 times slower than the fastest in vivo fusion reactions.[63] Second, promiscuous interaction of v-SNARE proteins with t-SNAREs of non cognate target membranes could be observed using recombinant proteins in vitro (example an endoplasmic reticulum v-SNARE interacting with a lysosomal t-SNARE), which should not happen if SNARE pairing would mediate any kind of specificity.[74] This was however largely contradicted by in vitro fusion assay[75,76] and coimmuno-precipitation experiments,[46,58] showing a rather important specificity of pairing. Third, NSF does not seem to play any function in SNARE dependent fusion assays in vitro.[77] Instead, SNAPs and NSF have been involved in the disassembly of the highly stable SNARE complex.[78,79] The requirement for NSF in the Golgi fusion assay may reflect the fact that v-SNARE recycles back to the donor membrane after fusion, thus pointing to a crucial role of SNARE complex disassembly to maintain homeostasis, although this has not yet been directly tested. Fourth, the yeast vacuole fusion assay suggested that SNARE proteins may operate for docking rather than for fusion. Indeed, this later step was found to depend on the presence of calmodulin on the V0 sector of the vacuolar ATPase, including the proteolipidic c-subunit, also called mediatophore.[80]

Altogether, these data point the possibility of complementary mechanisms beside SNAREs in membrane fusion.

Several studies have mentioned a fast track of exocytosis also called"kiss-and-run" (also for review see ref. 47). In this model, the synaptic vesicle does not fully collapse in the plasma membrane but its fusion pore flickers, thus allowing a partial release of the vesicular content. The question here is to reconcile the flickering observed with the high amount of energy required for lipid bilayers intermixing. Indeed, it seems unlikely that SNARE proteins would drive this flickering. The V0 sector of the V-ATPase mentioned above is proposed to form a "connexon-like" channel by trans pairing, thus forming a proteinaceous pore, then dilating if the fusion leads to full collapse of the vesicle in the plasma membrane.[81]

These important data may not contradict the SNARE model to a great extent as the c-subunit of the V-ATPase, if involved in membrane fusion, would only facilitate it. At least, this latter mechanism has not appeared as an alternative to the SNARE model which still seems to be the only one applicable to the largest range of membrane fusions in vivo, in which SNARE proteins supply the driving force to perform membrane fusion.

One of the key directions for the future is to complete the interactome of all SNARE proteins and to understand the regulatory networks in which they operate in different cell types. Another important issue to clarify is the role of TI-VAMP in neuronal cells and the possibility that the exocytosis mediated by this v-SNARE in mature neurons would be under completely different regulations than that seen in the case of synaptobrevin. Moreover, the findings that a mutation in SNAP29 promoter region is associated with schizophrenia,[82] and that a 1-bp deletion in SNAP29 coding region is responsible for a neurocutaneous syndrome characterized by cerebral dysgenesis, neuropathy, ichthyosis, and palmoplantar keratoderma[83] further suggests that the studies of human diseases is likely to uncover yet unexpected functions of SNARE proteins.

Acknowledgements

We are grateful to Rachel Rudge and Fabienne Paumet for their critical reading of the manuscript. This work was supported in part by grants from INSERM (Avenir Program), the European Commission ('Signalling and Traffic' STREP 503229), the Association pour la Recherche sur le Cancer (N°5873 and 4762), the Association Française contre les Myopathies, the Ministère de la Recherche (ACI-BDP), the Fondation pour la Recherche Médicale, the HFSP (RGY0027/2001-B101) and the Fondation pour la Recherche sur le Cerveau.

References

1. Söllner T, Bennett MK, Whiteheart SW et al. A protein assembly-disassembly pathway in vitro that may correspond to sequential steps of synaptic vesicle docking, activation, and fusion. Cell 1993; 75:409-418.
2. Balch WE, Dunphy WG, Braell WA et al. Reconstitution of the transport of protein between successive compartments of the Golgi measured by the coupled incorporation of N-acetylglucosamine. Cell 1984; 39(2 Pt 1):405-416.
3. Beckers CJ, Block MR, Glick BS et al. Vesicular transport between the endoplasmic reticulum and the Golgi stack requires the NEM-sensitive fusion protein. Nature 1989; 339:397-398.
4. Wilson DW, Wilcox CA, Flynn GC et al. A fusion protein required for vesicle-mediated transport in both mammalian cells and yeast. Nature 1989; 339:355-359.
5. Clary DO, Griff IC, Rothman JE. SNAPs, a family of NSF attachment proteins involved in intracellular membrane fusion in animals and yeast. Cell 1990; 61:709-721.
6. Söllner T, Whiteheart SW, Brunner M et al. SNAP receptors implicated in vesicle targeting and fusion. Nature 1993; 362:318-324.
7. Blasi J, Chapman ER, Link E et al. Botulinum neurotoxin A selectively cleaves the synaptic protein SNAP-25. Nature 1993; 365:160-163.
8. Schiavo G, Rossetto O, Catsicas S et al. Identification of the nerve terminal targets of botulinum neurotoxin serotypes A, D, and E. J Biol Chem 1993; 268:23784-23787.
9. Schiavo G, Santucci A, DasGupta BR et al. Botulinum neurotoxins serotypes A and E cleave SNAP-25 at distinct COOH-terminal peptide bonds. FEBS Lett 1993; 335:99-103.
10. Binz T, Blasi J, Yamasaki S et al. Proteolysis of SNAP-25 by types E and A botulinal neurotoxins. J Biol Chem 1994; 269:1617-1620.
11. Yamasaki S, Binz T, Hayashi T et al. Botulinum neurotoxin type G proteolyses the Ala[81]-Ala[82] bond of rat synaptobrevin 2. Biochem Biophys Res Commun 1994; 200:829-835.
12. Yamasaki S, Baumeister A, Binz T et al. Cleavage of members of the synaptobrevin/VAMP family by types D and F botulinal neurotoxins and tetanus toxin. J Biol Chem 1994; 269:12764-12772.
13. Niemann H, Blasi J, Jahn R. Clostridial neurotoxins: New tools for dissecting exocytosis. Trends Cell Biol 1994; 4:179-185.
14. Protopopov V, Govindan B, Novick P et al. Homologs of the synaptobrevin/VAMP family of synaptic vesicle proteins function on the late secretory pathway in S. cerevisiae. Cell 1993; 74:855-861.
15. Couve A, Gerst JE. Yeast Snc proteins complex with Sec9. Functional interactions between putative SNARE proteins. J Biol Chem 1994; 269(38):23391-23394.
16. Bennett MK, Scheller RH. The molecular machinery for secretion is conserved from yeast to neurons. Proc Natl Acad Sci USA 1993; 90:2559-2563.
17. Scales SJ, Chen YA, Yoo BY et al. SNAREs contribute to the specificity of membrane fusion. Neuron 2000; 26(2):457-464.
18. Weber T, Zemelman BV, McNew JA et al. SNAREpins: Minimal machinery for membrane fusion. Cell 1998; 92:759-772.
19. Schoch S, Deak F, Konigstorfer A et al. SNARE function analyzed in synaptobrevin/VAMP knockout mice. Science 2001; 294(5544):1117-1122.
20. Deitcher DL, Ueda A, Stewart BA et al. Distinct requirements for evoked and spontaneous release of neurotransmitter are revealed by mutations in the Drosophila gene neuronal-synaptobrevin. J Neurosci 1998; 18:2028-2039.
21. Sara Y, Virmani T, Deak F et al. An isolated pool of vesicles recycles at rest and drives spontaneous neurotransmission. Neuron 2005; 45(4):563-573.
22. Galli T, Zahraoui A, Vaidyanathan VV et al. A novel tetanus neurotoxin-insensitive vesicle-associated membrane protein in SNARE complexes of the apical plasma membrane of epithelial cells. Mol Biol Cell 1998; 9:1437-1448.
23. Martinez-Arca S, Coco S, Mainguy G et al. A common exocytotic mechanism mediates axonal and dendritic outgrowth. J Neurosci 2001; 21(11):3830-3838.
24. Alberts P, Rudge R, Hinners I et al. Cross talk between tetanus neurotoxin-insensitive vesicle-associated membrane protein-mediated transport and L1- mediated adhesion. Mol Biol Cell 2003; 14(10):4207-4220.

25. Osen-Sand A, Staple JK, Naldi E et al. Common and distinct fusion proteins in axonal growth and transmitter release. J Comp Neurol 1996; 367:222-234.
26. Borisovska M, Zhao Y, Tsytsyura Y et al. v-SNAREs control exocytosis of vesicles from priming to fusion. EMBO J 2005.
27. Randhawa VK, Thong FS, Lim DY et al. Insulin and hypertonicity recruit GLUT4 to the plasma membrane of muscle cells using NSF-dependent SNARE mechanisms but different v-SNAREs: Role of TI-VAMP. Mol Biol Cell 2004; 15(12):5565-5573.
28. Yang CM, Mora S, Ryder JW et al. VAMP3 null mice display normal constitutive, insulin- and exercise-regulated vesicle trafficking. Mol Cell Biol 2001; 21(5):1573-1580.
29. Allen LAH, Yang CM, Pessin JE. Rate and extent of phagocytosis in macrophages lacking vamp3. J Leukocyte Biol 2002; 72(1):217-221.
30. Schraw TD, Rutledge TW, Crawford GL et al. Granule stores from cellubrevin/VAMP-3 null mouse platelets exhibit normal stimulus-induced release. Blood 2003; 102(5):1716-1722.
31. Washbourne P, Thompson PM, Carta M et al. Genetic ablation of the t-SNARE SNAP-25 distinguishes mechanisms of neuroexocytosis. Nat Neurosci 2002; 5(1):19-26.
32. Vilinsky I, Stewart BA, Drummond J et al. A Drosophila SNAP-25 null mutant reveals context-dependent redundancy with SNAP-24 in neurotransmission. Genetics 2002; 162(1):259-271.
33. Schulze KL, Bellen HJ. Drosophila syntaxin is required for cell viability and may function in membrane formation and stabilization. Genetics 1996; 144:1713-1724.
34. Saifee O, Wei LP, Nonet ML. The Caenorhabditis elegans unc-64 locus encodes a syntaxin that interacts genetically with synaptobrevin. Mol Biol Cell 1998; 9:1235-1252.
35. Chae TH, Kim S, Marz KE et al. The hyh mutation uncovers roles for alphaSnap in apical protein localization and control of neural cell fate. Nat Genet 2004.
36. Hong HK, Chakravarti A, Takahashi JS. The gene for soluble N-ethylmaleimide sensitive factor attachment protein alpha is mutated in hydrocephaly with hop gait (hyh) mice. Proc Natl Acad Sci USA 2004; 101(6):1748-1753.
37. Proux-Gillardeaux V, Rudge R, Galli T. The tetanus neurotoxin-sensitive and insensitive routes to and from the plasma membrane: Fast and slow pathways? Traffic 2005; 6(5):366-373.
38. Chilcote TJ, Galli T, Mundigl O et al. Cellubrevin and synaptobrevins: Similar subcellular localization and biochemical properties in PC12 cells. J Cell Biol 1995; 129:219-231.
39. Advani RJ, Bae HR, Bock JB et al. Seven novel mammalian SNARE proteins localize to distinct membrane compartments. J Biol Chem 1998; 273:10317-10324.
40. Wang CC, Ng CP, Lu L et al. A role of VAMP8/endobrevin in regulated exocytosis of pancreatic acinar cells. Dev Cell 2004; 7(3):359-371.
41. Sutton RB, Fasshauer D, Jahn R et al. Crystal structure of a SNARE complex involved in synaptic exocytosis at 2.4 angstrom resolution. Nature 1998; 395:347-353.
42. Chen F, Foran P, Shone CC et al. Botulinum neurotoxin B inhibits insulin-stimulated glucose uptake into 3T3-L1 adipocytes and cleaves cellubrevin unlike type A toxin which failed to proteolyze the SNAP-23 present. Biochemistry 1997; 36:5719-5728.
43. Martinez-Arca S, Alberts P, Zahraoui A et al. Role of tetanus neurotoxin insensitive vesicle-associated membrane protein (TI-VAMP) in vesicular transport mediating neurite outgrowth. J Cell Biol 2000; 149(4):889-899.
44. Rao SK, Huynh C, Proux-Gillardeaux V et al. Identification of SNAREs involved in synaptotagmin VII-regulated lysosomal exocytosis. J Biol Chem 2004; 279(19):20471-20479.
45. Su Q, Mochida S, Tian JH et al. SNAP-29: A general SNARE protein that inhibits SNARE disassembly and is implicated in synaptic transmission. Proc Natl Acad Sci USA 2001; 98(24):14038-14043.
46. Martinez-Arca S, Rudge R, Vacca M et al. A dual mechanism controlling the localization and function of exocytic v-SNAREs. Proc Natl Acad Sci USA 2003; 100(15):9011-9016.
47. Galli T, Haucke V. Cycling of synaptic vesicles: How far? How fast! Sci STKE 2004; 264:19.
48. Prekeris R, Terrian DM. Brain myosin V is a synaptic vesicle-associated motor protein: Evidence for a Ca2+-dependent interaction with the synaptobrevin-synaptophysin complex. J Cell Biol 1997; 137:1589-1601.
49. Nevins AK, Thurmond DC. A direct interaction between Cdc42 and vesicle-associated membrane protein 2 regulates SNAREdependent insulin exocytosis. J Biol Chem 2005; 280(3):1944-1952.
50. Johnson DI. Cdc42: An essential Rho-type GTPase controlling eukaryotic cell polarity. Microbiol Mol Biol Rev 1999; 63(1):54-105.
51. Daniel S, Noda M, Cerione RA et al. A link between Cdc42 and syntaxin is involved in mastoparan-stimulated insulin release. Biochemistry 2002; 41(30):9663-9671.
52. Edelmann L, Hanson PI, Chapman ER et al. Synaptobrevin binding to synaptophysin: A potential mechanism for controlling the exocytotic fusion machine. EMBO J 1995; 14(2):224-231.
53. Galli T, McPherson PS, De Camilli P. The V0 sector of the V-ATPase, synaptobrevin and synaptophysin are associated on synaptic vesicles in a triton X-100 resistant, freeze-thawing sensitive complex. J Biol Chem 1996; 271:2193-2199.
54. Pennuto M, Bonanomi D, Benfenati F et al. Synaptophysin I controls the targeting of VAMP2/synaptobrevin II to synaptic vesicles. Mol Biol Cell 2003; 14(12):4909-4919.

55. Jahn R. Sec1/Munc18 proteins: Mediators of membrane fusion moving to center stage. Neuron 2000; 27(2):201-204.
56. Rowe J, Calegari F, Taverna E et al. Syntaxin 1a is delivered to the apical and basolateral domains of epithelial cells: The role of munc-18 proteins. J Cell Sci 2001; 114(18):3323-3332.
57. Rowe J, Corradi N, Malosio ML et al. Blockade of membrane transport and disassembly of the Golgi complex by expression of syntaxin 1A in neurosecretion-incompetent cells: Prevention by rbSEC1. J Cell Sci 1999; 112:1865-1877.
58. Martinez-Arca S, Proux-Gillardeaux V, Alberts P et al. Ectopic expression of syntaxin 1 in the ER redirects TI-VAMP- and cellubrevin-containing vesicles. J Cell Sci 2003; 116(Pt 13):2805-2816.
59. Junge HJ, Rhee JS, Jahn O et al. Calmodulin and Munc13 form a Ca2+ sensor/effector complex that controls short-term synaptic plasticity. Cell 2004; 118(3):389-401.
60. DeHaro L, Ferracci G, Opi S et al. Ca2+/calmodulin transfers the membrane-proximal lipid- binding domain of the v-SNARE synaptobrevin from cis to trans bilayers. Proc Nat Acad Sci USA 2004; 101(6):1578-1583.
61. Mahal LK, Sequeira SM, Gureasko JM et al. Calcium-independent stimulation of membrane fusion and SNAREpin formation by synaptotagmin I. J Cell Biol 2002; 158(2):273-282.
62. Bowen ME, Weninger K, Brunger AT et al. Single molecule observation of liposome-bilayer fusion thermally induced by soluble N-ethyl maleimide sensitive-factor attachment protein receptors (SNAREs). Biophys J 2004; 87(5):3569-3584.
63. Liu T, Tucker WC, Bhalla A et al. SNAREdriven, 25-millisecond vesicle fusion in vitro. Biophys J 2005; 89(4):2458-2472.
64. Fix M, Melia TJ, Jaiswal JK et al. Imaging single membrane fusion events mediated by SNARE proteins. Proc Nat Acad Sci USA 2004; 101(19):7311-7316.
65. Tucker WC, Weber T, Chapman ER. Reconstitution of Ca2+-regulated membrane fusion by synaptotagmin and SNAREs. Science 2004; 304(5669):435-438.
66. Seagar M, Leveque C, Charvin N et al. Interactions between proteins implicated in exocytosis and voltage-gated calcium channels. Phil Trans Roy Soc London B 1999; 354:289-297.
67. Koh TW, Bellen HJ. Synaptotagmin I, a Ca2+ sensor for neurotransmitter release. Trends Neurosci 2003; 26(8):413-422.
68. Jahn R, Grubmuller H. Membrane fusion. Curr Opin Cell Biol 2002; 14(4):488-495.
69. Chernomordik LV, Kozlov MM. Membrane hemifusion: Crossing a chasm in two leaps. Cell 2005; 123(3):375-382.
70. Pabst S, Margittai M, Vainius D et al. Rapid and selective binding to the synaptic SNARE complex suggests a modulatory role of complexins in neuroexocytosis. J Biol Chem 2002; 277(10):7838-7848.
71. Archer DA, Graham ME, Burgoyne RD. Complexin regulates the closure of the fusion pore during regulated vesicle exocytosis. J Biol Chem 2002; 277(21):18249-18252.
72. Proux-Gillardeaux V, Galli T, Callebaut I et al. D53 is a novel endosomal SNAREbinding protein that enhances interaction of syntaxin 1 with the synaptobrevin 2 complex in vitro. Biochem J 2003; 370(Pt 1):213-221.
73. Nickel W, Weber T, McNew JA et al. Content mixing and membrane integrity during membrane fusion driven by pairing of isolated v-SNAREs and t-SNAREs. Proc Natl Acad Sci USA 1999; 96:12571-12576, (MEDLINE).
74. Yang B, Gonzalez Jr L, Prekeris R et al. SNARE interactions are not selective. Implications for membrane fusion specificity. J Biol Chem 1999; 274(9):5649-5653.
75. McNew JA, Parlati F, Fukuda R et al. Compartmental specificity of cellular membrane fusion encoded in SNARE proteins. Nature 2000; 407(6801):153-159.
76. Paumet F, Rahimian V, Rothman JE. The specificity of SNAREdependent fusion is encoded in the SNARE motif. Proc Nat Acad Sci USA 2004; 101(10):3376-3380.
77. Weber T, Parlati F, McNew JA et al. SNAREpins are functionally resistant to disruption by NSF and alpha SNAP. J Cell Biol 2000; 149(5):1063-1072.
78. Otto H, Hanson PI, Jahn R. Assembly and disassembly of a ternary complex of synaptobrevin, syntaxin, and SNAP-25 in the membrane of synaptic vesicles. Proc Natl Acad Sci USA 1997; 94:6197-6201.
79. Hayashi T, Yamasaki S, Nauenburg S et al. Disassembly of the reconstituted synaptic vesicle membrane fusion complex in vitro. EMBO J 1995; 14:2317-2325.
80. Peters C, Bayer MJ, Buhler S et al. Trans-complex formation by proteolipid channels in the terminal phase of membrane fusion. Nature 2001; 409(6820):581-588.
81. Reese C, Heise F, Mayer A. Trans-SNARE pairing can precede a hemifusion intermediate in intracellular membrane fusion. Nature 2005; 436(7049):410-414.
82. Saito T, Guan F, Papolos DF et al. Polymorphism in SNAP29 gene promoter region associated with schizophrenia (Vol 6, pg 193, 2001). Mol Psychiatr 2001; 6(5):605.
83. Sprecher E, Ishida-Yamamoto A, Mizrahi-Koren M et al. A mutation in SNAP29, coding for a SNARE protein involved in intracellular trafficking, causes a novel neurocutaneous syndrome characterized by cerebral dysgenesis, neuropathy, ichthyosis, and palmoplantar keratoderma. Am J Hum Genet 2005; 77(2):242-251.

CHAPTER 2

Regulation of SNARE Complex Assembly by Second Messengers:
Roles of Phospholipases, Munc13 and Munc18

Alexander J.A. Groffen* and Matthijs Verhage

Abstract

Many specialized cell types are dedicated to the regulated exocytosis of biologically active substances. The secretory cargo is stored in vesicles, that have a low release probability in the absence of suitable secretory signals. Membrane fusion requires the assembly of soluble N-ethylmaleimide attachment protein receptor (SNARE) proteins into a trimeric SNARE complex, in mammals composed of syntaxin, SNAP-25/23 and synaptobrevin.[1] Although SNARE complex assembly is sufficient to drive fusion of liposomes in cell-free systems,[2,3] spatially and temporally controlled membrane fusion in living cells requires the activity of accessory proteins. Such accessory proteins have been identified, for instance Munc13 and Munc18, that are essential for regulated secretion.

Here we review the molecular mechanisms by which Munc13 and Munc18 may regulate SNARE complex assembly. Regulatory mechanisms can be invoked by external signals such as ligand-receptor interactions, as well as by the local production of second messengers due to high secretory activity. All these routes are convergently reflected by central signaling molecules such as intracellular Ca^{2+}, diacylglycerol, inositol-triphosphate or other phospholipase-generated products. Recently several novel pathways were described by which these molecules can provide regulatory feedback on the activity of Munc13 and Munc18 and hence, on SNARE-dependent exocytosis.

Modulation of Regulated Exocytosis

Regulated exocytosis, by definition, is not a constitutive process but requires a stimulus of some kind to trigger a secretory event. The possible routes to secretory activation can be categorized into three classes: (1) Ca^{2+} influx as a result of depolarization of excitable membranes, (2) G-protein activation, and (3) receptor tyrosine kinase activation (Fig. 1).

In general, the first class represents the major mechanism for secretory excitation, whereas the other pathways primarily act as modulators of exocytotic strength. Depolarizing stimuli are of vital importance in the nervous system as well as other excitable cell types. In synapses, vesicles dock and are processed to a fusion-ready state at the active zone, but their release probability remains low until Ca^{2+} enters the terminal as a result of an action potential that depolarizes the presynaptic membrane. The opening of voltage-dependent Ca^{2+} channels allows a strong inward Ca^{2+} current driven by the steep gradient in extracellular versus intracellular

*Corresponding Author: Alexander J.A. Groffen—Department of Functional Genomics, Center for Neurogenomics and Cognitive Research, Vrije Universiteit Amsterdam, De Boelelaan 1087, 1081 HV Amsterdam, The Netherlands. Email: sander.groffen@falw.vu.nl

Molecular Mechanisms of Exocytosis, edited by Romano Regazzi. ©2007 Landes Bioscience and Springer Science+Business Media.

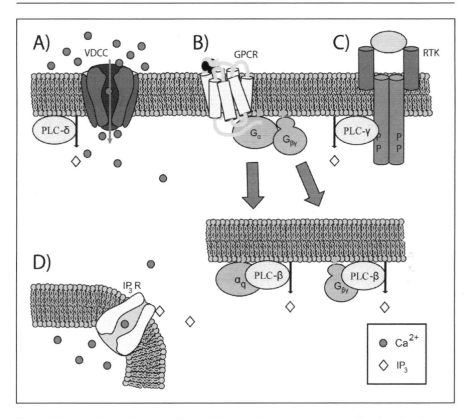

Figure 1. Alternative mechanisms for modulation of exocytosis are associated with activation of phosphoinositide-specific phospholipase C (PI-PLC). A) The major trigger for regulated exocytosis involves opening of voltage-dependent Ca^{2+} channels in response to membrane depolarization by an action potential. Ca^{2+}-sensitive phospholipases of the PLC-δ subfamily generate the second messenger inositol-triphosphate (IP_3). B) Binding of ligands to G-protein coupled receptors (GPCR) can induce downstream signaling by G protein α_q and $\beta\gamma$ subunits to PLC-β isoforms. C) Binding of ligands to receptor tyrosine kinases (RTK) induce their dimerization and activation of intracellular tyrosine kinase activity. Direct or indirect signal transuction pathways lead to activation of PLC-γ isozymes. D) IP_3 produced by phosphoinositide-specific PLC-β, -γ or -δ induces the opening of IP_3 receptor channels, causing the mobilization of Ca^{2+} ions from intracellular stores.

[Ca^{2+}]. This current causes a short local peak in intracellular [Ca^{2+}] near the entry site, which is strategically located in the vicinity of secretion-ready vesicles. The [Ca^{2+}] rise is coupled to vesicle exocytosis by an extremely rapid mechanism that is still only partly understood, but at least involves the proteins synaptotagmin[4,5] and complexin[6] besides the core components of the SNARE complex (for a comprehensive review, see ref. 7). Within milliseconds after the excitation event, the Ca^{2+} ions assume a more homogeneous distribution that is only mildly elevated.[8,9] This so-called 'residual' calcium concentration may be insufficient to evoke synaptic vesicle fusion directly, but it has many roles in regulating the secretory response to subsequent stimuli, i.e., synaptic plasticity.[10] A similar mechanism forms the major secretory signal in chromaffin cells, responsible for the secretion of catecholamines (adrenaline, noradrenaline) from the adrenal gland. Cholinergic input from the splanchnic nerve stimulates nicotinic receptors on the chromaffin cell surface, resulting in a depolarization that is amplified by voltage-sensitive Na^+ channels. Subsequently, voltage-dependent Ca^{2+} channels mediate Ca^{2+} influx and trigger exocytosis.

G-protein dependent signaling pathways (Fig. 1B) can modulate exocytotic strength in many cell types and, in some systems, can also evoke exocytosis.[11] The underlying mechanisms have been thoroughly studied in chromaffin cells, that show substantial exocytotis in response to ligands of G-protein coupled receptors (GPCRs); in particular histamine, muscarine and pituitary adenylate cyclase-activating polypeptide (PACAP). Downstream activities of G proteins affect many cellular signaling pathways including phospholipase activity and the release of Ca^{2+} from intracellular stores (see below). Exactly how these processes finally modulate exocytotic response is largely unknown (reviewed in ref. 10). The involvement of both $G\beta\gamma$ and $G\alpha$ subunits has been demonstrated, and their activities were shown to have regulatory effects on depolarization-induced secretion.[11]

Another mechanism of exocytotic modulation utilizes tyrosine kinase receptors (Fig. 1C). Generally, ligand binding causes the receptors to cluster and autophosphorylate, thus initiating a signaling cascade involving downstream activities of nonreceptor tyrosine kinases, phosphatidylinositide 3-kinase, and phospholipase C-γ. This finally results in the mobilization of intracellular Ca^{2+} accompanied by increased rates of exocytosis. Examples are mast cells that release inflammatory compounds upon binding of immunoglobulins to Fc receptors,[14] and adipocytes that incorporate a glucose transporter into the plasma membrane in response to .[12,13]

All these signals are collectively reflected by the concentrations of common second messenger molecules that coordinately adapt a variety of cellular processes to the activity state of secretory cell types. Among the essential messengers is Ca^{2+}, but in addition, many molecules that are produced by the turnover of phospholipids in the membrane play important roles in the modulation of regulated exocytosis. The pathways initiated by Ca^{2+} ions and phospholipid-derived second messengers are interdependent: Ca^{2+} increases can accelerate phospholipid turnover and vice versa, phospholipid-derived signaling molecules can increase cytoplasmic Ca^{2+} levels. Enzymes of the phospholipase superfamily are centrally involved in this event. Given their significant contribution of the modulation of regulated exocytosis, we describe in the next section how phospholipase activity is linked to secretory modulation.

The Activation of Phospholipases

Enhancement of exocytosis is usually accompanied by increased activities of phospholipases, that hydrolyze phospholipids to metabolites that induce signaling routes in the cell and at the membrane. Different families of phospholipases catalyze hydrolysis at different positions in the phospholipid molecule (Fig. 2), thereby producing essentially different products that either are released into the cytoplasm (e.g., IP3, arachnidonic acid) or remain in the hydrophobic environment of the membrane (e.g., diacylglycerol). The most relevant phospholipase families in the context of regulated exocytosis are the phosphoinositide-specific isoforms of phospholipase C (PI-PLC). In addition, the phospholipase A_2 (PLA$_2$) family deserves renewed attention.[15] Other phospholipases such as nonphosphoinositide-specific PLC, phospholipase D and phospholipase A_1 also contribute importantly, but are described elsewhere. Before elaborating on the signaling events induced by phospholipases, the molecular mechanisms responsible for their activation will first be described.

Activation of Phosphoinositide-Specific Phospholipase C

PI-PLC isozymes (EC 3.1.4.11) bind preferentially to phosphatidyl inositol-4,5-diphosphate (PIP$_2$). PIP$_2$ is a membrane component that is enriched in the inner leaflet of the bilayer, occurs in membrane subdomains[16] and is required for regulated exocytosis.[17] PI-PLC hydrolyses PIP$_2$ to produce inositol-triphosphate (IP$_3$) and diacylglycerol (DAG). Both products are important second messengers for downstream cellular signaling.

Since all isoforms lack a transmembrane segment, the recruitment of these enzymes from the cytosol to the inner plasma membrane surface constitutes a common step in their activation mechanism. On basis of their structural conservation, the PI-PLC isozymes were grouped

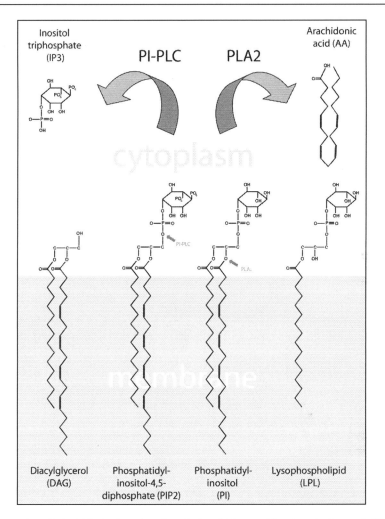

Figure 2. Phospholipid hydrolysis by phosphoinositide-specific phospholipase C (PI-PLC) and phospholipase A2 (PLA$_2$). PI-PLC converts the trace membrane component phosphatidylinositol-4,5-diphosphate (PIP$_2$) into diacylglycerol (DAG) and inositol-triphosphate (IP$_3$). PLA$_2$ hydrolyzes phospholipids such as, for example, phosphatidylinositol (PI) to produce lysophospholipid (LPL) and a free fatty acid such as arachidonic acid (AA). The water-soluble compounds IP$_3$ and AA diffuse into the cytoplasm (upper half), whereas the other hydrophobic molecules remain membrane-associated (lower half).

into four subfamilies: PLC-β, -γ, -δ and -ϵ. Each family has a characteristic domain composition, that provides possibilities for fundamentally distinct mechanisms of phospholipase activation. Taken together, each signaling route that is capable of inducing exocytosis (Fig. 1) is matched in the proteome by a stimulus-responsive PI-PLC subfamily. The spectrum of PI-PLC isoforms expressed in a particular cell type may thus encode different modes of secretory regulation.

PLC-β isozymes are primarily activated by G proteins. The dissociation of heterotrimeric G protein complexes can be triggered by binding of extracellular ligands to G protein-coupled receptors (GPCRs) on the membrane surface. The dissociation gives rise to a monomeric Gα

subunit and a dimeric $G\beta\gamma$ subunit. $G\alpha$ subunits that belong to the G_q family (i.e., subtypes α_q, α_{11}, α_{14} and α_{16}) induce PLC-β activity[18-20] by binding to its C-terminal domain.[21] This pathway is triggered by many GPCRs including the receptors for acetylcholine (muscarinic receptors m1, m3 and m5), angiotensin II, bombesin, bradykinin, glutamate (metabotropic receptors), histamine, thrombin, thromboxane A2, thyroid-stimulating hormone, vasopressin, and others. Alternatively, PLC-β can be activated by binding of the $G\beta\gamma$ subunit[22,23] to the N-terminal domain of PLC.[24] This activation route is relevant for the interleukin 8 and muscarinic acetylcholine (m2) receptor.

Activity of the PLC-γ subfamily (γ1, γ2) is evoked by receptors of the protein tyrosine kinase (PTK) type.[25] A well-characterized example of such a mechanism occurs in platelet-derived growth factor (PDGF) signal transduction[26] and mast cell degranulation.[14] PLC-γ interacts, after ligand-induced receptor autophosphorylation, with the cytoplasmic receptor subunits through its SH2 and SH3 domains (which are absent in other PLC subfamilies) and is thereby recruited to the plasma membrane. Additionally, PLC-γ is itself tyrosine-phosphorylated either directly by the receptor (e.g., the PDGF receptor)[27] or indirectly through a receptor-induced cascade of tyrosine phosphorylation events (e.g., the insulin receptor[28] or Fc receptors in hematopoietic cells).[14] At the same time activation of phosphatidylinositol 3-kinase (PI3K) occurs, resulting in an enhanced synthesis of phosphatidylinositol 3,4,5-trisphosphate (PIP$_3$) in the membrane, which further enhances PLC-γ activity. Besides by protein tyrosine kinases, PLC-γ can also be activated by the phospholipid metabolites phosphatidic acid (produced by PLD) or arachidonic acid (produced by cytosolic PLA$_2$).

Yet another activation mechanism prevails for the PLC-δ isozymes. These isozymes appear to be controlled by the concentration of intracellular Ca^{2+} (a second messenger that generally reflects the activity state of the cell). Ca^{2+} can also enhance the activity of other PI-PLC isozymes, but PLC-δ appears to be more sensitive to Ca^{2+}. Several domains within PLC-δ can bind Ca^{2+}, which enhances the membrane affinity as well as the catalytic function of the enzyme.[29-33] And finally, a more recently characterized PI-PLC subtype, PLC-ϵ, contains several domains that are absent in other PI-PLC isozymes. These domains, named CDC25, RA1 and RA2, together with a unique insert in the catalytic domain, mediate interactions of PLC-ϵ with small monomeric GTPase enzymes including members of the Ras, Rap (Rap1), and Rho family (RhoA, RhoB and RhoC). In analogy with the $G\alpha$ subunits of the G_q family that activate PLC-β isozymes, $G\alpha$ subunits of the G_{12} family can induce PLC-ϵ activity.[34] The latter mechanism may occur indirectly via Rho GTPase activation.[35] In conclusion, PI-PLC activity can be induced by multiple convergent pathways also known to modulate secretion.

Phospholipase A₂

Mammalian genomes encode many PLA$_2$ enzymes (EC 3.1.1.4) which can be either secreted or cytosolic, and either Ca^{2+}-dependent or -independent. Secreted PLA$_2$ enzymes are found in many digestive secretions (such as pancreatic juice), animal venoms (e.g., cobra venom), and in immune systems they play an important role in inflammation.[36] Cytosolic PLA$_2$ (cPLA$_2$) enzymes have found increasing interest in the context of constitutive and regulated exocytosis.[37] However, the cacophony of isozymes expressed in many different tissues makes it a complex task to dissect the individual functions of specific isoforms.

All PLA$_2$ enzymes hydrolyze the sn-2 ester bond of the phospholipid substrate (Fig. 2), resulting in a lysophospholipid (LPL) and a free fatty acid.[29] Different isozymes have a different preference for their phospholipid substrate. Due to the preference of cPLA$_2$ for phospholipids with polyunsaturated fatty acid chains, and the relative abundance arachidonate esters among these, the predominant fatty acid product is arachidonic acid (AA).[38] AA is a soluble second messenger that diffuses into the cytoplasm. In addition to the signaling properties of its own, AA can be metabolized to generate eicosanoids (prostaglandins, leukotrienes a.o.) that in the immune system have inflammatory properties and in the nervous system, signal transducing activities.[39-42] LPL may promote the fluidity and curvature of the membrane. Both LPL and AA produced by cPLA$_2$ positively affect regulated exocytosis.[43,44] Interestingly, AA also promotes the formation of syntaxin1-containing SNARE complexes, suggesting a direct involvement in secretory potentiation.[15]

The most common activation mechanism of cPLA$_2$ involves Ca^{2+} binding to a N-terminally located C2 domain motif present in the α, β and δ isoforms.[38,45] Structural studies suggested that membrane recruitment of the enzyme is accompanied by opening of the catalytic center, referred to as 'interfacial activation'.[46] The cPLA$_2$-γ isoform lacks the C2 domain, but as an alternative membrane-binding mechanism it is post-translationally modified with a farnesyl residue.[47] PLA$_2$ activity can also be provoked by phosphorylation (e.g., by microtubule-associate protein kinases or calmodulin-dependent protein kinase II), protein-protein interactions (e.g., the intermediate filament vimentin), or synthesis of PIP$_2$ or PIP$_3$ in the membrane.[48,49]

How Central Signaling Molecules Modulate Exocytosis

As outlined above, extensive cross-talk exists between events in regulated exocytosis and regulation of various phospholipases. With little inaccuracy, these intricate signaling pathways can be simplified by the rule of thumb that the second messengers [Ca^{2+}]$_i$, DAG and IP$_3$ (and to a lesser extent, AA and PA) are indicators for the activity state of secretory cells. The activity of these central signaling molecules diverges onto a multitude of cellular processes, coordinately required to adapt the cellular logistics according to the biological demands. While ignoring a vast amount of signaling cascades that extend beyond regulated exocytosis, the following survey describes several key processes that provide means to modulate secretory strength.

IP$_3$ Mobilizes Ca^{2+} from Intracellular Stores

PI-PLC activity at the membrane produces inositol-triphosphate (IP$_3$) and diacylglycerol (DAG) from PIP$_2$-rich regions in the plasma membrane. The small hydrophilic molecule IP$_3$ diffuses into the cytoplasm and to the surface of intracellular Ca^{2+} stores. Here it recognizes an IP$_3$-gated Ca^{2+} channel, resulting in an ion current into the cytosol. The Ca^{2+} elevations induced by IP$_3$-generating agonists have been imaged in many exocrine cell types, and showed a characteristic oscillatory behavior.[50,51] This phenomenon is explained by a biphasic feedback mechanism that regulates the channel opening probability of the IP$_3$ receptor.[52] The IP$_3$-induced increase in [Ca^{2+}]$_i$ further enhances opening of the IP$_3$ receptor at low physiological Ca^{2+} levels (100-300 nM), but inhibits its opening at higher levels.[53] The presence of Ca^{2+} binding sites in regions outside the channel pore suggests that the IP$_3$ receptor may be directly regulated by Ca^{2+} ions.[54] However, a contribution of the Ca^{2+} binding protein calmodulin also exists.[55,56] Thus, IP$_3$ production by PI-PLC finally results in an elevation of the cytosolic Ca^{2+} concentration that can be local[57] and oscillatory[50,51] in nature. The cytosolic Ca^{2+} concentration subsequently acts as a potent modulator of exocytotic strength by mechanisms that will be described later in this chapter.

Divergent Signaling Activities of Diacylglycerol

While IP$_3$ is liberated from the plasma membrane by the activity of PI-PLC, the hydrophobic molecule DAG remains associated with the membrane. Besides acting as a second messenger on its target proteins, local DAG accumulation in the membrane also has a direct effect on the physical properties of the phospholipid bilayer. These changes include a partial dehydration of the membrane surface and increased curvature of phospholipid monolayers (leaflets),[58,59] and thereby enhances membrane fusion.[60]

The activity of DAG can be mimicked by synthetic phorbolesters such as 12-O-tetradecanoylphorbol-13-acetate (TPA), phorbol 12-myristate 13-acetate (PMA), or phorbol 12,13-dibutyrate (PDBU).[61] This has facilitated the characterization of the cellular processes that DAG induces. Besides processes that reorganize the cytoskeleton (e.g., ref. 48) , it is well established that DAG increases secretory strength.[49,50,51] Multiple DAG effector proteins include at least six protein families: protein kinase C (PKC), protein kinase D (PKD), DAG kinase, chimaerin, Ras guanyl nucleotide-releasing proteins and Munc13.[66]

An important property shared by all target proteins is the presence of a C1 domain (named after its initial identification as the first conserved domain in members of the PKC family).[67] The C1 domain physically interacts with DAG, an interaction that has been described to the level of molecular structures for rat PKC-γ[68] and mouse PKC-δ.[69] These studies indicate that DAG binding causes the C1 domain to be buried into the phospholipid bilayer. Several other structures of DAG effector domains in their unbound state indicate that the interaction must be associated with drastic conformational changes that are predicted to modulate the protein's activity.[70,71]

Divergent Signaling Activities of Intracellular Ca²⁺ Ions

Elevated concentrations of intracellular Ca^{2+} ions, originating either from the extracellular environment through Ca^{2+} channels in the plasma membrane, or from intracellular stores through channels on intracellular stores, play a key role in modulating secretory strength (as well as many other processes). A dynamic redistribution of Ca^{2+} ions occurs after their entry into the cytoplasm, governed by diffusion, buffering and active transport processes. In contrast to the extremely rapid event of synchronous vesicle fusion, regulatory activities of Ca^{2+} take place on a somewhat slower timescale (from seconds to hours or even longer) and are mostly responsive to Ca^{2+} fluctuations in the range of 100-1000 nM.[8,9,72,73] In this stage, the 'residual' Ca^{2+} concentration has assumed an essentially homogenous distribution in the cytosol.[10]

An extensive repertoire of Ca^{2+}-dependent proteins populates the mammalian proteome, and our current insight in their individual contributions to exocytotic regulation is by no means complete. The Ca^{2+} dependency is often mediated by structurally related domains within the proteins, the two most abundant being the EF-hand and the C2 domain. EF-hands are short helix-loop-helix motifs that usually function in pairs to bind Ca^{2+} ions, in many cases resulting in a conformational change that initiates protein interactions. The human genome encodes at least 230 annotated EF-hand proteins, including the archetypal protein calmodulin, many S100 proteins, calbindin D_{9k}, parvalbumin, recoverin, neuronal calcium sensor 1, and calcineurin.[77,78] Some functions of calmodulin will be touched upon below.

The C2 domain motif, unlike the EF-hand, plays a vital role in Ca^{2+}-regulated processes at membrane interfaces. Structural and functional studies of C2 domains in well-characterized Ca^{2+}-regulated proteins (protein kinase C-α, synaptotagmin I, cPLA2-α, PLCδ-1, and others) have suggested that stereotype C2 domains function as a Ca^{2+}-dependent switch for membrane binding.[74] In the case of phospholipases, the C2 domain-mediated activation of these enzymes is responsible for the increased production of DAG and IP₃, as detailed above. Other C2 domains are independent of Ca^{2+}, and have other activities at the phospholipid interface (e.g., PIP_2 binding or protein-protein interactions).[75,76] The Ca^{2+}- and DAG-dependent activity of protein kinase C plays an important role in the regulation of exocytosis, is extremely well documented, and will briefly be addressed below.

Central Role of Protein Kinase C in Divergent Signaling by DAG and Ca²⁺

As the first characterized DAG effector, protein kinase C (PKC) has been the subject of extensive investigation for over 25 years.[79] The mammalian PKC isoforms were grouped into the 'classical' (PKC-α, -βI, -βII and -γ), 'novel' (PKC-δ, -ϵ, -η, -θ) and 'atypical' isoforms (PKC-λ and -ζ). All PKC isoforms contain a C2 domain that is thought to mediate phospholipid interactions but only the classical isoforms are regulated by Ca^{2+}, presumably due to a different structural organization of the C2 domains.[80] Both the classical and novel PKC isoforms contain a C1 domain and are regulated by DAG, whereas the atypical isoforms lack C1 domains. Upon activation either by Ca^{2+}, DAG, or both, PKC is recruited to the membrane and undergoes a conformational change that involves the exposure of the kinase domain, allowing the phosphorylation of serine or threonine residues of substrate proteins.

The relevance of PKC has historically been somewhat overestimated by the assumption that it is the only effector activated by DAG / phorbolester, in combination with the use of pharmaca with limited specificity.[66,81-84] Still, it remains undisputed that PKC modulates secretory strength through many bonafide signaling processes, as confirmed by the generation of substrate proteins with phosphomimetic or phosphorylation-deficient mutations. Target phosphorylation by PKC is involved in the regulation of secretion from adrenal chromaffin cells,[85,86] GLUT4 vesicle exocytosis in adipocytes[87] and the neuron-specific isoform PKC-γ contributes essentially to long-term plasticity processes in the hippocampus and cerebellum.[88] Among the PKC substrates are several proteins with established roles in SNARE-dependent exocytosis including Munc18, SNAP-25, $Ca_v2.1$ and $Ca_v2.2$, synaptotagmin I and MARCKS / GAP43.

Central Role of Calmodulin in Ca^{2+} Signaling

One Ca^{2+} sensor protein of the EF-hand type that has been intensively studied for many years is calmodulin (CaM).[89] There is ample evidence that CaM - Ca^{2+} signaling has important effects on regulated exocytosis.[90,91] As a recurring theme, the divergent nature of calmodulin's signaling activities makes it impossible to ascribe these observations to a single molecular mechanism. Many calmodulin effectors are involved in different parts of the exocytotic pathway; recent attention has been paid to Munc13, synaptobrevin, synaptotagmin, Rab3 and ion channels.[92] Direct binding of Ca^{2+}/CaM to synaptobrevin, the vesicle SNARE, favors the interaction of synaptobrevin with 'trans' phospholipids on the target membrane, at the expense of 'cis' phospholipids on the vesicular membrane.[93,94] The Ca^{2+}-dependent interaction of CaM with Rab3 was found to be essential for the secretion-inhibiting capacity of an overexpressed, GTP-locked form of the GTPase.[95] In association with ion channels, CaM can function as a Ca^{2+}-dependent subunit to regulate Ca^{2+} and K^+ currents. This applies to Ca^{2+} and K^+ channels in excitable membranes,[96-98] as well as channels on intracellular Ca^{2+} stores (the IP_3 and ryanodine receptors).[99,100] Furthermore, the involvement of calmodulin-dependent kinases (CaMK) is of continued interest as they are required for the induction of long-term plasticity in neurons.[101-103] Introduction of calmodulin-dependent kinase induces synaptic potentiation by phosphorylating synaptic vesicle-associated proteins (e.g., synapsins, rabphilin) as well as other targets (e.g., the synprint site of voltage-dependent Ca^{2+} channels).

Role of Munc13 in Regulated Exocytosis

Munc13 is a mammalian protein family orthologous to UNC-13 in *C. elegans*. It binds directly to syntaxin,[104,105] and is essential for neurotransmission.[106-108] It is well established that Munc13 supports the priming step that converts immature vesicles to a fusion-ready state.[108,109] The predominant isoform Munc13-1 is expressed throughout the brain[110] and is essential for glutamatergic (excitatory) neurotransmission, but dispensable for GABA-ergic (inhibitory) neurotransmission.[108] Munc13-1 is also implicated in the secretion of acetylcholine from motor neurons,[111] insulin from pancreatic beta cells,[112] and catecholamines from chromaffin cells[109] (the latter upon Munc13-1 overexpression). Munc13-2 is coexpressed with Munc13-1 in some brain regions including the hippocampus, and required for GABA-ergic neurotransmission.[113] Within a single hippocampal neuron, a small amount of glutamatergic synapses are Munc13-1 independent. These synapses employ Munc13-2 and show a different type of short-term plasticity, ascribed to differences in Ca^{2+}- and PLC-dependent pathways.[114] A splice variant named ubMunc13-2 is expressed also outside the central nervous system. The third isoform, Munc13-3, is expressed together with Munc13-1 in the cerebellum. At synapses between granule cells and Purkinje cells in the cerebellum, Munc13-3 deficient mice showed a partial impairment in glutamatergic neurotransmission characterized by increased paired-pulse facilitation.[115] Munc13-4 is expressed in several peripheral tissues including spleen and thymus, and is required for the degranulation of lymphocytes, mast cells and platelets.[116-119] A structurally similar protein named BAP3 is regarded as a fifth member of the Munc13 family.[120] The brain-expressed isoforms Munc13-1/2/3 are localized to presynaptic active zones by interactions with RIM,[121] and thus participate in a large complex containing Munc13, RIM, CAST, bassoon and piccolo.[122,123] Both Rab3 and RIM are involved in long-term potentiation of mossy fiber synapses.[124] An indirect interaction with secretory vesicles through RIM and

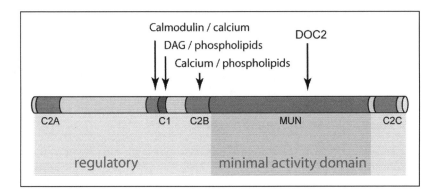

Figure 3. Domain organization of the vesicle priming factor Munc13-1. Colors indicate C1 domain (blue), C2 domains (red), MUN domain (green), and Ca^{2+} / calmodulin binding region (dark grey). The MUN domain represents an autonomously folding domain that can complement vesicle priming in Munc13-deficient cells. Interactions with putative regulatory molecules are indicated by arrows. The boundaries of the DOC2-interacting domain (Did) are nearly identical to that of the MUN domain. A color version of this figure is available online at www.Eurekah.com.

Rab3 may be involved in the acceptance of new synaptic vesicles for priming.[119,125] In hemapoietic cells, Munc13-4 interacts directly with the GTPase Rab27 on the secretory granules.[118,119]

Munc13 contains four elements that may provide sensitivity towards DAG and Ca^{2+}: a C1 domain, a DOC2 ('double C2')-interacting domain, a calmodulin binding domain, and two to three C2 domains (Fig. 3). The C1 domain is present in all three brain-expressed Munc13 isoforms, but not in Munc13-4, and serves as a target region for activation by DAG.[116] Although Munc13 proteins were initially neglected as a target of DAG signaling, a recent study showed convincingly that DAG-Munc13 signaling is essential for life.[81] Cultured neurons from knock-in mice that express a mutant Munc13-1 protein insensitive for DAG were capable of neurotransmission, but showed a defect in presynaptic potentiation induced by phorbolesters. This defect was associated with an accelerated synaptic depression during high frequency stimulation.[81] Upon DAG binding the C1 domain undergoes a conformational change expected to be critical for its function in the priming of secretory vesicles,[70] and enhances the affinity of Munc13 for the plasma membrane.[126]

At the same time, DAG induces binding of Munc13 to DOC2. By this mechanism, DOC2 cotranslocates with Munc13 to the membrane upon DAG synthesis.[127] DOC2 also acts as a Ca^{2+} sensor: the C2 domains of DOC2 isoforms α and β associate rapidly and reversibly with the membrane during intracellular Ca^{2+} elevation.[135] Overexpression of DOC2α increases exocytotic strength in chromaffin-like PC12 cells,[129] and mice lacking DOC2α showed abnormal short-term plasticity in hippocampal neurons.[130] Several studies have investigated the relevance of the Munc13-DOC2 interaction[131-133] and found that it contributes importantly to Ca^{2+}-dependent exocytosis from chromaffin-like PC12 cells,[133] acetylcholine release from cultured superior cervical ganglion neurons,[132] and phorbolester-induced potentiation in the giant synapse at the calyx of Held.[131]

The minimal domain required for the DOC2 interaction was defined as Did (for Doc2α-interacting domain).[133] This region is located between the second and third C2 domain. Similar, the region in the N-terminal domain of DOC2 that is required for Munc13 binding was named Mid (for Munc13-interacting domain). The Did domain in Munc13 contains significant sequence conservation. A fragment very similar to the Did domain (amino acids 851-1461 of Munc13-1), termed the MUN domain (a.a. 859-1531), could be expressed as a highly soluble recombinant protein in bacteria presumably because it folds autonomously from other domains.[134] This raised the question if it can also function autonomously. Confirming this, the MUN domain represented the minimal Munc13-1 fragment that can rescue vesicle priming activity in Munc13-1 and Munc13-2 double-knockout neurons.[134] In another study, a Munc13-1 fragment comprising amino acids 1100-1735 was found to be sufficient for priming activity in chromaffin cells.[135] Thus, the priming activity of Munc13-1 is confined to the C-terminal half, whereas the N-terminal half provides various modes of modulation, including Ca^{2+}- and DAG-sensitive interactions located in the C1 and C2 domains. DOC2 binding to the functional domain of Munc13 may further modulate its priming activity.[131]

Adding further to Munc13's Ca^{2+} sensitivity, a calmodulin-interacting domain was recently characterized that is located in the N-terminal (regulatory) half of the molecule.[136] The Munc13-calmodulin interaction is Ca^{2+} dependent. Mutant variants of Munc13-1 and -2 that specifically lack calmodulin binding were able to rescue neurotransmission in Munc13-1 and -2 double-deficient neurons.[136] At high stimulation frequencies however, the neurons expressing mutant Munc13 showed a severe defect in short-term augmentation, suggesting that the Ca^{2+}/calmodulin interaction is involved in Ca^{2+}-dependent replenishment of the vesicle pool. Finally, additional Ca^{2+} sensitivity of Munc13 proteins is suggested by the presence of three or two conserved C2 domains. Experimental evidence for their functionality is not yet available however. To summarize, Munc13 functions in vesicle priming, and is therefore essential for exocytosis. In addition it provides several handles for modulation by Ca^{2+} and DAG, thereby providing activity-dependent control over vesicle pool replenishment rates.

Role of Munc18 in Regulated Exocytosis

Sec1/Munc18 proteins, together referred to as the SM proteins, are syntaxin-binding proteins that are required for secretion at a step upstream from the exocytotic event. All orthologs within this family play important roles in regulated exocytosis: Sec1p in yeast, UNC18 in nematodes, Rop in flies, and Munc18 in mammals.[137] *Sec1* mutant yeast cells were isolated from a forward genetic screen aimed to identify genes with functions in the secretory pathway.[138] In yeast, *sec1* mutants show a temperature-sensitive secretion defect characterized by an accumulation of secretory vesicles. A secretion defect also occurs in neuromuscular synapses of *unc-18* mutant nematodes, that express the ortholog in the nervous system.[139] Although the synaptic terminals in mutant worms contain high numbers of synaptic vesicles, the amount of docked and primed vesicles appears to be reduced.[140] In mice lacking Munc18-1 neurotransmission is completely blocked, including both evoked and spontaneous fusion events, despite the presence of synaptic vesicles in nerve terminals.[141] Adrenal chromaffin cells also showed a severe secretion defect (a 10-fold reduction), combined with a substantial decrease in the amount of morphologically docked vesicles.[142] Besides Munc18-1, the mammalian genome also encodes other isoforms: Munc18-2, expressed in polarized epithelia, and Munc18c, expressed in pancreatic beta cells and adipocytes. In addition, the mouse SM proteins mVPS45, mVPS33 and mSly1 are more distant paralogs to Munc18-1.[137] Mice lacking Munc18c were early embryonic lethal, but heterozygous animals showed glucose intolerance when they received a high-fat diet. This phenotype presumably results from impaired exocytosis of insulin granules (in pancreatic β-cells after stimulation by glucose) and GLUT4 storage vesicles (in adipocytes after stimulation by insulin).[143] Combining the available data, it is clear that Munc18 functions upstream of Munc13-dependent vesicle priming, perhaps in vesicle docking. As soluble cytoplasmic proteins, the subcellular localization of SM proteins relies on interactions with proteins enriched at secretion sites: syntaxins and Mint/X11 proteins (for Munc18-interacting). Mint proteins have PDZ domains that bind to secretory specializations through interactions with neurexins and CASK.[144]

The best characterized molecular property of SM proteins is their high-affinity interaction with syntaxin. Interactions were observed only in particular combinations of Munc18 and syntaxin isoforms (e.g., Munc18-1 binds syntaxin-1, -2 or -3; Munc18c binds syntaxin-2 or -4). Munc18 is thought to function as a chaperone that is required for syntaxin stability, and may prevent premature SNARE pairing at improper locations in the membrane trafficking route. In free solution, syntaxin-1 can cycle between 'closed' and 'open' conformations by virtue of a flexible linker that separates the membrane-anchored SNARE domain from a three-helical H3 domain.[145] In the dimer with Munc18-1 syntaxin-1 is in its closed conformation, which precludes SNARE pairing with its partners, SNAP-25 and synaptobrevin-2.[146] This phenomenon has led to the hypothesis that the dissociation of Munc18-1 from syntaxin1 is necessary to make syntaxin's SNARE domain available for entry into the SNARE complex (Fig. 4).

Dissociation of the Munc18/syntaxin dimer is likely an active process that requires additional factors such as Munc13,[147] arachidonic acid,[15] tomosyn,[148] and/or protein kinases,[86,149] four possibilities that we will briefly explain below. In support of a role of Munc13 in disassembly of the Munc18/syntaxin complex, a mutant syntaxin variant that is forced into an open conformation in worms[145] rescues the paralyzed phenotype of *unc-13 (s69)* worms (deficient for the nematode ortholog of Munc13).[150] However the paralyzed phenotype of *unc-18* null worms could not be rescued by open syntaxin.[140] This result may relate to multiple independent observations that Munc18 also required for a second, syntaxin-independent process.[142,151-153] A role of arachidonic acid was recently suggested by the finding that this fatty acid promotes the transition of Munc18-dimerized syntaxin-1 to SNARE complexes in vitro.[15] Possibly, the presence of soluble poly-unsaturated fatty acids promotes the transient opening of syntaxin-1, which is subsequently stabilized by events that inhibit reentry into the Munc18-syntaxin dimer.[15] An additional factor that was reported to dissociate the Munc18-1/syntaxin-1 dimer is tomosyn.[148,154] Since tomosyn competes with synaptobrevin in SNARE-like complexes it seems unlikely however, that this activity of tomosyn contributes positively to SNARE fusion.[155] Rather, it was suggested by several independent studies that tomosyn has an inhibitory role in regulated exocytosis, perhaps by occupying t-SNARE complexes that are incorrect targets for vesicle delivery.[156] The fourth mode of Munc18/syntaxin dissociation may involve PKC-mediated phosphorylation of Munc18, that occurs in vitro on residues Ser[306] and Ser[313].[149] In situ Ser[313] phosphorylation was observed in response to phorbolester and other stimuli (histamine, intracellular Ca^{2+} increase, membrane depolarization) in chromaffin cells and synaptic terminals.[86,157,158] After PKC phosphorylation, Munc18-1 loses its affinity for syntaxin.[86,149] The same effect was accomplished by phosphomimetic mutations at Ser[306] and Ser[313]. Since syntaxin binding causes Munc18-1 to associate with the plasma membrane,[159] phosphorylation dissociated Munc18 from the membrane, an effect that was also observed for the syntaxin-4 binding isoform Munc18-3 (Munc18c) expressed in acinar cells.[160,161] Due to

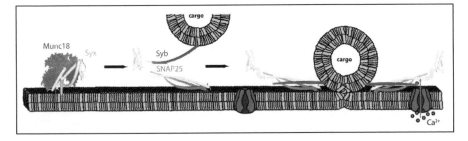

Figure 4. A working hypothesis of Munc18 and Munc13 function in regulated assembly of the neuronal SNARE complex. Munc18 functions upstream of vesicle docking / priming by stabilizing syntaxin (Syx) in its closed conformation. Dissociation of the Munc18-syntaxin dimer occurs by an unknown mechanism, in which arachidonic acid (AA), Munc13 and protein kinase C (PKC) may be involved. The SNARE domain of syntaxin then becomes available for entry into the SNARE complex, also containing SNAP25 and synaptobrevin (Syb). Fusion-ready vesicles are tethered in the vicinity of voltage-dependent Ca^{2+} channels (VDCC), where their release probability remains low until secretion is triggered by the entry of Ca^{2+} ions. The Ca^{2+}-sensing apparatus that mediates the fusion event is not depicted.

high protein phosphatase activities in the cytosol the phosphorylation is only transient.[157,158] This suggests that Munc18 cycles between a phosphorylated and nonphosphorylated state to exert its docking / priming activity.

Ultrastructural analyses in central neurons and chromaffin cells lacking Munc18-1 have produced different results with respect to the number of morphologically docked vesicles.[141,142] Central neurons analyzed at stage E16 completely lacked synaptic vesicle release, but their terminals showed normal numbers of morphologically docked vesicles.[141] In contrast, adrenal chromaffin cells lacking Munc18-1 showed a strong defect in vesicle docking.[142] A large reduction of large dense-cored vesicle (LDCV) docking was also observed in peptidergic cells of the pituitary gland,[153] suggesting a mechanistic difference between the docking and secretion of synaptic vesicles and LDCVs. It is not yet clear if the apparent function of Munc18 in vesicle docking represents a second, syntaxin-independent function of the protein, e.g., via its interaction with Mint,[159] or should be viewed as a consequence of its syntaxin-binding activity. To illustrate the latter possibility, the tethering of vesicles to their target membrane could require the presence of Munc18/syntaxin dimers, or the formation of intermediate protein complexes prior to ternary SNARE complex assembly (e.g., a syntaxin/SNAP25 pair).[162]

Taken together, Munc18 contributes essentially to regulated exocytosis by controlling a step upstream from SNARE-dependent membrane fusion. The Munc18/syntaxin complex is likely centrally involved in its function, in conjunction with the activities of Munc13, fatty acids, protein kinases, Mint, and perhaps other factors. While the function of SM proteins is most accurately described in the context of classical neurotransmitter release, there may be substantial variation in the molecular mechanisms that govern peptidergic secretion from LDCVs. The mechanism may also differ between organisms. For example, Sec1p in yeast can also bind syntaxin when it resides in the SNARE complex,[163] and in vertebrates, the principle of syntaxin opening and closing appears to be incompletely conserved among paralogs.[164]

Final Remarks

As the molecular mechanism for excitation-secretion coupling becomes increasingly well understood, it is also evident that a significant proportion of the exocytotic protein machinery is dedicated to the strict modulation of this process. Whereas the ternary SNARE complex is sufficient to drive membrane fusion of liposomes in solution, secretion from living cells requires additional proteins including Munc18 and Munc13, that may cooperatively regulate SNARE complex assembly. Recent findings support the possibility that Munc13 and Munc18 are regulated by Ca^{2+}, diacylglycerol, calmodulin, DOC2, arachidonic acid, and protein kinases. Future investigations should address the question how the current model for the controlled assembly of neuronal SNARE complex can be extrapolated to other types of regulated exocytosis. Furthermore, it should be of primary interest to test which regulatory mechanisms are most biologically relevant for neuro-endocrine function in an organismal context.

References

1. Chen YA, Scales SJ, Patel SM et al. SNARE complex formation is triggered by Ca2+ and drives membrane fusion. Cell 1999; 97:165-174.
2. Weber T, Zemelman BV, McNew JA et al. SNARE-pins: Minimal machinery for membrane fusion. Cell 1998; 92:759-772.
3. Nickel W, Weber T, McNew JA et al. Content mixing and membrane integrity during membrane fusion driven by pairing of isolated v-SNAREs and t-SNAREs. Proc Natl Acad Sci USA 1999; 96:12571-12576.
4. Fernandez-Chacon R, Konigstorfer A, Gerber SH et al. Synaptotagmin I functions as a calcium regulator of release probability. Nature 2001; 410:41-49.
5. Brose N, Petrenko AG, Sudhof TC et al. Synaptotagmin: A calcium sensor on the synaptic vesicle surface. Science 1992; 256:1021-1025.
6. Reim K, Mansour M, Varoqueaux F et al. Complexins regulate a late step in Ca2+-dependent neurotransmitter release. Cell 2001; 104:71-81.
7. Sudhof TC. The synaptic vesicle cycle. Annu Rev Neurosci 2004; 27:509-547.

8. Korogod N, Lou X, Schneggenburger R. Presynaptic Ca2+ requirements and developmental regulation of posttetanic potentiation at the calyx of Held. J Neurosci 2005; 25:5127-5137.
9. Meinrenken CJ, Borst JG, Sakmann B. Local routes revisited: The space and time dependence of the Ca2+ signal for phasic transmitter release at the rat calyx of Held. J Physiol 2003; 547:665-689.
10. Zucker RS. Calcium- and activity-dependent synaptic plasticity. Curr Opin Neurobiol 1999; 9:305-313.
11. Lang J, Nishimoto I, Okamoto T et al. Direct control of exocytosis by receptor-mediated activation of the heterotrimeric GTPases Gi and G(o) or by the expression of their active G alpha subunits. EMBO J 1995; 14:3635-3644.
12. Marley PD. Mechanisms in histamine-mediated secretion from adrenal chromaffin cells. Pharmacol Ther 2003; 98:1-34.
13. Chen XK, Wang LC, Zhou Y et al. Activation of GPCRs modulates quantal size in chromaffin cells through G(betagamma) and PKC. Nat Neurosci 2005; 8:1160-1168.
14. Wilde JI, Watson SP. Regulation of phospholipase C gamma isoforms in haematopoietic cells: Why one, not the other? Cell Signal 2001; 13:691-701.
15. Rickman C, Davletov B. Arachidonic acid allows SNARE complex formation in the presence of Munc18. Chem Biol 2005; 12:545-553.
16. De Matteis MA, Godi A. PI-loting membrane traffic. Nat Cell Biol 2004; 6:487-492.
17. Aikawa Y, Martin TF. ARF6 regulates a plasma membrane pool of phosphatidylinositol(4,5) bisphosphate required for regulated exocytosis. J Cell Biol 2003; 162:647-659.
18. Smrcka AV, Hepler JR, Brown KO et al. Regulation of polyphosphoinositide-specific phospholipase C activity by purified Gq. Science 1991; 251:804-807.
19. Taylor SJ, Chae HZ, Rhee SG et al. Activation of the beta 1 isozyme of phospholipase C by alpha subunits of the Gq class of G proteins. Nature 1991; 350:516-518.
20. Lee CH, Park D, Wu D et al. Members of the Gq alpha subunit gene family activate phospholipase C beta isozymes. J Biol Chem 1992; 267:16044-16047.
21. Lee SB, Shin SH, Hepler JR et al. Activation of phospholipase C-beta 2 mutants by G protein alpha q and beta gamma subunits. J Biol Chem 1993; 268:25952-25957.
22. Park D, Jhon DY, Lee CW et al. Activation of phospholipase C isozymes by G protein beta gamma subunits. J Biol Chem 1993; 268:4573-4576.
23. Camps M, Carozzi A, Schnabel P et al. Isozyme-selective stimulation of phospholipase C-beta 2 by G protein beta gamma-subunits. Nature 1992; 360:684-686.
24. Wang T, Pentyala S, Rebecchi MJ et al. Differential association of the pleckstrin homology domains of phospholipases C-beta 1, C-beta 2, and C-delta 1 with lipid bilayers and the beta gamma subunits of heterotrimeric G proteins. Biochemistry 1999; 38:1517-1524.
25. Rhee SG. Regulation of phosphoinositide-specific phospholipase C. Annu Rev Biochem 2001; 70:281-312.
26. Kim HK, Kim JW, Zilberstein A et al. PDGF stimulation of inositol phospholipid hydrolysis requires PLC-gamma 1 phosphorylation on tyrosine residues 783 and 1254. Cell 1991; 65:435-441.
27. Tallquist M, Kazlauskas A. PDGF signaling in cells and mice. Cytokine Growth Factor Rev 2004; 15:205-213.
28. Watson RT, Kanzaki M, Pessin JE. Regulated membrane trafficking of the insulin-responsive glucose transporter 4 in adipocytes. Endocr Rev 2004; 25:177-204.
29. Grobler JA, Essen LO, Williams RL et al. C2 domain conformational changes in phospholipase C-delta 1. Nat Struct Biol 1996; 3:788-795.
30. Grobler JA, Hurley JH. Catalysis by phospholipase C delta1 requires that Ca2+ bind to the catalytic domain, but not the C2 domain. Biochemistry 1998; 37:5020-5028.
31. Yamamoto T, Takeuchi H, Kanematsu T et al. Involvement of EF hand motifs in the Ca(2+)-dependent binding of the pleckstrin homology domain to phosphoinositides. Eur J Biochem 1999; 265:481-490.
32. Lomasney JW, Cheng HF, Roffler SR et al. Activation of phospholipase C delta1 through C2 domain by a Ca(2+)-enzyme-phosphatidylserine ternary complex. J Biol Chem 1999; 274:21995-22001.
33. Essen LO, Perisic O, Cheung R et al. Crystal structure of a mammalian phosphoinositide-specific phospholipase C delta. Nature 1996; 380:595-602.
34. Lopez I, Mak EC, Ding J et al. A novel bifunctional phospholipase c that is regulated by Galpha 12 and stimulates the Ras/mitogen-activated protein kinase pathway. J Biol Chem 2001; 276:2758-2765.
35. Wing MR, Bourdon DM, Harden TK. PLC-epsilon: A shared effector protein in Ras-, Rho-, and G alpha beta gamma-mediated signaling. Mol Interv 2003; 3:273-280.

36. Balsinde J, Winstead MV, Dennis EA. Phospholipase A(2) regulation of arachidonic acid mobilization. FEBS Lett 2002; 531:2-6.
37. Brown WJ, Chambers K, Doody A. Phospholipase A2 (PLA2) enzymes in membrane trafficking: Mediators of membrane shape and function. Traffic 2003; 4:214-221.
38. Clark JD, Lin LL, Kriz RW et al. A novel arachidonic acid-selective cytosolic PLA2 contains a Ca(2+)-dependent translocation domain with homology to PKC and GAP. Cell 1991; 65:1043-1051.
39. Malik KU, Sehic E. Prostaglandins and the release of the adrenergic transmitter. Ann NY Acad Sci 1990; 604:222-236.
40. Matsumoto Y, Yamaguchi T, Watanabe S et al. Involvement of arachidonic acid cascade in working memory impairment induced by interleukin-1 beta. Neuropharmacology 2004; 46:1195-1200.
41. Sublette ME, Russ MJ, Smith GS. Evidence for a role of the arachidonic acid cascade in affective disorders: A review. Bipolar Disord 2004; 6:95-105.
42. Farooqui AA, Horrocks LA. Brain phospholipases A2: A perspective on the history. Prostaglandins Leukot Essent Fatty Acids 2004; 71:161-169.
43. Juhl K, Hoy M, Olsen HL et al. cPLA2alpha-evoked formation of arachidonic acid and lysophospholipids is required for exocytosis in mouse pancreatic beta-cells. Am J Physiol Endocrinol Metab 2003; 285:E73-81.
44. Farooqui AA, Yang HC, Rosenberger TA et al. Phospholipase A2 and its role in brain tissue. J Neurochem 1997; 69:889-901.
45. Nalefski EA, Slazas MM, Falke JJ. Ca2+-signaling cycle of a membrane-docking C2 domain. Biochemistry 1997; 36:12011-12018.
46. Dessen A, Tang J, Schmidt H et al. Crystal structure of human cytosolic phospholipase A2 reveals a novel topology and catalytic mechanism. Cell 1999; 97:349-360.
47. Murakami M, Masuda S, Kudo I. Arachidonate release and prostaglandin production by group IVC phospholipase A2 (cytosolic phospholipase A2gamma). Biochem J 2003; 372:695-702.
48. Six DA, Dennis EA. Essential Ca(2+)-independent role of the group IVA cytosolic phospholipase A(2) C2 domain for interfacial activity. J Biol Chem 2003; 278:23842-23850.
49. Hirabayashi T, Murayama T, Shimizu T. Regulatory mechanism and physiological role of cytosolic phospholipase A2. Biol Pharm Bull 2004; 27:1168-1173.
50. Thorn P, Lawrie AM, Smith PM et al. Local and global cytosolic Ca2+ oscillations in exocrine cells evoked by agonists and inositol trisphosphate. Cell 1993; 74:661-668.
51. Kasai H, Li YX, Miyashita Y. Subcellular distribution of Ca2+ release channels underlying Ca2+ waves and oscillations in exocrine pancreas. Cell 1993; 74:669-677.
52. Bezprozvanny I, Watras J, Ehrlich BE. Bell-shaped calcium-response curves of Ins(1,4,5)P3- and calcium-gated channels from endoplasmic reticulum of cerebellum. Nature 1991; 351:751-754.
53. Patterson RL, Boehning D, Snyder SH. Inositol 1,4,5-trisphosphate receptors as signal integrators. Annu Rev Biochem 2004; 73:437-465.
54. Tu H, Nosyreva E, Miyakawa T et al. Functional and biochemical analysis of the type 1 inositol (1,4,5)-trisphosphate receptor calcium sensor. Biophys J 2003; 85:290-299.
55. Kelly PT, Mackinnon IInd RL, Dietz RV et al. Postsynaptic IP3 receptor-mediated Ca2+ release modulates synaptic transmission in hippocampal neurons. Brain Res Mol Brain Res 2005; 135:232-248.
56. Taylor CW, Laude AJ. IP3 receptors and their regulation by calmodulin and cytosolic Ca2+. Cell Calcium 2002; 32:321-334.
57. Miyata M, Finch EA, Khiroug L et al. Local calcium release in dendritic spines required for long-term synaptic depression. Neuron 2000; 28:233-244.
58. Goni FM, Alonso A. Structure and functional properties of diacylglycerols in membranes. Prog Lipid Res 1999; 38:1-48.
59. Hindenes JO, Nerdal W, Guo W et al. Physical properties of the transmembrane signal molecule, sn-1-stearoyl 2-arachidonoylglycerol. Acyl chain segregation and its biochemical implications. J Biol Chem 2000; 275:6857-6867.
60. Sanchez-Migallon MP, Aranda FJ, Gomez-Fernandez JC. Role of phosphatidylserine and diacylglycerol in the fusion of chromaffin granules with target membranes. Arch Biochem Biophys 1994; 314:205-216.
61. Castagna M, Takai Y, Kaibuchi K et al. Direct activation of calcium-activated, phospholipid-dependent protein kinase by tumor-promoting phorbol esters. J Biol Chem 1982; 257:7847-7851.
62. Shariff A, Luna EJ. Diacylglycerol-stimulated formation of actin nucleation sites at plasma membranes. Science 1992; 256:245-247.
63. Zoukhri D, Sergheraert C, Rossignol B. Phorbol ester-stimulated exocytosis in lacrimal gland: PKC might not be the sole effector. Am J Physiol 1993; 264:C1045-1050.

64. Rose SD, Lejen T, Zhang L et al. Chromaffin cell F-actin disassembly and potentiation of cat-echolamine release in response to protein kinase C activation by phorbol esters is mediated through myristoylated alanine-rich C kinase substrate phosphorylation. J Biol Chem 2001; 276:36757-36763.
65. Lou X, Scheuss V, Schneggenburger R. Allosteric modulation of the presynaptic Ca2+ sensor for vesicle fusion. Nature 2005; 435:497-501.
66. Brose N, Betz A, Wegmeyer H. Divergent and convergent signaling by the diacylglycerol second messenger pathway in mammals. Curr Opin Neurobiol 2004; 14:328-340.
67. Hurley JH, Newton AC, Parker PJ et al. Taxonomy and function of C1 protein kinase C homol-ogy domains. Protein Sci 1997; 6:477-480.
68. Xu RX, Pawelczyk T, Xia TH et al. NMR structure of a protein kinase C-gamma phorbol-binding domain and study of protein-lipid micelle interactions. Biochemistry 1997; 36:10709-10717.
69. Zhang G, Kazanietz MG, Blumberg PM et al. Crystal structure of the cys2 activator-binding do-main of protein kinase C delta in complex with phorbol ester. Cell 1995; 81:917-924.
70. Shen N, Guryev O, Rizo J. Intramolecular occlusion of the diacylglycerol-binding site in the C1 domain of munc13-1. Biochemistry 2005; 44:1089-1096.
71. Canagarajah B, Leskow FC, Ho JY et al. Structural mechanism for lipid activation of the Rac-specific GAP, beta2-chimaerin. Cell 2004; 119:407-418.
72. Ales E, Fuentealba J, Garcia AG et al. Depolarization evokes different patterns of calcium signals and exocytosis in bovine and mouse chromaffin cells: The role of mitochondria. Eur J Neurosci 2005; 21:142-150.
73. Villalobos C, Nunez L, Montero M et al. Redistribution of Ca2+ among cytosol and organella during stimulation of bovine chromaffin cells. Faseb J 2002; 16:343-353.
74. Rizo J, Sudhof TC. C2-domains, structure and function of a universal Ca2+-binding domain. J Biol Chem 1998; 273:15879-15882.
75. Bai J, Tucker WC, Chapman ER. PIP2 increases the speed of response of synaptotagmin and steers its membrane-penetration activity toward the plasma membrane. Nat Struct Mol Biol 2004; 11:36-44.
76. Tsuboi T, Fukuda M. The C2B domain of rabphilin directly interacts with SNAP-25 and regu-lates the docking step of dense-core vesicle exocytosis in PC12 cells. J Biol Chem 2005.
77. Haiech J, Moulhaye SB, Kilhoffer MC. The EF-Handome: Combining comparative genomic study using FamDBtool, a new bioinformatics tool, and the network of expertise of the European Cal-cium Society. Biochim Biophys Acta 2004; 1742:179-183.
78. Lewit-Bentley A, Rety S. EF-hand calcium-binding proteins. Curr Opin Struct Biol 2000; 10:637-643.
79. Takai Y, Kishimoto A, Iwasa Y et al. Calcium-dependent activation of a multifunctional protein kinase by membrane phospholipids. J Biol Chem 1979; 254:3692-3695.
80. Ochoa WF, Garcia-Garcia J, Fita I et al. Structure of the C2 domain from novel protein kinase Cepsilon. A membrane binding model for Ca(2+)-independent C2 domains. J Mol Biol 2001; 311:837-849.
81. Rhee JS, Betz A, Pyott S et al. Beta phorbol ester- and diacylglycerol-induced augmentation of transmitter release is mediated by Munc13s and not by PKCs. Cell 2002; 108:121-133.
82. Brose N, Rosenmund C. Move over protein kinase C, you've got company: Alternative cellular effectors of diacylglycerol and phorbol esters. J Cell Sci 2002; 115:4399-4411.
83. Silinsky EM, Searl TJ. Phorbol esters and neurotransmitter release: More than just protein kinase C? Br J Pharmacol 2003; 138:1191-1201.
84. Yang C, Kazanietz MG. Divergence and complexities in DAG signaling: Looking beyond PKC. Trends Pharmacol Sci 2003; 24:602-608.
85. Nagy G, Matti U, Nehring RB et al. Protein kinase C-dependent phosphorylation of synaptosome-associated protein of 25 kDa at Ser187 potentiates vesicle recruitment. J Neurosci 2002; 22:9278-9286.
86. Barclay JW, Craig TJ, Fisher RJ et al. Phosphorylation of Munc18 by protein kinase C regulates the kinetics of exocytosis. J Biol Chem 2003; 278:10538-10545.
87. Mendez CF, Leibiger IB, Leibiger B et al. Rapid association of protein kinase C-epsilon with insulin granules is essential for insulin exocytosis. J Biol Chem 2003; 278:44753-44757.
88. Abeliovich A, Paylor R, Chen C et al. PKC gamma mutant mice exhibit mild deficits in spatial and contextual learning. Cell 1993; 75:1263-1271.
89. Schulman H, Greengard P. Stimulation of brain membrane protein phosphorylation by calcium and an endogenous heat-stable protein. Nature 1978; 271:478-479.
90. Steinhardt RA, Alderton JM. Calmodulin confers calcium sensitivity on secretory exocytosis. Na-ture 1982; 295:154-155.

91. Chen YA, Duvvuri V, Schulman H et al. Calmodulin and protein kinase C increase Ca(2+)-stimulated secretion by modulating membrane-attached exocytic machinery. J Biol Chem 1999; 274:26469-26476.
92. Burgoyne RD, Clague MJ. Calcium and calmodulin in membrane fusion. Biochim Biophys Acta 2003; 1641:137-143.
93. Quetglas S, Leveque C, Miquelis R et al. Ca2+-dependent regulation of synaptic SNARE complex assembly via a calmodulin- and phospholipid-binding domain of synaptobrevin. Proc Natl Acad Sci USA 2000; 97:9695-9700.
94. de Haro L, Ferracci G, Opi S et al. Ca2+/calmodulin transfers the membrane-proximal lipid-binding domain of the v-SNARE synaptobrevin from cis to trans bilayers. Proc Natl Acad Sci USA 2004; 101:1578-1583.
95. Coppola T, Perret-Menoud V, Luthi S et al. Disruption of Rab3-calmodulin interaction, but not other effector interactions, prevents Rab3 inhibition of exocytosis. EMBO J 1999; 18:5885-5891.
96. Schumacher MA, Rivard AF, Bachinger HP et al. Structure of the gating domain of a Ca2+-activated K+ channel complexed with Ca2+/calmodulin. Nature 2001; 410:1120-1124.
97. Lee A, Zhou H, Scheuer T et al. Molecular determinants of Ca(2+)/calmodulin-dependent regulation of Ca(v)2.1 channels. Proc Natl Acad Sci USA 2003; 100:16059-16064.
98. Mori MX, Erickson MG, Yue DT. Functional stoichiometry and local enrichment of calmodulin interacting with Ca2+ channels. Science 2004; 304:432-435.
99. Moreau B, Straube S, Fisher RJ et al. Ca2+-calmodulin-dependent facilitation and Ca2+ inactivation of Ca2+ release-activated Ca2+ channels. J Biol Chem 2005; 280:8776-8783.
100. Zhu X, Ghanta J, Walker JW et al. The calmodulin binding region of the skeletal ryanodine receptor acts as a self-modulatory domain. Cell Calcium 2004; 35:165-177.
101. Wei F, Qiu CS, Liauw J et al. Calcium calmodulin-dependent protein kinase IV is required for fear memory. Nat Neurosci 2002; 5:573-579.
102. Ninan I, Arancio O. Presynaptic CaMKII is necessary for synaptic plasticity in cultured hippocampal neurons. Neuron 2004; 42:129-141.
103. Hinds HL, Goussakov I, Nakazawa K et al. Essential function of alpha-calcium/calmodulin-dependent protein kinase II in neurotransmitter release at a glutamatergic central synapse. Proc Natl Acad Sci USA 2003; 100:4275-4280.
104. Garcia EP, Gatti E, Butler M et al. A rat brain Sec1 homologue related to Rop and UNC18 interacts with syntaxin. Proc Natl Acad Sci USA 1994; 91:2003-2007.
105. Betz A, Okamoto M, Benseler F et al. Direct interaction of the rat unc-13 homologue Munc13-1 with the N terminus of syntaxin. J Biol Chem 1997; 272:2520-2526.
106. Richmond JE, Davis WS, Jorgensen EM. UNC-13 is required for synaptic vesicle fusion in C. elegans. Nat Neurosci 1999; 2:959-964.
107. Aravamudan B, Fergestad T, Davis WS et al. Drosophila UNC-13 is essential for synaptic transmission. Nat Neurosci 1999; 2:965-971.
108. Augustin I, Rosenmund C, Sudhof TC et al. Munc13-1 is essential for fusion competence of glutamatergic synaptic vesicles. Nature 1999; 400:457-461.
109. Ashery U, Varoqueaux F, Voets T et al. Munc13-1 acts as a priming factor for large dense-core vesicles in bovine chromaffin cells. EMBO J 2000; 19:3586-3596.
110. Augustin I, Betz A, Herrmann C et al. Differential expression of two novel Munc13 proteins in rat brain. Biochem J 1999; 337(Pt 3):363-371.
111. Varoqueaux F, Sons MS, Plomp JJ et al. Aberrant morphology and residual transmitter release at the Munc13-deficient mouse neuromuscular synapse. Mol Cell Biol 2005; 25:5973-5984.
112. Sheu L, Pasyk EA, Ji J et al. Regulation of insulin exocytosis by Munc13-1. J Biol Chem 2003; 278:27556-27563.
113. Varoqueaux F, Sigler A, Rhee JS et al. Total arrest of spontaneous and evoked synaptic transmission but normal synaptogenesis in the absence of Munc13-mediated vesicle priming. Proc Natl Acad Sci USA 2002; 99:9037-9042.
114. Rosenmund C, Sigler A, Augustin I et al. Differential control of vesicle priming and short-term plasticity by Munc13 isoforms. Neuron 2002; 33:411-424.
115. Augustin I, Korte S, Rickmann M et al. The cerebellum-specific Munc13 isoform Munc13-3 regulates cerebellar synaptic transmission and motor learning in mice. J Neurosci 2001; 21:10-17.
116. Koch H, Hofmann K, Brose N. Definition of Munc13-homology-domains and characterization of a novel ubiquitously expressed Munc13 isoform. Biochem J 2000; 349:247-253.
117. Feldmann J, Callebaut I, Raposo G et al. Munc13-4 is essential for cytolytic granules fusion and is mutated in a form of familial hemophagocytic lymphohistiocytosis (FHL3). Cell 2003; 115:461-473.
118. Neeft M, Wieffer M, de Jong AS et al. Munc13-4 is an effector of rab27a and controls secretion of lysosomes in hematopoietic cells. Mol Biol Cell 2005; 16:731-741.

119. Shirakawa R, Higashi T, Tabuchi A et al. Munc13-4 is a GTP-Rab27-binding protein regulating dense core granule secretion in platelets. J Biol Chem 2004; 279:10730-10737.
120. Shiratsuchi T, Oda K, Nishimori H et al. Cloning and characterization of BAP3 (BAI-associated protein 3), a C2 domain-containing protein that interacts with BAI1. Biochem Biophys Res Commun 1998; 251:158-165.
121. Betz A, Thakur P, Junge HJ et al. Functional interaction of the active zone proteins Munc13-1 and RIM1 in synaptic vesicle priming. Neuron 2001; 30:183-196.
122. Dresbach T, Qualmann B, Kessels MM et al. The presynaptic cytomatrix of brain synapses. Cell Mol Life Sci 2001; 58:94-116.
123. Takao-Rikitsu E, Mochida S, Inoue E et al. Physical and functional interaction of the active zone proteins, CAST, RIM1, and Bassoon, in neurotransmitter release. J Cell Biol 2004; 164:301-311.
124. Castillo PE, Schoch S, Schmitz F et al. RIM1alpha is required for presynaptic long-term potentiation. Nature 2002; 415:327-330.
125. Dulubova I, Lou X, Lu J et al. A Munc13/RIM/Rab3 tripartite complex: From priming to plasticity? EMBO J 2005; 24:2839-2850.
126. Betz A, Ashery U, Rickmann M et al. Munc13-1 is a presynaptic phorbol ester receptor that enhances neurotransmitter release. Neuron 1998; 21:123-136.
127. Duncan RR, Betz A, Shipston MJ et al. Transient, phorbol ester-induced DOC2-Munc13 interactions in vivo. J Biol Chem 1999; 274:27347-27350.
128. Groffen AJ, Brian EC, Dudok JJ et al. Ca(2+)-induced recruitment of the secretory vesicle protein DOC2B to the target membrane. J Biol Chem 2004; 279:23740-23747.
129. Orita S, Sasaki T, Komuro R et al. Doc2 enhances Ca2+-dependent exocytosis from PC12 cells. J Biol Chem 1996; 271:7257-7260.
130. Sakaguchi G, Manabe T, Kobayashi K et al. Doc2alpha is an activity-dependent modulator of excitatory synaptic transmission. Eur J Neurosci 1999; 11:4262-4268.
131. Hori T, Takai Y, Takahashi T. Presynaptic mechanism for phorbol ester-induced synaptic potentiation. J Neurosci 1999; 19:7262-7267.
132. Mochida S, Orita S, Sakaguchi G et al. Role of the Doc2 alpha-Munc13-1 interaction in the neurotransmitter release process. Proc Natl Acad Sci USA 1998; 95:11418-11422.
133. Orita S, Naito A, Sakaguchi G et al. Physical and functional interactions of Doc2 and Munc13 in Ca2+-dependent exocytotic machinery. J Biol Chem 1997; 272:16081-16084.
134. Basu J, Shen N, Dulubova I et al. A minimal domain responsible for Munc13 activity. Nat Struct Mol Biol 2005.
135. Stevens DR, Wu ZX, Matti U et al. Identification of the minimal protein domain required for priming activity of munc13-1. Curr Biol 2005; 15:2243-2248.
136. Junge HJ, Rhee JS, Jahn O et al. Calmodulin and Munc13 form a Ca2+ sensor/effector complex that controls short-term synaptic plasticity. Cell 2004; 118:389-401.
137. Toonen RF, Verhage M. Vesicle trafficking: Pleasure and pain from SM genes. Trends Cell Biol 2003; 13:177-186.
138. Novick P, Field C, Schekman R. Identification of 23 complementation groups required for post-translational events in the yeast secretory pathway. Cell 1980; 21:205-215.
139. Hosono R, Sassa T, Kuno S. Mutations affecting acetylcholine levels in the nematode Caenorhabditis elegans. J Neurochem 1987; 49:1820-1823.
140. Weimer RM, Richmond JE, Davis WS et al. Defects in synaptic vesicle docking in unc-18 mutants. Nat Neurosci 2003; 6:1023-1030.
141. Verhage M, Maia AS, Plomp JJ et al. Synaptic assembly of the brain in the absence of neurotransmitter secretion. Science 2000; 287:864-869.
142. Voets T, Toonen RF, Brian EC et al. Munc18-1 promotes large dense-core vesicle docking. Neuron 2001; 31:581-591.
143. Oh E, Spurlin BA, Pessin JE et al. Munc18c heterozygous knockout mice display increased susceptibility for severe glucose intolerance. Diabetes 2005; 54:638-647.
144. Biederer T, Sudhof TC. Mints as adaptors. Direct binding to neurexins and recruitment of munc18. J Biol Chem 2000; 275:39803-39806.
145. Dulubova I, Sugita S, Hill S et al. A conformational switch in syntaxin during exocytosis: Role of munc18. EMBO J 1999; 18:4372-4382.
146. Misura KM, Scheller RH, Weis WI. Three-dimensional structure of the neuronal-Sec1-syntaxin 1a complex. Nature 2000; 404:355-362.
147. Sassa T, Harada S, Ogawa H et al. Regulation of the UNC-18-Caenorhabditis elegans syntaxin complex by UNC-13. J Neurosci 1999; 19:4772-4777.
148. Fujita Y, Shirataki H, Sakisaka T et al. Tomosyn: A syntaxin-1-binding protein that forms a novel complex in the neurotransmitter release process. Neuron 1998; 20:905-915.

149. Fujita Y, Sasaki T, Fukui K et al. Phosphorylation of Munc-18/n-Sec1/rbSec1 by protein kinase C: Its implication in regulating the interaction of Munc-18/n-Sec1/rbSec1 with syntaxin. J Biol Chem 1996; 271:7265-7268.
150. Richmond JE, Weimer RM, Jorgensen EM. An open form of syntaxin bypasses the requirement for UNC-13 in vesicle priming. Nature 2001; 412:338-341.
151. Ciufo LF, Barclay JW, Burgoyne RD et al. Munc18-1 regulates early and late stages of exocytosis via syntaxin-independent protein interactions. Mol Biol Cell 2005; 16:470-482.
152. Weimer RM, Richmond JE. Synaptic vesicle docking: A putative role for the Munc18/Sec1 protein family. Curr Top Dev Biol 2005; 65:83-113.
153. Korteweg N, Maia AS, Thompson B et al. The role of Munc18-1 in docking and exocytosis of peptide hormone vesicles in the anterior pituitary. Biol Cell 2005; 97:445-455.
154. Groffen AJ, Jacobsen L, Schut D et al. Two distinct genes drive expression of seven tomosyn isoforms in the mammalian brain, sharing a conserved structure with a unique variable domain. J Neurochem 2005; 92:554-568.
155. Pobbati AV, Razeto A, Boddener M et al. Structural basis for the inhibitory role of tomosyn in exocytosis. J Biol Chem 2004; 279:47192-47200.
156. Sakisaka T, Baba T, Tanaka S et al. Regulation of SNAREs by tomosyn and ROCK: Implication in extension and retraction of neurites. J Cell Biol 2004; 166:17-25.
157. de Vries KJ, Geijtenbeek A, Brian EC et al. Dynamics of munc18-1 phosphorylation/dephosphorylation in rat brain nerve terminals. Eur J Neurosci 2000; 12:385-390.
158. Craig TJ, Evans GJ, Morgan A. Physiological regulation of Munc18/nSec1 phosphorylation on serine-313. J Neurochem 2003; 86:1450-1457.
159. Schutz SF, Lang T et al. A dual function for Munc-18 in exocytosis of PC12 cells. Eur J Neurosci 2005; 21:2419-2432.
160. Gaisano HY, Lutz MP, Leser J et al. Supramaximal cholecystokinin displaces Munc18c from the pancreatic acinar basal surface, redirecting apical exocytosis to the basal membrane. J Clin Invest 2001; 108:1597-1611.
161. Imai A, Nashida T, Shimomura H. Roles of Munc18-3 in amylase release from rat parotid acinar cells. Arch Biochem Biophys 2004; 422:175-182.
162. Fasshauer D, Margittai M. A transient N-terminal interaction of SNAP-25 and syntaxin nucleates SNARE assembly. J Biol Chem 2004; 279:7613-7621.
163. Carr CM, Grote E, Munson M et al. Sec1p binds to SNARE complexes and concentrates at sites of secretion. J Cell Biol 1999; 146:333-344.
164. Dulubova I, Yamaguchi T, Arac D et al. Convergence and divergence in the mechanism of SNARE binding by Sec1/Munc18-like proteins. Proc Natl Acad Sci USA 2003; 100:32-37.

CHAPTER 3

Rab GTPases and Their Role in the Control of Exocytosis

Romano Regazzi*

Abstract

A large number of Rab GTPases are distributed between the different cellular compartments of eukaryotes and orchestrate intracellular vesicular transport. A subset of these proteins, including Rab3 and Rab27, are assigned to the control of the final steps of the regulated secretory pathway involving docking and fusion of secretory organelles with the plasma membrane. These GTPases function by recruiting on the vesicle surface a variety of protein effectors that, once bound to the Rabs, link the organelles to other components of the secretory machinery, cytoskeletal elements or membrane phospholipids. The purpose of this chapter is to highlight recent progresses in the understanding of the mode of action of Rab3 and Rab27 and in their possible involvement in human pathologies.

Introduction

Rab GTPases are monomeric, small molecular mass proteins (Mr~21-27kDa) that share both structural and functional homologies with the Ras oncogene product (Fig. 1). More than 60 genes encoding for members of this large protein family have been identified in the human genome and 11 in yeast. Several Rab proteins appear to be ubiquitously expressed but others are restricted to cells with specialized functions. Although the precise role of many family members remains to be explored, a wealth of genetic and biochemical data indicate that Rab GTPases are key regulators of intracellular vesicular trafficking in the secretory and endocytic pathways. First evidences for a role of Rab family members in vesicular trafficking came in 1988 in yeast. Indeed, a mutation in the yeast Rab GTPase Ypt-1 was discovered to cause a secretory-defective phenotype[1] and, a couple of months later, one of the genes involved in the final stages of the yeast secretory pathway was found to encode the Rab GTPase Sec4.[2] After these initial findings it became rapidly clear that the mammalian counterparts of yeast GTPases called Rab (*ras* genes from rat brain) were playing similar roles in vesicular transport.[3,4] Rabs participate in all aspects of vesicular trafficking; they facilitate vesicle formation from donor compartments, contribute to transport by recruiting motor proteins and regulate membrane fusion. Consistent with their function in distinct vesicular transport processes Rabs are localized to specific cellular compartments (for review see ref. 5).

Rab Properties and Functions

As is the case for other GTP-binding proteins, Rabs act as molecular switches, cycling between "inactive" GDP-bound and "active" GTP-bound states. Taking advantage of the

*Corresponding Author: Romano Regazzi—Department of Cell Biology and Morphology, Rue du Bugnon 9, 1005 Lausanne, Switzerland. Email: Romano.Regazzi@unil.ch

Molecular Mechanisms of Exocytosis, edited by Romano Regazzi. ©2007 Landes Bioscience and Springer Science+Business Media.

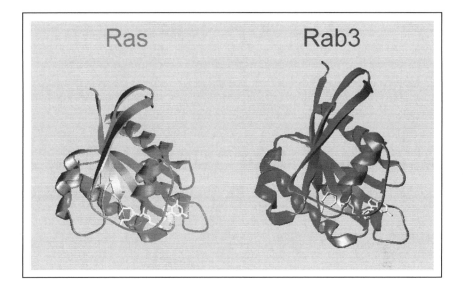

Figure 1. Comparison of the three dimensional structure of Ras and Rab3A.

structural similarities of Rabs with the Ras oncogene product mutants locked in the active or inactive state can be easily generated. Activity changes of Rab GTPases are coordinated by several regulatory factors and are coupled to a reversible association with cellular membranes allowing both spatial and temporal control of Rab function (Fig. 2). Some of the regulatory factors are shared by many Rabs while others are devoted to the control of only one or few Rabs. The association with cellular membranes is rendered possible by post-translational addition of prenyl residues to two cysteines located close to the carboxy-terminal end.[5] In most cases, these cysteine residues are arranged in double-cysteine prenylation motifs (e.g., XXCC, XCXC, CCXX, CCXXX or XCCX, where X can be any amino acid). Newly synthesized Rab proteins are relatively hydrophilic and bind to Rab Escort Protein (REP). This protein complex is recognized by a Rab geranylgeranyl transferase (Rab GGTase) that catalyzes the prenylation of the cystein residues of the GTPase. The prenylated Rab is then chaperoned by a protein called Rab GDP Dissociation Inhibitor (Rab GDI) and delivered to cellular membranes. The human genome encodes two GDI proteins that show tissue-specific expression.[6] The GDIα isoform is enriched in brain while GDIβ is ubiquitously expressed. Both GDI isoforms display high affinity binding (K_d = 20-500nM) to all different Rabs. This promiscuity is explained by the fact that GDIs interact with a domain that is highly conserved among the members of the Rab family called switch region. When Rabs are bound to GDIs their prenylated moieties are accommodated in a GDI pocket preventing the association of the protein complex to cellular membranes.[6] The mechanism permitting delivery of Rabs to specific membrane compartments is not yet fully understood. GDI Displacement Factors (GDFs) are thought to favor the attachment of Rabs to the appropriate membranes by disassembling the complex with GDIs (for review see ref. 7). Yip3 (Ypt-interacting protein 3) also called PRA1 (Prenylated Rab Acceptor 1) has recently been shown to dissociate Rab9-GDI complexes and to possess GDF activity. Indeed, silencing of Yip3/PRA1 by RNA interference or introduction of antibodies directed against this protein prevent the recruitment of Rab9 onto endosomes.[8] Yip3/PRA1 is part of a relatively large Yip protein family (16 genes in humans). At present, it is still unclear whether the other members of this family have also GDF activities.

Once anchored at the membrane, the Rabs are activated by specific Guanine Exchange Factors (GEFs) that favor the exchange of GDP for GTP. In the active GTP-bound state the

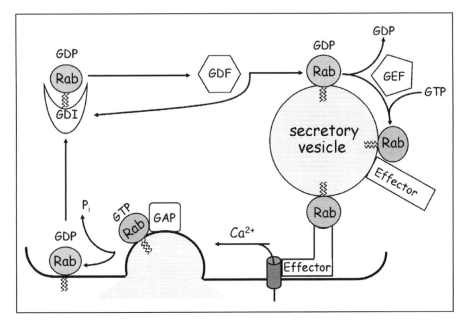

Figure 2. The Rab cycle. Membrane association and activation of Rabs are controlled by several regulatory factors. GDI: GDP Dissociation Inhibitor; GDF: GDI Displacement Factor; GEF: GDP/ GTP Exchange Factor; GAP: GTPase Activating Protein. In the active GTP-bound state the Rabs interact with different effector molecules allowing their recruitment on the vesicle surface. In some cases the Rab effector is localized on the target membrane. This configuration may contribute to the targeting of the vesicles to specific domains of the acceptor membrane.

Rabs are able to recruit several effector molecules (see below) that regulate various aspects of intracellular trafficking including vesicle formation, vesicle movement and membrane fusion. To switch off the signal and inactivate the Rabs the GTP molecule attached to the protein is hydrolysed and converted to GDP. This occurs thanks to the intrinsic GTPase activity of Rabs that is boosted by the interaction with specific GTPase Activating Proteins (GAP). Although GTP hydrolysis is known to be coupled to exocytosis[9] it is still unclear whether GTP hydrolysis occurs immediately before during or after the fusion event. There is evidence that some Rab-GEFs and -GAPs associate in multiprotein complexes.[10] This is likely to ensure spatial and functional coordination of the activities of these factors and to permit tight control of the pool of activated Rabs present in the cells. The Rabs in the GDP-bound form are extracted by the membranes by RabGDI and can then be reused for another cycle.

Rab Subfamilies

Rabs can be grouped according to segregating patterns in phylogenetic trees.[11] These groups reflect similarities of sequence, localization and/or functions. Thus, the Rabs belonging to group V are involved in endosome trafficking while members of group III are generally localized to secretory granules and are likely to govern the final steps of the secretory pathway.[11] Within the large Rab GTPase family, a subgroup of Rabs including the Rab3 and Rab27 subfamilies is devoted to the regulation of the final steps in the regulated secretory pathway. Rab26, Rab33 and Rab37 are also targeted to secretory organelles[7,11] and may participate to the control of exocytosis in specific cell systems but the information available on these Rabs is still limited and they will not be discussed in this review.

The Rab3 Subfamily

Rab3 proteins are encoded by four different genes that drive the expression of Rab3A, -B, -C and -D, respectively. These closely related isoforms share a largely conserved central region and differ significantly only in two small domains located at the N- and C-terminal portion of the molecule. Rab3 proteins are involved in the control of exocytosis in many cells with regulated secretory functions.[5,11-13] Consistent with this role, Rab3 proteins are expressed at highest level in brain and endocrine tissues and to a lesser extent in exocrine glands, adipocytes and other peripheral tissues.[14] Rab3A and Rab3C are the predominant isoforms in neurons and neuroendocrine cells, while Rab3D appears to play a role in the control of exocytosis outside the nervous system.[15] Many different functions have been attributed to Rab3 GTPases, ranging from vesicle biogenesis to docking and fusion of the vesicles with the plasma membrane. The role of Rab3 proteins was initially assessed by introducing in different secretory systems a series of mutants locked in the GDP or GTP-bound conformation. Overexpression of GTPase deficient mutants of the four Rab3 isoforms in catecholamine and insulin-secreting cell lines were found to inhibit stimulus-induced secretion suggesting a negative modulatory role for Rab3 proteins in the exocytotic process.[16-18] The characterization of the phenotype of Rab3A knockout mice led to observations in agreement with this model. Indeed, the absence of Rab3A in hippocampal CA1 pyramidal cells results in an increase in synaptic depression after trains of repetitive stimuli.[19] This effect was initially attributed to a defect in synaptic vesicle recruitment[19] but detailed analysis of the properties of Rab3A-deficient synapses revealed that the rundown of synaptic transmission is not due to changes in the size of the readily releasable vesicle pool but to an increase in the number of Ca^{2+}-evoked vesicle fusions.[20] Despite these concordant findings a series of observations in different cell systems are difficult to reconcile with an inhibitory role of Rab3 proteins and would rather support a positive action of these GTPases on exocytosis. Thus, inhibition of Rab3B expression in pituitary cells using antisense oligonucleotides results in a decrease in Ca^{2+}-induced exocytosis[21] and Rab3A null mice develop fasting hyperglycemia and glucose intolerance because of a defect in glucose-induced insulin secretion.[22] In addition, in transgenic mice overexpressing Rab3D in pancreatic acinar cells secretagogue-stimulated amylase release is enhanced and not diminished as would be expected if Rab3D exerts an inhibitory control on exocytosis.[23] In the absence of all Rab3 isoforms no obvious developmental or structural abnormalities were detected in the brain but Ca^{2+}-evoked synaptic vesicle exocytosis was decreased by about 30%.[14] Consistent with these data mice lacking Rab3-GEF that is required for Rab3 activation exhibit a similar decrease in Ca^{2+} triggered neurotransmitter release.[24] Part of the difficulty in assigning a precise role to Rab3 isoforms is likely to reside in the functional redundancy of Rab3 proteins. Deletion of individual Rab3 isoforms or any combination of two Rab3 proteins does not affect mice survival. In contrast, deletion of three Rab3 isoforms is deleterious when one of the tree deleted proteins is Rab3A and quadruple knockouts are lethal due respiratory failure.[14] These results suggest at least partial redundancy in the function of Rab3 proteins.

Despite many efforts it has been difficult to attribute a precise function to Rab3 proteins. The observations gathered in more than ten years of research indicate that Rab3 GTPases are not essential for the exocytotic process but can modulate the response to stimuli in either a positive or negative fashion. This regulatory function can be exerted by acting on different steps of the secretory pathway including vesicle biogenesis, transport, docking and fusion. It is possible that the regulatory effect of Rab3 proteins varies from cell to cell, according to the presence of specific combinations of isoforms and effectors.

The Rab27 Subfamily

Rab27A and Rab27B are structurally and evolutionarily related to Rab3 proteins. Their involvement in the regulation of the final steps of the secretory pathway was recognized only much later than for Rab3. Mutations in the RAB27A gene were identified as a cause of Griscelli syndrome, a human genetic disease characterized by pigmentation disorders (pigmentary dilution of the skin and hair) and immunodeficiency. These symptoms are due to impaired melanosomes transport in melanocytes[25] and defects in lytic granules release from T-lymphocytes.[26] Patients suffering from Griscelli syndrome can present with different symptoms. This heterogeneity is the consequence of mutations in different genes. Interestingly, some forms of the Griscelli syndrome are caused by mutations of Melanophilin and Myosin Va, two proteins that associate in a functional complex with Rab27A (see below). Rab27A is expressed at high level not only in melanocytes and T-lymphocytes but also in several endocrine tissues such as pancreatic islets, pituitary gland and chromaffin cells.[27-29] Rab27A is poorly expressed in brain and in neuronal tissues where it is probably replaced by Rab27B, an isoform possessing partially overlapping functions.[30] Several recent observations have demonstrated a central role for Rab27A in neuroendocrine exocytosis. Thus, in β-cells overexpression of Rab27A enhances secretion while selective reduction of Rab27A expression by RNA interference, lead to reduction in insulin release triggered by secretagogues.[31,32] The function of Rab27A has been studied also in vivo by investigating the phenotype of Ashen mice, which exhibit a mutation of Rab27A mimicking the Griscelli syndrome. Ashen mice display glucose intolerance and loss of both phases of insulin secretion resulting in abnormal in vivo insulin levels.[33] Surprisingly, glucose-induced insulin secretion is impaired, but the response to other secretagogues is unaffected. Thus, the loss of Rab27A function results in a defect that is reminiscent to that of Rab3A knockout mice suggesting that the two proteins may share partially redundant functions.

Rab Effectors

The function of Rab GTPases depends on the nature and properties of their effectors. Many potential Rab3 and Rab27 effectors displaying tissue specific distributions have been identified. Since each secretory cell appears to express a characteristic set of effectors the proportion of Rab3 and Rab27 associated with a given partner may vary according to the cellular context. In addition, because of the structural similarities between Rab3 and Rab27, it is not always easy to discriminate between "true" Rab3 effectors, "true" Rab27 effectors and proteins that can potentially mediate the action of both GTPases. Indeed in protein/protein interaction studies several Rab3 partners can bind to Rab27 partners and vice versa and, in some cases, also to more distantly related Rabs.[34,35]

Rabphilin-3A

Rabphilin-3A was the first putative Rab3 effector isolated from brain because of its interaction with Rab3A.[36] The observation that the levels of Rabphilin-3A are reduced in Rab3A knockout mice confirmed a possible functional link between this effector and Rab3A.[19] However, systematic analysis using a large set of Rabs revealed that the protein displays high affinity binding also to Rab27A. Indeed, only the interaction with Rab27 appears to be conserved through evolution and, in PC12 cells, Rabphilin-3A is recruited to dense-core vesicle in a Rab27A dependent manner.[37] The phenotype of Rabphilin-3A knockout mice is also not fully compatible with a major role for this effector in mediating Rab3A action. In fact, Rabphilin-3A knockout mice do not display the defects in neurotransmitters release observed in mice lacking Rab3A.[38] Thus, at present it is still unclear whether Rabphilin-3A is an effector of Rab3, Rab27 or both. This Rab effector contains a zinc-finger domain which confers the capacity to bind Rab3 and Rab27 and two C2 domains homologous to the calcium/phospholipid binding modules of PKC. Different studies have demonstrated that regulated exocytosis is enhanced after overexpression of Rabphilin-3A.[39,40] However, this effect on exocytosis is at least partially independent from the capacity to bind Rab3. Indeed, some Rabphilin-3A mutants with impaired binding to Rab3 are still capable to increase the secretory response suggesting the involvement of other partners.[40]

Despite the fact that Rabphilin-3A was identified more than thirteen years ago the mode of action of this protein is only beginning to be understood. Rabphilin-3A knockout mice are fertile and viable and display no abnormalities in synaptic transmission.[38] In addition, the synaptic properties that are impaired in Rab3A knockout mice are unchanged in mice lacking Rabphilin-3A indicating that this effector is not essential for the regulatory functions of Rab3A. At the molecular level Rabphilin-3A was found to interact with α-actinin and to facilitate the association of dense-core granules of chromaffin cells with actin cytoskeleton.[41,42] Moreover, overexpression of Rabphilin-3A was reported to increase the number of docked vesicles in PC12 cells.[43] This effect was mediated by a direct interaction of one of the C2 domains of Rabphilin-3A (C2B) with the plasma membrane associated SNARE protein SNAP25. These findings suggest that the role of Rabphilin-3A may be to chaperon the secretory vesicles through the cortical actin cytoskeleton and to favor their docking to the plasma membrane. There are other Rab effectors suspected to accomplish analogous functions (see below). It is possible that, if necessary, these proteins can compensate for the absence of Rabphilin-3A possibly explaining the apparently normal phenotype of Rabphilin-3A knockout mice.[38]

NOC2

NOC2, for NO C2 domains, has been discovered in a screen for Rabphilin-3A homologues.[44] This protein is abundantly expressed in pancreatic β-cells and in other endocrine cells.[44] NOC2 has been shown to bind to both Rab3 and Rab27.[34,35,45] In neuroendocrine cells, overexpression of NOC2 enhanced exocytosis in one study[44] and was reported to inhibit secretion in another.[45] Recent studies using RNA interference to knock down the expression of NOC2 have contributed to shed some light on the possible function of this Rab effector. Thus, silencing of NOC2 by RNA interference in insulin-secreting cell lines led to a strong reduction in stimulus-induced secretion.[34] This observation was confirmed and extended by the characterization of NOC2 knockout mice.[46] If these mice are kept under normal conditions, blood glucose levels and insulin release are not significantly different from wild type animals. In contrast, under stress conditions NOC2$^{-/-}$ mice display glucose intolerance and impaired insulin secretion. Pancreatic β-cells lacking NOC2 were found to be more sensitive to the inhibition of insulin secretion caused by adrenaline agonists. These findings suggest that under stress conditions NOC2 exerts a positive effect on insulin secretion by attenuating the inhibitory action of the adrenergic signaling. The characterization of the phenotype of NOC2 knockout mice revealed also a previously unsuspected involvement of NOC2, in the secretory process of exocrine glands. Thus, in mice lacking NOC2 amylase secretion from pancreatic acinar cells is abolished and the size and number of zymogen granules is greatly altered.[46] Similar changes in secretory granule morphology were observed in several other exocrine glands. The secretory phenotype of NOC2$^{-/-}$ mice was corrected by reintroducing wild type NOC2 but not a point mutant unable to bind Rab3. These observations led to the conclusion that the interaction with Rab3 is necessary for the action of NOC2.[46] However, we now know that the NOC2 mutant used in this particular study is unable to interact also with Rab27,[34] rendering the interpretation of the data less straight forward.

The RIM Family

The first member of the RIM (Rab3 Interacting Molecule) family was discovered in a Yeast Two-hybrid screen using constitutively active Rab3C as bait.[47] RIM proteins are encoded by one gene in *C. elegans* (Unc10) and four genes in mammals (RIM1-4).[48] These genes drive the expression of a large number of splicing variants of different sizes and lacking one or more functional domains. RIM proteins are mainly found in neuronal cells, but RIM2 is expressed at high level in endocrine cells, such as β-cells.[49] RIM1α and RIM2α contain a zinc-finger domain responsible for the binding of Rab3 GTPases, a PDZ domain that is thought to mediate interactions with components of synaptic active zone and two C2 domains, (C2A and C2B). Many different RIMα variants displaying variable distances between the functional modules of the protein can be generated by insertion of alternative spliced sequences. In contrast to the other Rab3 and Rab27 effectors RIM1α and RIM2α are not associated with secretory vesicles but are located at the plasma membrane.[47] The β and γ isoforms lack the amino terminal domain required for Rab3 binding. The role of these proteins in exocytosis is unclear but, if any, it is most probably unrelated to Rab3 GTPases.

Many observations point to an involvement of RIM proteins in secretion. Unc10 mutants in *C. elegans* exhibit a severe defect in neurotransmitter release associated with impaired synaptic vesicle priming.[50] In mammals, the phenotype of RIM1α knockout mice is milder, probably because of partial redundancy with RIM2α, but defects in neurotransmitter release and its regulation are also observed.[51,52] RIM proteins play also an important role in the control of the final steps of exocytosis of pancreatic β-cells. In particular, RIM2α was shown to be involved in a Protein kinase A independent pathway that triggers insulin exocytosis in response to cAMP-raising agents.[49]

The different domains of RIMα isoforms have been implicated in multiple protein/protein interactions. The N-terminal domain is not only responsible for Rab3 binding, as previously mentioned, but is also involved in the interaction with 14-3-3[53] and with Munc13-1,[54,55] an active zone protein thought to play an essential role in synaptic vesicle priming. Interestingly, Rab3A and Munc13-1 compete each other for the amino terminus of RIM1 suggesting that their bindings are mutually exclusive. The switch from the RIM1/Rab3A complex to the RIM1/Munc13-1 complex was proposed to occur just after docking of synaptic vesicles to the plasma membrane insuring optimal coordination between the docking and priming processes. The PDZ and the two C2 domains of RIM proteins, are also involved in both Ca^{2+}-dependent and -independent protein/protein interactions. The potential partners of RIM include Synaptotagmin-I,[52,56] the pore-forming subunit of N-type Ca^{2+} channels,[56] SNAP-25,[56] Liprins,[52] RBP (RIM binding proteins)[57] and CAST (CAZ-associated Structural protein).[58] Taken together these results suggest that RIM acts as scaffold protein that modulates sequential steps in synaptic vesicle exocytosis through serial protein-protein interactions.

Synaptotagmin-Like Proteins (Slps)

Synaptotagmin-like proteins (Slps) constitute a family of five Rab effectors (Slp1-5) characterized by an N-terminal Slp homology domain (SHD), a linker region of varying length and two C-terminally located C2 domains.[28] For most Slps alternative splicing isoforms lacking one or more domains have been described. Below, I briefly summarize the properties and function of these proteins. Further details on the role of Slps in the secretory pathway can be found in another chapter of the book ("The role Synaptotagmin and Synaptotagmin-like protein (Slp) in regulated exocytosis").

Slps have been discovered in 2001, because of their structural similarities to the Ca^{2+} sensor proteins of the Synaptotagmin family.[59] However, the characterization of the properties and functions of these proteins revealed that they differ significantly from Synaptotagmins. Thus, Slps are not endowed with transmembrane stretches (most often present in Synaptotagmins) and possess an N-terminal domain that confers high affinity binding to Rab27 isoforms. Because of homologies with other Rab effectors some authors have proposed to include Slps as members of a new family of proteins involved in exocytosis called Exophilins (Exocytosis associated rabphilin3/granuphilin-like proteins).[29] To avoid confusion, here only the original nomenclature is given.

Slp proteins are peripheral membrane proteins and are recruited on secretory vesicles by Rab27. Several studies in which their cellular level was increased or reduced by RNA interference have demonstrated that Slps regulate secretion by controlling docking of secretory vesicles to the plasma membrane. Slps accomplish their function by simultaneously interacting with Rab27 located on the vesicle surface and with proteins or lipids associated with the plasma membrane. Thus, Slp1 interacts with phosphatidylinositol 3,4,5-triphosphate,[60] Slp2 with phosphatidylserine[61] and Slp4 with Munc18/syntaxin complexes.[62,63] Slp4 that is also called Granuphilin has some peculiar properties compared to the other members of the Slp family. This protein was first identified in a screening for genes differentially expressed in pancreatic β- and α-cells.[64] Granuphilin/Slp4 is specifically expressed in pancreatic β-cells, in the pituitary gland and to a much lower extent in chromaffin cells. Contrary to all other Slps, overexpression of Granuphilin/Slp4 in β-cells and PC12 results in a strong inhibitory effect on exocytosis.[32,63] Moreover, glucose-induced insulin release in Granuphilin/Slp4 knockout mice is augmented, further confirming that this Rab27 effector exerts an inhibitory effect on the exocytotic process. Surprisingly, in Granuphilin/Slp4-deficient β-cells in spite of an increase in insulin release the number of secretory granules morphologically docked to the plasma membrane is markedly reduced. A possible interpretation of these observations is that Granuphilin/Slp4 is required to ensure docking of insulin granules at specific plasma membrane locations permitting a tight control of the exocytotic rate. In the absence of Granuphilin part of the secretory vesicles may circumvent this regulatory step leading to an exaggerated number of fusion events in reponse to external stimuli.

Proteins Related to Slps

Slac2 proteins (Slac2-a/Melanophilin, Slac2-b and Slac2-c/MyRIP) possess an N-terminal Rab27 binding domain homologous to that of Slps but they lack C2 domains (hence the name Slp homologue lacking C2 domains). Despite the absence of C2 domains multiple interactions between Slac2 and other cellular components, in particular with cytoskeletal elements, have been reported.

Melanophilin/Slac2a

A mutation in the *Melanophilin/Slac2-a* gene was discovered to be responsible for an alteration in mice coat color.[65] Later on mutations of Rab27A and Myosin Va genes were reported to lead to a similar phenotype, suggesting a functional relationship between Melanophilin and these two proteins.[65] Indeed, a number of results obtained in melanocyte cultures demonstrate that these three proteins form a functional complex that is essential for melanosome transport. If any of the components of this tripartite complex is lacking melanosomes instead of being transported to the cell periphery and distributed to the surrounding cells accumulate in the perinuclear region of melanocytes. Melanophilin possesses a domain enabling the interaction with Myosin Va.[66] Rab27A present on melanosomes recruits Melanophilin; once anchored at the surface of the pigment-containing organelle the effector binds to Myosin Va, linking melanosomes to actin filaments and contributing to their delivery and retention at the cell periphery. The melanosome is a specialized secretory vesicle that is equivalent to secretory lysosomes.[5] At present, it is not known whether Melanophilin controls vesicular trafficking and exocytosis through a similar mechanism in other Rab27-expressing cells.

MyRIP/Slac2-c

MyRIP (Myosin Rab Interacting Protein) has been identified thanks to its ability to interact with Myosin VIIa and to Rab27.[67] Later on, MyRIP was found to bind also to Myosin Va and, via its C-terminal domain, to associate to actin.[68] Functional studies on MyRIP have been performed in chromaffin cells, in pancreatic β-cells and in parotid gland acinar cells.[31,69,70] In all these cells MyRIP is localized on secretory granules. A selective decrease of MyRIP by RNA interference impairs insulin secretion in response to secretagogues.[31] Studies performed using deletion mutants revealed that both the Rab binding motif and the actin-binding domain of MyRIP are involved in the regulation of insulin secretion. These data, together with experiments in which wild type or dominant negative mutants of MyRIP were transfected in PC12 cells, led to the conclusion that the Rab27/MyRIP complex creates a functional link between the secretory granules and the actin skeleton, possibly contributing to guide the secretory organelles toward their release sites.

Calmodulin

Rab3 has been shown to associate with the Ca^{2+}-binding protein Calmodulin (CaM) in a Ca^{2+} and nucleotide dependent manner.[71] The interaction with CaM does not occur via the effector loop (Fig. 3) and necessitates the presence of two arginine residues found in Rab3 but not in other Rab proteins that are not involved in regulated exocytosis (Fig. 4). Replacement of these two amino acids was found to prevent the inhibition of exocytosis caused by constitutively active Rab3A in insulin-secreting cell lines and in PC12 cells.[72] However, these findings were not reproduced in another independent study.[73] Rab3A affinity for CaM is relatively low compared to other CaM-binding proteins such as Ca^{2+}/CAM kinase-II. These findings led to the proposal that transient Rab3A/CaM interactions could allow the recruitment of CaM to the cytoplasmic face of secretory vesicles. Subsequent transfer of the Ca^{2+}-binding protein to other vesicle-associated CaM-binding proteins would then permit the adaptation of the secretory machinery to local rises in the Ca^{2+} concentration. This mechanism was suggested to operate to couple the activation of granule-associated Ca^{2+}/CaM dependent phosphatase, Calcineurin, to an increase in intracellular Ca^{2+}.[74,75] Activated Calcineurin would then dephosphorylate the motor protein Kinesin and favor the transport of secretory granules at the plasma membrane to join a readily releasable pool of vesicles committed to undergo exocytosis. According to this model, in the absence of Rab3A, activation of Calcineurin and Kinesin would not occur and granule transport would be less efficient. This would lead to a reduction in the number of granules in the readily releasable pool and insulin exocytosis would be impaired.

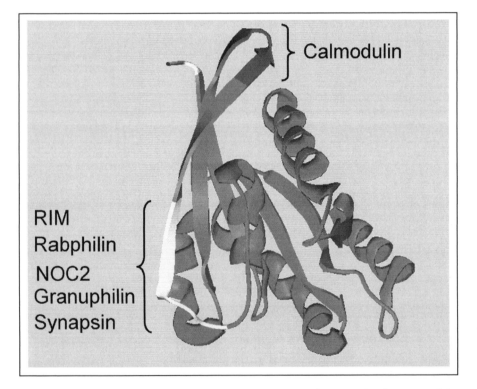

Figure 3. Domains of Rab3 involved in the interaction with effector molecules. Most of the partners of Rab3A interact with the so-called effector loop that undergoes a conformational change upon GTP binding. In contrast, Calmodulin binds to a domain of Rab3 that is located at a short distance from the C-terminus of the GTPase.

		66		70			
Rab3A	KTIY	R	NDK	R	IKLQI	WDTAGQE	RYR
Rab1A	RTIE	L	DGK	T	IKLQI	WDTAGQE	RFR
Rab2A	RMIT	I	DGK	Q	IKLQI	WDTAGQE	SFR
Rab4A	KIIN	V	GGK	Y	VKLQI	WDTAGQE	RFR
Rab5A	QTVC	L	DDT	T	VKFEI	WDTAGQE	GYH
Rab6A	KTMY	L	EDR	T	VRLQL	WDTAGQE	RFR
Rab7	KEVM	V	DDR	L	VTMQI	WDTAGQE	RFQ
Sec4	KTVD	I	NGK	K	IKLQI	WDTAGQE	RFR

Figure 4. Alignment of the Calmodulin-binding domain of Rab3A with the corresponding segments of other Rab proteins. The two arginines (Arg66 and Arg70) essential for Calmodulin binding are indicated by the boxes.

Synapsin I

Synapsin I is the most recently identified partner of Rab3.[76,77] Synapsins belong to a family of neuron specific proteins, which bridge synaptic vesicles and actin cytoskeleton.[78] Their role in neurotransmitter release is discussed in detail in another chapter of the book. Synapsin I is also expressed in insulin-secreting cell lines and has been reported to be associated with secretory granules[79] but this finding is controversial.[80] The interaction between Synapsin I and Rab3A was discovered using an unbiased phage display approach and was then confirmed using recombinant proteins and chemical cross-linking in intact nerve terminals. A functional link between the two proteins is also suggested by the observation that in Synapsin knockout mice the amount of Rab3A associated with synaptic vesicles is decreased.[76] Synapsin I associates preferentially with the GTP-bound form of Rab3A and this interaction has functional consequences for the biological activities of both proteins. On one hand, the presence of Synapsin I stimulates GTP binding and GTPase activity of Rab3A. On the other hand, Rab3A inhibits the association of Synapsin I to F-actin. In view of the involvement of Rab3A and Synapsin I in exocytosis this interaction may play an important role in the control of neurotransmitter and, possibly, also in insulin release.

Munc13-4

Munc13-4 is a distantly related member of the Munc13 family discovered thanks to the presence of a Munc13 homology domain.[81] In contrast to the other members of the Munc13 family, Munc13-4 has recently been shown to be a novel potential partner of Rab27. Thus, Munc13-4 was found to bind Rab27 in a GTP-dependent manner, suggesting a possible involvement of this Munc13 isoform in Rab27-regulated events in platelets.[82] The expression of Munc13-4 has been demonstrated in many cells of hematopoietic origin.[83] In these cells, the complex formed by Rab27A and Munc13-4 appears to be an essential regulator of membrane fusion and controls lysosome secretion.[83]

Conclusion and Perspectives

The data gathered in the last decade have highlighted the role of Rab27 and Rab3 GTPases in the fine tuning of exocytosis in several cell systems that use the regulated secretory pathway to release biologically active peptides, hormones or neurotransmitters. The mode of action of each of these Rabs is complex and evidence for an involvement in a variety of important steps in the secretory pathway ranging from vesicle biogenesis to vesicle fusion has been provided. A number of burning questions remain unsolved and awaits clarification. Future investigations will need to assess whether Rab3 and Rab27 control different steps or accomplish partially redundant functions. They should also determine the relative contribution of each Rab and Rab effector in different cell systems. The advent of new powerful tools such as RNA interference and the use of sophisticated imaging techniques that allow tracking of individual secretory vesicles in living cells[84] are opening new avenues in the field and promise to furnish satisfactory answers to these central questions in the near future.

Acknowledgments

Work in the author's laboratory is supported by the Swiss National Science Foundation.

References

1. Segev N, Mulholland J, Botstein D. The yeast GTP-binding YPT1 protein and a mammalian counterpart are associated with the secretion machinery. Cell 1988; 52(6):915-924.
2. Goud B, Salminen A, Walworth NC et al. A GTP-binding protein required for secretion rapidly associates with secretory vesicles and the plasma membrane in yeast. Cell 1988; 53(5):753-768.
3. Touchot N, Chardin P, Tavitian A. Four additional members of the ras gene superfamily isolated by an oligonucleotide strategy: Molecular cloning of YPT-related cDNAs from a rat brain library. Proc Natl Acad Sci USA 1987; 84(23):8210-8214.
4. Zahraoui A, Touchot N, Chardin P et al. The human Rab genes encode a family of GTP-binding proteins related to yeast YPT1 and SEC4 products involved in secretion. J Biol Chem 1989; 264(21):12394-12401.
5. Seabra MC, Mules EH, Hume AN. Rab GTPases, intracellular traffic and disease. Trends Mol Med 2002; 8(1):23-30.
6. Pfeffer S, Aivazian D. Targeting Rab GTPases to distinct membrane compartments. Nat Rev Mol Cell Biol 2004; 5(11):886-896.
7. Seabra MC, Wasmeier C. Controlling the location and activation of Rab GTPases. Curr Opin Cell Biol 2004; 16(4):451-457.
8. Sivars U, Aivazian D, Pfeffer SR. Yip3 catalyses the dissociation of endosomal Rab-GDI complexes. Nature 2003; 425(6960):856-859.
9. Stahl B, von Mollard GF, Walch-Solimena C et al. GTP cleavage by the small GTP-binding protein Rab3A is associated with exocytosis of synaptic vesicles induced by alpha-latrotoxin. J Biol Chem 1994; 269(40):24770-24776.
10. Nagano F, Kawabe H, Nakanishi H et al. Rabconnectin-3, a novel protein that binds both GDP/GTP exchange protein and GTPase-activating protein for Rab3 small G protein family. J Biol Chem 2002; 277(12):9629-9632.
11. Pereira-Leal JB, Seabra MC. Evolution of the Rab family of small GTP-binding proteins. J Mol Biol 2001; 313(4):889-901.
12. Fischer von Mollard G, Stahl B, Li C et al. Rab proteins in regulated exocytosis. Trends Biochem Sci 1994; 19(4):164-168.
13. Lledo PM, Johannes L, Vernier P et al. Rab3 proteins: Key players in the control of exocytosis. Trends Neurosci 1994; 17(10):426-432.
14. Schluter OM, Schmitz F, Jahn R et al. A complete genetic analysis of neuronal Rab3 function. J Neurosci 2004; 24(29):6629-6637.
15. Millar AL, Pavios NJ, Xu J et al. Rab3D: A regulator of exocytosis in nonneuronal cells. Histol Histopathol 2002; 17(3):929-936.
16. Holz RW, Brondyk WH, Senter RA et al. Evidence for the involvement of Rab3A in Ca(2+)-dependent exocytosis from adrenal chromaffin cells. J Biol Chem 1994; 269(14):10229-10234.
17. Iezzi M, Escher G, Meda P et al. Subcellular distribution and function of Rab3A, B, C, and D isoforms in insulin-secreting cells. Mol Endocrinol 1999; 13(2):202-212.
18. Johannes L, Lledo PM, Roa M et al. The GTPase Rab3a negatively controls calcium-dependent exocytosis in neuroendocrine cells. EMBO J 1994; 13(9):2029-2037.

19. Geppert M, Bolshakov VY, Siegelbaum SA et al. The role of Rab3A in neurotransmitter release. Nature 1994; 369(6480):493-497.
20. Geppert M, Goda Y, Stevens CF et al. The small GTP-binding protein Rab3A regulates a late step in synaptic vesicle fusion. Nature 1997; 387(6635):810-814.
21. Lledo PM, Vernier P, Vincent JD et al. Inhibition of Rab3B expression attenuates Ca(2+)-dependent exocytosis in rat anterior pituitary cells. Nature 1993; 364(6437):540-544.
22. Yaekura K, Julyan R, Wicksteed BL et al. Insulin secretory deficiency and glucose intolerance in Rab3A null mice. J Biol Chem 2003; 278(11):9715-9721.
23. Ohnishi H, Samuelson LC, Yule DI et al. Overexpression of Rab3D enhances regulated amylase secretion from pancreatic acini of transgenic mice. J Clin Invest 1997; 100(12):3044-3052.
24. Yamaguchi K, Tanaka M, Mizoguchi A et al. A GDP/GTP exchange protein for the Rab3 small G protein family upregulates a postdocking step of synaptic exocytosis in central synapses. Proc Natl Acad Sci USA 2002; 99(22):14536-14541.
25. Hume AN, Collinson LM, Rapak A et al. Rab27a regulates the peripheral distribution of melanosomes in melanocytes. J Cell Biol 2001; 152(4):795-808.
26. Menasche G, Pastural E, Feldmann J et al. Mutations in RAB27A cause Griscelli syndrome associated with haemophagocytic syndrome. Nat Genet 2000; 25(2):173-176.
27. Cheviet S, Waselle L, Regazzi R. Noc-king out exocrine and endocrine secretion. Trends Cell Biol 2004; 14(10):525-528.
28. Fukuda M. Versatile role of Rab27 in membrane trafficking: Focus on the Rab27 effector families. J Biochem (Tokyo) 2005; 137(1):9-16.
29. Izumi T, Gomi H, Kasai K et al. The roles of Rab27 and its effectors in the regulated secretory pathways. Cell Struct Funct 2003; 28(5):465-474.
30. Barral DC, Ramalho JS, Anders R et al. Functional redundancy of Rab27 proteins and the pathogenesis of Griscelli syndrome. J Clin Invest 2002; 110(2):247-257.
31. Waselle L, Coppola T, Fukuda M et al. Involvement of the Rab27 binding protein Slac2c/MyRIP in insulin exocytosis. Mol Biol Cell 2003; 14(10):4103-4113.
32. Yi Z, Yokota H, Torii S et al. The Rab27a/granuphilin complex regulates the exocytosis of insulin-containing dense-core granules. Mol Cell Biol 2002; 22(6):1858-1867.
33. Kasai K, Ohara-Imaizumi M, Takahashi N et al. Rab27a mediates the tight docking of insulin granules onto the plasma membrane during glucose stimulation. J Clin Invest 2005; 115(2):388-396.
34. Cheviet S, Coppola T, Haynes LP et al. The Rab-binding protein Noc2 is associated with insulin-containing secretory granules and is essential for pancreatic beta-cell exocytosis. Mol Endocrinol 2004; 18(1):117-126.
35. Fukuda M. Distinct Rab binding specificity of Rim1, Rim2, rabphilin, and Noc2. Identification of a critical determinant of Rab3A/Rab27A recognition by Rim2. J Biol Chem 2003; 278(17):15373-15380.
36. Shirataki H, Kaibuchi K, Yamaguchi T et al. A possible target protein for smg-25A/rab3A small GTP-binding protein. J Biol Chem 1992; 267(16):10946-10949.
37. Fukuda M, Kanno E, Yamamoto A. Rabphilin and Noc2 are recruited to dense-core vesicles through specific interaction with Rab27A in PC12 cells. J Biol Chem 2004; 279(13):13065-13075.
38. Schluter OM, Schnell E, Verhage M et al. Rabphilin knock-out mice reveal that rabphilin is not required for rab3 function in regulating neurotransmitter release. J Neurosci 1999; 19(14):5834-5846.
39. Chung SH, Takai Y, Holz RW. Evidence that the Rab3a-binding protein, rabphilin3a, enhances regulated secretion. Studies in adrenal chromaffin cells. J Biol Chem 1995; 270(28):16714-16718.
40. Joberty G, Stabila PF, Coppola T et al. High affinity Rab3 binding is dispensable for Rabphilin-dependent potentiation of stimulated secretion. J Cell Sci 1999; 112(Pt 20):3579-3587.
41. Baldini G, Martelli AM, Tabellini G et al. Rabphilin localizes with the cell actin cytoskeleton and stimulates association of granules with F-actin cross-linked by {alpha}-actinin. J Biol Chem 2005; 280(41):34974-34984.
42. Kato M, Sasaki T, Ohya T et al. Physical and functional interaction of rabphilin-3A with alpha-actinin. J Biol Chem 1996; 271(50):31775-31778.
43. Tsuboi T, Fukuda M. The C2B domain of rabphilin directly interacts with SNAP-25 and regulates the docking step of dense core vesicle exocytosis in PC12 cells. J Biol Chem 2005; 280(47):39253-39259.
44. Kotake K, Ozaki N, Mizuta M et al. Noc2, a putative zinc finger protein involved in exocytosis in endocrine cells. J Biol Chem 1997; 272(47):29407-29410.
45. Haynes LP, Evans GJ, Morgan A et al. A direct inhibitory role for the Rab3-specific effector, Noc2, in Ca2+-regulated exocytosis in neuroendocrine cells. J Biol Chem 2001; 276(13):9726-9732.
46. Matsumoto M, Miki T, Shibasaki T et al. Noc2 is essential in normal regulation of exocytosis in endocrine and exocrine cells. Proc Natl Acad Sci USA 2004; 101(22):8313-8318.

47. Wang Y, Okamoto M, Schmitz F et al. Rim is a putative Rab3 effector in regulating synaptic-vesicle fusion. Nature 1997; 388(6642):593-598.
48. Wang Y, Sudhof TC. Genomic definition of RIM proteins: Evolutionary amplification of a family of synaptic regulatory proteins (small star, filled). Genomics 2003; 81(2):126-137.
49. Ozaki N, Shibasaki T, Kashima Y et al. cAMP-GEFII is a direct target of cAMP in regulated exocytosis. Nat Cell Biol 2000; 2(11):805-811.
50. Koushika SP, Richmond JE, Hadwiger G et al. A post-docking role for active zone protein Rim. Nat Neurosci 2001; 4(10):997-1005.
51. Calakos N, Schoch S, Sudhof TC et al. Multiple roles for the active zone protein RIM1alpha in late stages of neurotransmitter release. Neuron 2004; 42(6):889-896.
52. Schoch S, Castillo PE, Jo T et al. RIM1alpha forms a protein scaffold for regulating neurotransmitter release at the active zone. Nature 2002; 415(6869):321-326.
53. Sun L, Bittner MA, Holz RW. Rim, a component of the presynaptic active zone and modulator of exocytosis, binds 14-3-3 through its N terminus. J Biol Chem 2003; 278(40):38301-38309.
54. Betz A, Thakur P, Junge HJ et al. Functional interaction of the active zone proteins Munc13-1 and RIM1 in synaptic vesicle priming. Neuron 2001; 30(1):183-196.
55. Dulubova I, Lou X, Lu J et al. A Munc13/RIM/Rab3 tripartite complex: From priming to plasticity? EMBO J 2005; 24(16):2839-2850.
56. Coppola T, Magnin-Luthi S, Perret-Menoud V et al. Direct interaction of the Rab3 effector RIM with Ca2+ channels, SNAP-25, and synaptotagmin. J Biol Chem 2001; 276(35):32756-32762.
57. Hibino H, Pironkova R, Onwumere O et al. RIM binding proteins (RBPs) couple Rab3-interacting molecules (RIMs) to voltage-gated Ca(2+) channels. Neuron 2002; 34(3):411-423.
58. Ohtsuka T, Takao-Rikitsu E, Inoue E et al. Cast: A novel protein of the cytomatrix at the active zone of synapses that forms a ternary complex with RIM1 and munc13-1. J Cell Biol 2002; 158(3):577-590.
59. Fukuda M, Mikoshiba K. Synaptotagmin-like protein 1-3: A novel family of C-terminal-type tandem C2 proteins. Biochem Biophys Res Commun 2001; 281(5):1226-1233.
60. Catz SD, Johnson JL, Babior BM. The C2A domain of JFC1 binds to 3'-phosphorylated phosphoinositides and directs plasma membrane association in living cells. Proc Natl Acad Sci USA 2002; 99(18):11652-11657.
61. Kuroda TS, Fukuda M. Rab27A-binding protein Slp2-a is required for peripheral melanosome distribution and elongated cell shape in melanocytes. Nat Cell Biol 2004; 6(12):1195-1203.
62. Torii S, Takeuchi T, Nagamatsu S et al. Rab27 effector granuphilin promotes the plasma membrane targeting of insulin granules via interaction with syntaxin 1a. J Biol Chem 2004; 279(21):22532-22538.
63. Coppola T, Frantz C, Perret-Menoud V et al. Pancreatic beta-cell protein granuphilin binds Rab3 and Munc-18 and controls exocytosis. Mol Biol Cell 2002; 13(6):1906-1915.
64. Wang J, Takeuchi T, Yokota H et al. Novel rabphilin-3-like protein associates with insulin-containing granules in pancreatic beta cells. J Biol Chem 1999; 274(40):28542-28548.
65. Matesic LE, Yip R, Reuss AE et al. Mutations in Mlph, encoding a member of the Rab effector family, cause the melanosome transport defects observed in leaden mice. Proc Natl Acad Sci USA 2001; 98(18):10238-10243.
66. Strom M, Hume AN, Tarafder AK et al. A family of Rab27-binding proteins. Melanophilin links Rab27a and myosin Va function in melanosome transport. J Biol Chem 2002; 277(28):25423-25430.
67. El-Amraoui A, Schonn JS, Kussel-Andermann P et al. MyRIP, a novel Rab effector, enables myosin VIIa recruitment to retinal melanosomes. EMBO Rep 2002; 3(5):463-470.
68. Fukuda M, Kuroda TS. Slac2-c (synaptotagmin-like protein homologue lacking C2 domains-c), a novel linker protein that interacts with Rab27, myosin Va/VIIa, and actin. J Biol Chem 2002; 277(45):43096-43103.
69. Desnos C, Schonn JS, Huet S et al. Rab27A and its effector MyRIP link secretory granules to F-actin and control their motion towards release sites. J Cell Biol 2003; 163(3):559-570.
70. Imai A, Yoshie S, Nashida T et al. The small GTPase Rab27B regulates amylase release from rat parotid acinar cells. J Cell Sci 2004; 117(Pt 10):1945-1953.
71. Park JB, Farnsworth CC, Glomset JA. Ca2+/calmodulin causes Rab3A to dissociate from synaptic membranes. J Biol Chem 1997; 272(33):20857-20865.
72. Coppola T, Perret-Menoud V, Luthi S et al. Disruption of Rab3-calmodulin interaction, but not other effector interactions, prevents Rab3 inhibition of exocytosis. EMBO J 1999; 18(21):5885-5891.
73. Schluter OM, Khvotchev M, Jahn R et al. Localization versus function of Rab3 proteins. Evidence for a common regulatory role in controlling fusion. J Biol Chem 2002; 277(43):40919-40929.

74. Donelan MJ, Morfini G, Julyan R et al. Ca2+-dependent dephosphorylation of kinesin heavy chain on beta-granules in pancreatic beta-cells. Implications for regulated beta-granule transport and insulin exocytosis. J Biol Chem 2002; 277(27):24232-24242.
75. Kajio H, Olszewski S, Rosner PJ et al. A low-affinity Ca2+-dependent association of calmodulin with the Rab3A effector domain inversely correlates with insulin exocytosis. Diabetes 2001; 50(9):2029-2039.
76. Giovedi S, Darchen F, Valtorta F et al. Synapsin is a novel Rab3 effector protein on small synaptic vesicles. II. Functional effects of the Rab3A-synapsin I interaction. J Biol Chem 2004; 279(42):43769-43779.
77. Giovedi S, Vaccaro P, Valtorta F et al. Synapsin is a novel Rab3 effector protein on small synaptic vesicles. I. Identification and characterization of the synapsin I-Rab3 interactions in vitro and in intact nerve terminals. J Biol Chem 2004; 279(42):43760-43768.
78. Greengard P, Benfenati F, Valtorta F. Synapsin I, an actin-binding protein regulating synaptic vesicle traffic in the nerve terminal. Adv Second Messenger Phosphoprotein Res 1994; 29:31-45.
79. Matsumoto K, Ebihara K, Yamamoto H et al. Cloning from insulinoma cells of synapsin I associated with insulin secretory granules. J Biol Chem 1999; 274(4):2053-2059.
80. Krueger KA, Ings EI, Brun AM et al. Site-specific phosphorylation of synapsin I by Ca2+/calmodulin-dependent protein kinase II in pancreatic betaTC3 cells: Synapsin I is not associated with insulin secretory granules. Diabetes 1999; 48(3):499-506.
81. Koch H, Hofmann K, Brose N. Definition of Munc13-homology-domains and characterization of a novel ubiquitously expressed Munc13 isoform. Biochem J 2000; 349(Pt 1):247-253.
82. Shirakawa R, Higashi T, Tabuchi A et al. Munc13-4 is a GTP-Rab27-binding protein regulating dense core granule secretion in platelets. J Biol Chem 2004; 279(11):10730-10737.
83. Neeft M, Wieffer M, de Jong AS et al. Munc13-4 is an effector of rab27a and controls secretion of lysosomes in hematopoietic cells. Mol Biol Cell 2005; 16(2):731-741.
84. Rink J, Ghigo E, Kalaidzidis Y et al. Rab conversion as a mechanism of progression from early to late endosomes. Cell 2005; 122(5):735-749.

CHAPTER 4

The Role of Synaptotagmin and Synaptotagmin-Like Protein (Slp) in Regulated Exocytosis

Mitsunori Fukuda*

Abstract

Cells secrete a variety of substances by "regulated exocytosis" (i.e., fusion of secretory vesicles with the plasma membrane) in response to extracellular stimuli. Since two membranes must be apposed with each other prior to the secretion event, phospholipid-binding domains are often found in exocytotic proteins on secretory vesicles, and one of the best characterized phospholipid-binding domains involved in regulated exocytosis is the C2 domain. C2 domains are found in tandem in the C-terminal region of C-type tandem C2 proteins, including synaptotagmins (Syts), synaptotagmin-like proteins (Slps), rabphilin, and Doc2s. In this chapter I provide an overview of the structure and function of the Syt and Slp families, especially focusing on recent advances in research on the molecular mechanisms of secretory vesicle trafficking (e.g., docking, fusion, and recycling of vesicles) mediated by Syt or Slp proteins.

Introduction

Secretion of neurotransmitters, hormones, and enzymes is a fundamental biological activity of the cell, and is achieved by vesicular exocytosis, namely, fusion of secretory vesicles with the plasma membrane. Vesicular exocytosis generally consists of at least three distinct steps: docking/tethering of a transport vesicle to the plasma membrane, ATP-dependent priming of the vesicle, and actual fusion of the vesicle to the plasma membrane, which is often triggered by extracellular stimuli (e.g., increase in intracellular Ca^{2+} concentrations).[1] Since two membranes (i.e., vesicle membrane and plasma membrane) must be apposed with each other prior to the secretion event, phospholipid-binding domains are often found in exocytotic proteins on secretory vesicles. Among the various phospholipid-binding domains that have been identified, the C2 domain, a putative Ca^{2+}- and phospholipid-binding motif originally identified in Ca^{2+}-dependent protein kinase C,[2] is of particular interest, because two C2 domains are often found in tandem in the C-terminal portion of a group of exocytotic proteins (named the C-type tandem C2 protein family).[3,4] Tandem C2 domains are also found in the N-terminal portion of GAP1 family proteins and copine family proteins (named the N-type tandem C2 protein family),[3] but whether they are involved in secretory vesicle exocytosis is completely unknown. Thus far, 27 different C-type tandem C2 proteins have been identified in mice and

*Mitsunori Fukuda—Fukuda Initiative Research Unit, RIKEN (The Institute of Physical and Chemical Research), 2-1 Hirosawa, Wako, Saitama 351-0198, Japan.
Email: mnfukuda@brain.riken.go.jp.

Molecular Mechanisms of Exocytosis, edited by Romano Regazzi. ©2007 Landes Bioscience and Springer Science+Business Media.

humans, and they have been classified into three distinct groups based on their N-terminal structures.[3,4] The first, and best characterized group is the synaptotagmin (Syt) family.[4-6] Syt is defined as a protein with a single N-terminal transmembrane domain and tandem C-terminal cytoplasmic C2 domains (referred to as the C2A domain and the C2B domain; Fig. 1). The second group consists of rabphilin and Doc2s, both of which share highly homologous tandem C2 domains, although their N-terminal structures are completely different: rabphilin contains an N-terminal Rab-binding domain (RBD),[7] whereas Doc2 contains an N-terminal Munc13-1-interacting domain (MID) (Fig. 1).[8,9] The final group is the recently identified synaptotagmin-like protein (Slp) family.[10-12] Slp family members contain an N-terminal Rab27A/B-binding domain (also called Slp homology domain [SHD] or RBD27),[12-14] and have been suggested to control a variety of secretion events.[12] B/K,[15,16] Strep14 (Syt XIV-related protein),[17] and Tac2-N (tandem C2 protein in nucleus)[18] also contain tandem C2 domains at the C terminus, but they do not fall into any of the above groups, because they lack a specific N-terminal sequence (Fig. 1). This chapter describes recent studies (specifically after 2001) on the role of the C-type tandem C2 protein family, particularly focusing on the function of the Syt and Slp families in secretory vesicle trafficking.

Role of the Synaptotagmin Family Members in Regulated Exocytosis

Syt forms the largest branch in the phylogenetic tree of the C-type tandem C2 protein family (Fig. 1) and is found in a variety of species in different phyla.[17,19,20] In principle, Syt family members consist of five different domains, a short extracellular domain (from 0 to less than 70 amino acids), "a single transmembrane domain", a spacer domain of varying length, and tandem C2 domains, but some Syts lack one or several domains as a result of alternative splicing.[4,19] All Syt members reported thus far lack a signal peptide sequence, but they are believed to display type I membrane topology (i.e., tandem C2 domains are present in the cytoplasm). To date, 15 distinct *syt* genes (*syts I-XV* or *syts 1-15*) have been identified in mice, rats, and humans (Table 1), and several *syt* genes have been found in invertebrates.[17,19] Mammalian Syt isoforms are further classified into several subfamilies based on their sequence similarities, the Syt I/II/IX subfamily, Syt III/V/VI/X subfamily, and Syt IV/XI subfamily (dotted circles in Fig. 1), and there are no clear sequence similarities between the N-terminal domains of the different Syt subfamilies. Two additional C-type tandem C2 proteins, Strep14 and B/K are sometimes referred to as Syt XVI and Syt XVII, respectively, because they are present in the Syt branch in the phylogenetic tree (Fig. 1). At present, however, there is no evidence that either protein contains an N-terminal transmembrane domain at the protein level or mRNA level,[19] indicating that they fall outside the Syt category. Land plants also possess several Syts (e.g., At-Syts A-D), but they from a branch of the phylogenetic tree that is completely distinct from the animal Syt branch (Fig. 1), indicating that the plant Syts evolved from a different source, possibly from yeast Tricalbin proteins[21] (three C2 domains), as a result of losing their third C2 domain (Fig. 1).[17] In contrast to the animal Syts, the plant Syts lacks a putative fatty-acylation site just downstream of the transmembrane domain that may be responsible for the stable oligomerization,[22] and at present nothing is known about the function of the plant Syts in Ca^{2+}-regulated exocytosis. This section describes recent advances in our understanding of the function of mammalian Syt isoforms (or their invertebrate orthologues) in regulated exocytosis.

Syt I, II, and IX Subfamily

Role of Syt I (or II) in Synaptic Vesicle Exocytosis and Endocytosis in Neurons

Syts I, II, and IX are an evolutionarily conserved subfamily of Ca^{2+}-dependent Syts that are involved in the control of exocytosis of secretory vesicles (e.g., synaptic vesicles in neurons and/ or dense-core vesicles in neuroendocrine cells).[23-25] Both Syt I and Syt II are present on synaptic vesicles, with Syt I being predominant in the rostral region of the brain and Syt II predominating in the caudal region of the brain. The best characterized Syt isoform, Syt I, is now

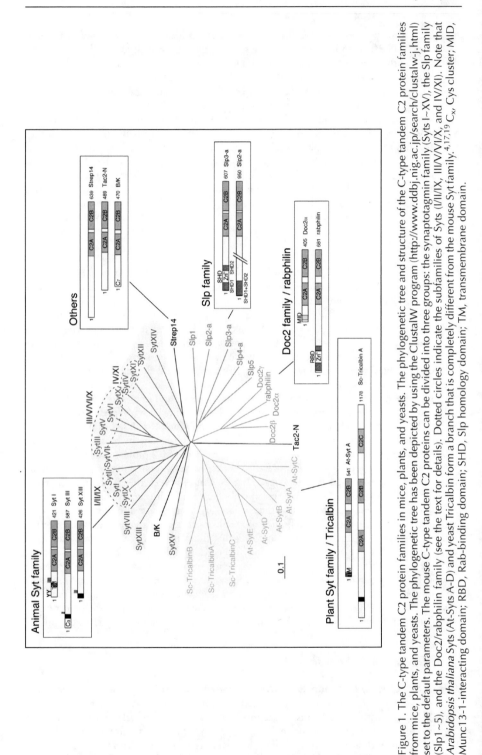

Figure 1. The C-type tandem C2 protein families in mice, plants, and yeasts. The phylogenetic tree and structure of the C-type tandem C2 protein families from mice, plants, and yeasts. The phylogenetic tree has been depicted by using the ClustalW program (http://www.ddbj.nig.ac.jp/search/clustalw-j.html) set to the default parameters. The mouse C-type tandem C2 proteins can be divided into three groups: the synaptotagmin family (Syts I~XV), the Slp family (Slp1~5), and the Doc2/rabphilin family (see the text for details). Dotted circles indicate the subfamilies of Syts (I/II/IX, III/V/VI/X, and IV/XI). Note that *Arabidopsis thaliana* Syts (At-Syts A-D) and yeast Tricalbin form a branch that is completely different from the mouse Syt family.[4,17,19] C_x Cys cluster; MID, Munc13-1-interacting domain; RBD, Rab-binding domain; SHD, Slp homology domain; TM, transmembrane domain.

widely believed to be the major low affinity Ca^{2+}-sensor for neurotransmitter release.[26-29] Syt I is not a simple Ca^{2+}-sensor, and presumably regulates synaptic vesicle exocytosis itself by modulating the synaptic vesicle docking step,[30,31] fusion pore expansion dynamics,[32,33] and endocytosis.[34-36] A variety of potential ligands of the C2A (or C2B) domain of Syt I (or II), including proteins, divalent cations, phospholipids, and soluble inositol polyphosphates, have been reported thus far (for a review see refs. 4-6), and it has been hypothesized that their binding to the Syt I C2 domains regulates synaptic vesicle exocytosis and endocytosis. For example, Ca^{2+}-dependent binding of Syt I with SNARE (soluble *N*-ethylmaleimide-sensitive factor attachment protein receptor) complex composed of syntaxin-1a, SNAP-25, and VAMP-2/synaptobrevin-2 or PIP_2 (phosphatidylinositol 4,5-bisphosphate) has been hypothesized to promote Ca^{2+}-dependent synaptic vesicle fusion.[37-39] Ca^{2+}-independent binding of Syt I with t-SNARE heterodimers to be involved in synaptic vesicle exocytosis.[40] Ca^{2+}-dependent self-oligomerization of Syt I to control fusion pore dynamics;[41] and Ca^{2+}-independent binding of the Syt I C2B domain with clathrin assembly protein AP-2 and/or stonin-2 to control synaptic vesicle endocytosis.[42]

How does Syt I control distinct steps of synaptic vesicle trafficking? Although the precise mechanism remains to be elucidated, the presence of the multiple ligand binding sites in the Syt I C2 domains may ensure binding to a variety of molecules and the control of distinct steps in synaptic vesicle trafficking.[4] As shown in Figure 2, both C2 domains are composed of an eight-stranded anti-parallel β-sandwich structure (β1-β8 strands), and two and three Ca^{2+} ions bind the loop regions formed at the top of the C2A and C2B β-sandwich structure, respectively, and the C2B domain contains an additional α-helix between the β7 and β8 strands.[43] The Ca^{2+}-binding loops of the C2A domain and C2B domain are presumed to face with each other, and redundant Ca^{2+}-binding sites that mediate binding of t-SNAREs are thought to be formed (Fig. 2A).[44] In contrast to the C2A domain, the C2B domain contains two additional ligand-binding sites (Fig. 2A,B). A polybasic sequence in the β4 strand (known as the "C2B effector domain") is required for binding of a variety of molecules, including AP-2 and inositol polyphosphates,[45] and has been suggested to regulate synaptic vesicle exocytosis and endocytosis. The polybasic sequence is also required for suppression of spontaneous neurotransmitter release.[46] The conserved WHXL motif in the β8 strand is required for binding of a plasma membrane protein, neurexin Iα, in vitro, and it has been suggested to be involved in the synaptic vesicle docking to the plasma membrane.[30] The WHXL motif is also important for the maintenance of the C2B structure,[47] and C-terminal fusion of fluorescent protein (e.g., GFP and CFP) has been shown to impair Syt I function, presumably by abrogating the function of the WHXL motif.[48] These three ligand-binding sites in the C2B domain are not totally independent, and ligand-binding of one site presumably affects the function of the other ligand-binding sites.

Syts I and II undergo several posttranslational modifications (e.g., phosphorylation, *N*- and *O*-glycosylation at the extracellular domain,[49] and fatty-acylation just downstream of the transmembrane domain),[22] and the glycosylation or palmitoylation of Syt I has recently been shown to be required for efficient targeting of the Syt I molecule to secretory vesicles.[50-52]

Role of Syt IX in Dense-Core Vesicle Exocytosis in Neuroendocrine Cells

Although Syt I is also expressed on the dense-core vesicles in certain neuroendocrine cells (e.g., chromaffin cells and pituitary cells) and involved in the control of their exocytosis,[53,54] Syt I is dispensable for dense-core vesicle exocytosis by PC12 cells, because Syt I-deficient PC12 cells display normal hormone secretion activity,[55,56] indicating the presence of an alternate Ca^{2+}-sensor on dense-core vesicles in PC12 cells. The most likely candidate for the alternate Ca^{2+}-sensor for dense-core vesicle exocytosis is Syt IX (originally described as Syt V),[57,58] which is abundantly expressed on the Syt I-containing dense-core vesicles in PC12 cells.[23,38,56] Syt IX is also expressed on insulin-containing vesicles in pancreatic β-cell lines[59] and islet β-cells[60] which do not endogenously express Syt I. Functional ablation of Syt IX has been shown to reduce Ca^{2+}-dependent hormone secretion by PC12 cells and pancreatic

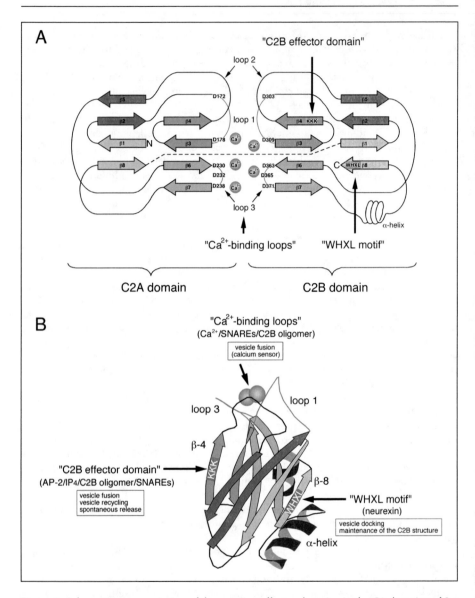

Figure 2. Schematic representation of the putative effector domains in the C2 domains of Syt I. A) Schematic representation of the eight-stranded β-sandwich structure of the C2A domain and C2B domain of Syt I. Three Ca^{2+} ions are bound to the top loops 1 and 3 of the C2A domain and two Ca^{2+} ions are bound to the top loops 1 and 3 of the C2B domain. Five Asp residues (D172, D178, D230, D232, and D328 in the C2A domain; and D303, D309, D363, D365, and D371 in the C2B domain) are thought to be essential for Ca^{2+}-binding. Two C2 domains are presumed to face with each other, and redundant Ca^{2+}-binding sites seem to form between them. An additional α-helix is present between the β7 and β8 strands of the C2B domain.[43] The polybasic sequence (KKK) in the β4 strand of the C2B domain (named the C2B effector domain) is involved in binding a variety of ligands.[4-6] The WHXL motif in the β8 strand is involved in binding of neurexin in vitro,[30] and is required for proper folding of the C2B domain.[47] B) Schematic representation of three ligand-binding sites of the Syt I C2B domain and their proposed function in synaptic vesicle trafficking (see the text for details).

β-cells.[23,56,59,60] Interestingly, Syt IX is also expressed in a rat basophilic leukemia mast cell line (RBL-2H3) and regulates transport from the perinuclear endocytic recycling compartment (ERC) to the cell surface via interaction with microtubules.[61,62]

Syt III, V, VI, and X Subfamily

Syts III, V, VI, and X exhibit high sequence similarities throughout the entire proteins, and this subfamily of Syts is characterized by the presence of an N-terminal conserved Cys motif, which is essential for homo-dimer formation through disulfide bonding.[63] All members of this subfamily exhibit Ca^{2+}/phospholipid binding activity,[25,64,65] and have been suggested to regulate Ca^{2+}-dependent exocytosis of secretory vesicles. This subfamily of Syts has been retained in vertebrates alone, and is not found in invertebrates.[4,19]

Role of Syt III in Regulated Exocytosis: Plasma Membrane or Vesicular Ca^{2+}-Sensor?

Syt III is abundantly expressed in all brain regions, and its expression increases in parallel with synaptogenesis during postnatal development of the mouse brain. Unlike Syts I and II, Syt III protein is mainly present on the presynaptic plasma membrane, rather than on synaptic vesicles,[66] and it has been suggested to function as a plasma membrane high-affinity Ca^{2+}-sensor for neurotransmitter release.[64] At present, however, nothing is known about the functional involvement of endogenous Syt III in synaptic vesicle exocytosis, and the expression and subcellular localization of Syt III in neuroendocrine cells (particularly in pancreatic β-cells) are still matters of controversy (dense-core vesicles versus plasma membrane). Recent work indicates that Syt III protein is present just beneath the plasma membrane of pancreatic polypeptide cells, but that it is not present in α-, β-, or δ-cells.[60] Syt III is also expressed in RBL-2H3 mast cells and is involved in the formation and delivery of cargo to the perinuclear endocytic recycling compartment.[67]

Role of Syt V in Dense-Core Vesicle Exocytosis in Specific Types of Neuroendocrine Cells

Syt V has been proposed to function as a Ca^{2+}-sensor for exocytosis of specific populations of dense-core vesicles.[68] Syt V is enriched in the dense-core vesicle fraction of specific regions of mouse brain, and it is also expressed in pancreatic islet α-cells that secrete glucagon, but not in β- or δ-cells.[59,68]

Role of Syt VI in the Acrosome Reaction of Sperm

An alternative splicing isoform of Syt VI lacking a transmembrane domain (named Syt VIΔTM)[69] is predominantly expressed in selected regions of mouse brain (e.g., olfactory bulb), and it is present in the various membrane fractions.[66] Although the function of Syt VI (or Syt VIΔTM) in regulated exocytosis in the brain remains completely unknown, Syt VI regulates acrosome reaction, a unique Ca^{2+}-regulated exocytosis, in sperm.[70,71] Phosphorylation of the polybasic region of the Syt VI C2 domains, which corresponds to the C2B effector domain of Syt I (see above and Fig. 2), modulates Syt VI function in acrosomal exocytosis in human sperm.[72]

Syt X Is an Immediate Early Gene Product in Brain Whose Expression Is Induced by Kainic Acid-Induced Seizures

Syt X has been identified as an immediate early gene product in rat brain by differential display between brain exposed and unexposed to kainic acid.[73] However, the tissue distribution and subcellular localization of Syt X protein remain completely unknown.

Syt IV and XI Subfamily

The Syt IV and XI subfamily is evolutionarily conserved from *Caenorhabditis elegans* to humans,[4,19,74] and shares a point mutation of one of the conserved acidic residues (e.g., Ser-244 of mouse Syt IV, which corresponds to the Asp-232 of mouse Syt I; see Fig. 2A) in the putative Ca^{2+}-binding loop3 of the C2A domain. As a result of this mutation, the isolated C2A domain of Syts IV and XI lacks Ca^{2+}/phospholipid binding activity. Recent structural analysis of the mouse Syt IV C2B domain has shown that it is unlikely to bind Ca^{2+} ions despite possessing the five conserved acidic residues in the putative Ca^{2+}-binding loops.[75] Moreover, "overexpression studies" have suggested that Syt IV functions as a negative regulator of regulated exocytosis.[76,77] Interestingly, however, more recent analysis of "endogenous" Syt IV protein has indicated that Syt IV functions as a positive regulator of some forms of regulated exocytosis (e.g., glutamate release from astrocytes).[78,79]

Subcellular Localization and Possible Function of Syt IV Protein in Brain and PC12 Cells

The Syt IV isoform has also been identified as an immediate early gene product induced by membrane depolarization (e.g., high-KCl, forskolin, and kainic acid) in brain and PC12 cells.[80-82] Syt IV expression is found in all brain regions, but, unlike other Syt isoforms, its protein expression levels in mouse brain are highest during the 1st week of postnatal development.[81] Syt IV is mainly present in the Golgi and in the tips of growing neurites in developing neurons, and the signals in the tips of axons and dendrites almost disappear in mature neurons.[81] Syt IV is not a synaptic vesicle protein in either mice or *Drosophila*, and it is mainly localized on certain vesicle/organelle structures in dendrites.[74,83] Although the specific cargo of Syt IV-containing vesicles is unknown, it is highly possible that they contain molecules that are directly involved in changes in synaptic plasticity, because Syt IV knockout mice exhibit abnormalities of some forms of memory related to the hippocampus.[80] In contrast to neurons, in which Syts I and II are abundantly expressed, the Syt IV isoform, but not Syt I, is specifically expressed in astrocytes, and Syt IV is required for Ca^{2+}-induced glutamate release from astrocytes, strongly indicating that Syt IV is a positive regulator of exocytosis.[79] Although Syt IV itself is not a synaptic vesicle protein, if Syt IV is ectopically expressed on synaptic vesicles in neurons, it can rescue the impairment of neurotransmitter release in the Syt I null mutant fly.[84]

In nerve growth factor (NGF)-differentiated PC12 cells, Syt IV is also present in the Golgi and distal portion of neurites, where dense-core vesicles are accumulated.[78] In undifferentiated PC12 cells, however, Syt IV is present in the Golgi and immature secretory vesicles, rather than mature dense-core vesicles. It is very interesting that Syt IV is sorted to newly formed, mature dense-core vesicles that undergo Ca^{2+}-dependent exocytosis in response to NGF.[78] The spacer domain of Sty IV, which is not conserved in other Syt isoforms, primarily determines the Golgi localization of Syt IV.[85]

Syt XI Interaction with Parkin

Syt XI was originally identified as a closely related isoform of Syt IV that is abundantly expressed in brain. The same as Syt IV, Syt XI expressed in PC12 cells is mainly targeted to the BFA (brefeldin A)-sensitive perinuclear region (presumably the Golgi),[86] although almost nothing is known about the function and localization of endogenous Syt XI in brain. The only information available is that parkin, an autosomal recessive juvenile Parkinson disease gene product, interacts with and ubiquitinates Syt XI.[87]

Syt VII

Syt VII is a ubiquitous Syt isoform that is evolutionarily conserved from *C. elegans* to humans. There are several alternative splicing isoforms of mammalian Syt VII (e.g., Syt VIIα-γ or Syt VIIa-e),[88,89] but the shortest form (named Syt VIIα) is predominant in most mouse, rat, and human tissues.[89] The mammalian Syt VII isoform binds Ca^{2+}/phopholipids with higher affinity than Syt I does, suggesting that Syt VII functions as a high affinity Ca^{2+}-sensor.[64] In addition, both C2 domains of Syt VII contribute to the formation of Ca^{2+}-dependent multimers and hetero-oligomers with other Syt isoforms.[90,91]

The subcellular localization of Syt VII protein is a matter of some controversy, and it has been shown to be present in three different compartments: presynaptic plasma membranes in neurons,[88] dense-core vesicles in endocrine cells,[92,93] and lysosomes in fibroblasts.[94] Syt VII regulates Ca^{2+}-dependent lysosomal exocytosis in fibroblasts, which contributes to plasma membrane repair.[94-96] Syt VII knockout mice also exhibit impaired plasma membrane repair as well as autoimmune myositis, but no neurological abnormalities have been observed.[97] While transiently overexpressed Syt VII in PC12 cells is mainly targeted to the plasma membrane, stably expressed Syt VII in PC12 cells is mainly targeted to dense-core vesicles and regulates their exocytosis, suggesting that Syt VII functions as a vesicular Ca^{2+}-sensor, rather than as a plasma membrane Ca^{2+}-sensor.[92,93]

Other Syt Isoforms (VIII, XII-XV)

The C2 domains of all other Syt isoforms (VIII, XII-XV)[17,20,86] lack Ca^{2+}/phospholipid binding activity.[98] Syt XII (also known as Syt-related gene 1, Srg1) is specifically expressed in brain, and its expression level is regulated by thyroid hormone.[99] Syt VIII is a ubiquitous Syt isoform and is localized at the acrosomal crescent of sperm cells.[100] Syt VIII has been proposed to regulate the acrosome reaction through Ca^{2+}-dependent interaction with syntaxin-2.[71] Nothing is known about the subcellular localization and function of other Ca^{2+}-independent Syt isoforms (XIII-XV).

Role of Slp Family Members in Rab27-Dependent Membrane Trafficking

Slp (Syt-like protein) is defined as a protein that consists of an N-terminal "SHD (Slp homology domain)", a linker domain of varying length, and C-terminal tandem C2 domains.[3,10-12,101] Five different isoforms (Slp1~5)[101,102] are present in mice, rats, and humans, a single isoform (dm-Slp/Btsz)[103] in *Drosophila*, and none in *C. elegans* (Table 1).[12,17] Several alternative splicing isoforms that lack one or several domains have been reported in most Slp members.[101] The SHD consists of two potential α-helical regions (named SHD1 and SHD2; see Fig. 1), which are often separated by two zinc finger motifs. The SHD shows weak similarity to the Rab3A-binding domain of rabphilin, and it is now widely believed to function as a specific effector domain for Rab27A (or Rab27B), one of the small GTPase Rab proteins.[12-14,104-106] The SHD is also found in the N-terminal region of members of the other protein family, named the Slac2 family (Slac2-a/melanophilin, Slac2-b, and Slac2-c/MyRIP) (for a review see ref. 12). The first α-helical SHD1 alone, and not the SHD2, is capable of interacting with GTP-Rab27A in vitro, and a single point mutation in the SHD1 (e.g., E14A in Slac2-a) completely abrogates Rab27A/B-binding activity.[14] In vivo, however, both domains are absolutely required for targeting the SHD to endogenous Rab27. For example, the entire SHD of Slac2-a or Slp4-a, but not the Slac2-a-SHD1, Slac2-a-SHD2, or the Slp4-a-ΔSHD1, is recruited to Rab27A on dense-core vesicles in the neurites of NGF-differentiated PC12 cells (Fig. 3). By contrast, zinc finger domains are not required for Rab27A/B binding activity, and they are presumably involved in the stability of the SHD structure.[14] This section describes the recent discovery of the Rab27 effector function of Slps in Rab27-dependent membrane trafficking, with particular emphasis on the docking of Rab27-containing organelles to the plasma membrane.

Role of Slp2-a in Melanosome Transport in Melanocytes

Slp2-a is abundantly expressed on the melanosomes of cultured melanocytes and is involved in intracellular melanosome transport.[13] Slp2-a simultaneously interacts with Rab27A on the melanosome via the N-terminal SHD and with phosphatidylserine (PS) in the plasma

Table 1. C-type tandem C2 proteins in humans, mice, rats, and invertebrates

Name of Protein	Human Chromosome	Mouse Chromosome	Rat Chromosome	Drosophila (Dm) C. elegans (Ce)	Tissue Distribution and Subcellular Localization	Function and Other Properties
Syt I	12q21	10	7q13	Dm/Ce	Rostral brain; SVs and growth cone vesicles. Endocrine cells; DCVs and SLMVs.	Putative low affinity Ca^{2+}-sensor for neurotransmitter release. Neurite outgrowth; axonal repair; endocrine exocytosis.
Syt II	1q31.1	1	13q13		Caudal brain; SVs.	Putative Ca^{2+}-sensor for neurotransmitter release in caudal brain.
Syt III	19q13.33	7	1q21		Mast cells; lysosomes. Brain; presynaptic PM. Pancreatic β-, δ-, or polypeptide cells; SGs or PM. Mast cells; perinuclear ERC.	Lysosomal exocytosis in mast cells. Putative Ca^{2+}-sensor for secretory granule exocytosis? Control of secretory granule size in mast cells.
Syt IV	18q12.3	18	18q12.1	Dm/Ce	Brain; Golgi and unknown vesicles in axons/dendrites. Astrocytes; glutamate-containing vesicles. PC12 cells; Golgi, immature SVs, and DCVs	Immediate early gene; Syt IV protein level rapidly changes in response to various extracellular stimuli. Putative Ca^{2+}-sensor for glutamate release from astrocytes.
Syt V[a]	19q	7 E3	1q12		Brain and pancreatic α-cells; DCVs.	Putative Ca^{2+}-sensor for DCV exocytosis.
Syt VI	1p12	3 F3	2q34		Brain; various membrane fractions (Syt VIΔTM). Sperm head; outer acrosomal membrane.	Putative Ca^{2+}-sensor for acrosome reaction in sperm.
Syt VII	11q12.2	19 B	1q41	Dm/Ce	Brain; presynaptic PM. Fibroblasts; lysosomes. Endocrine cells; DCVs.	Plasma membrane repair and lysosomal exocytosis in fibroblasts. Putative Ca^{2+}-sensor for DCV exocytosis.
Syt VIII	11p15.5	7 F5	1q37		Sperm head; acrosomal crescent. Renal tubule epithelial cells; unknown.	Putative Ca^{2+}-sensor for acrosome reaction in sperm.
Syt IX (Syt V) [a]	11p15.4	7 A1	1q33		Endocrine cells; DCVs.	Putative major Ca^{2+}-sensor for DCV exocytosis in endocrine cells.

Table continued on next page

Table 1. Continued

Name of Protein	Human Chromosome	Mouse Chromosome	Rat Chromosome	Drosophila (Dm) C. elegans (Ce)	Tissue Distribution and Subcellular Localization	Function and Other Properties
Syt X	12p11.1	15 E3	7q34		Neuronal and non-neuronal tissues; unknown.	Syt X mRNA level is rapidly increased by kainic acid seizures.
Syt XI	1q21.2	3 F1	2q34		Brain; unknown.	Function unknown.
Syt XII/Srg1	11q13.1	19 A	1q41	Dm	Brain; unknown.	Syt XII protein level is rapidly increased by thyroid hormone.
Syt XIII	11p12-p11	2 E1	3q24	Dm	Abundant in brain. Non-neuronal tissues; unknown.	Function unknown.
Syt XIV	1q32.2	1 H6	ND[d]	Dm	Non-neuronal tissues; unknown.	Function unknown.
Syt XV	10q11.1	14 B	16p15		Non-neuronal tissues; unknown.	Function unknown.
Strep14 (Syt XVI)[b]	14q23.2	12 C3	6q24		Non-neuronal tissues; unknown.	Function unknown. Syt IV-related protein without TM.
Slp1/JFC1	1p36.11	4 D2.3	5q36	Dm?[c]	Prostate epithelial cells; prostate-specific antigen-containing vesicles.	PIP$_3$- and NADPH oxidase-binding protein. Putative Rab27 effector. Phosphorylation of JFC1 at serine 241 by Akt.
Slp2	11q14	7 D3	1q32		Melanocytes; melanosomes.	Anchoring of melanosomes to the PM through interaction with PS.
Slp3	6q25.3	17 A1	1q11		Unknown.	Enhancement of DCV exocytosis in PC12 cells. Putative Rab27 effector.
Slp4/ granuphilin	Xq21.33	X E3	Xq34		Pancreatic β-cells and PC12 cells; DCVs. Parotid gland; amylase-containing granules.	Docking of DCVs to the PM through interactionwith Munc18-1/2 and/or syntaxin-1~3. Rab3/8/27-binding protein.
Slp5	Xq21.1	X A1.1	Xq13		Pancreatic β-cells; unknown.	Enhancement of DCV exocytosis in PC12 cells. Putative Rab27 effector.

Table continued on next page

Table 1. Continued

Name of Protein	Human Chromosome	Mouse Chromosome	Rat Chromosome	Drosophila (Dm) C. elegans (Ce)	Tissue Distribution and Subcellular Localization	Function and Other Properties
rabphilin	12q24.13	5 F	12q16	Dm/Ce	Brain; SVs. Certain endocrine cells; DCVs.	Enhancement of SG exocytosis. Rab3/Rab27 effector. Docking of DCVs to the PM through interaction with SNAP-25.
Doc2α	16p11.2	7 F4	1q36		Brain; SVs. Certain endocrine cells; DCVs.	Enhancement of SG exocytosis. Munc13-1- and Munc18-1-binding protein.
Doc2β	17p13.3	11 B5	10q24		Brain; SVs. Non-neuronal tissues; unknown.	Enhancement of SG exocytosis? Munc13-1- and Munc18-1-binding protein.
Doc2γ	11q13.2	19 A	1q42		Certain non-neuronal tissues; nucleus?	Function unknown. Targeted to nucleus via the NLS in the C2A domain.
B/K (Syt XVII)[b]	16p12.3	7	1q35		Brain and kidney; ER.	B/K protein expression level is increased by ER stress. B/K is targeted to TGN in PC12 cells via the N-terminal Cys cluster.
Tac2-N	14q32.12	12 F1	6q32		Certain non-neuronal tissues; nucleus?	Function unknown. Targeted to nucleus via the NLS in the C2B domain.

[a] Two different Syt V sequences were reported at the same time, but the Syt V identified by Craxton and Goedert[57] and by Hudson and Birnbaum[58] is usually referred to as Syt IX (or Syt 9) in the literature to distinguish it from the Syt V identified by Li et al.[133] [b] Although the official gene symbols Syt XVI and Syt XVII refer to Strep14 and B/K, respectively, Strep14 and B/K proteins do not contain a putative transmembrane domain and thus fall outside the definition of a Syt category (i.e., a protein with a single transmembrane domain and tandem C2 domains). [c] Drosophila Slp homologue bitesize protein normally lacks a SHD, even though the putative exon 2 encodes a typical type I SHD, and dm-Slp/Btsz shows the highest homology to Slp2-a in the phylogenetic tree.[4] [d] ND: not determined; DCV: dense-core vesicle; ER: endoplasmic reticulum; ERC: endocytic recycling compartment; NLS: nuclear localization signal; PM: plasma membrane; PS: phosphatidylserine; SG: secretory granule; SLMV: synaptic-like microvesicle; SV: synaptic vesicle; TGN: trans-Golgi network; TM: transmembrane domain.

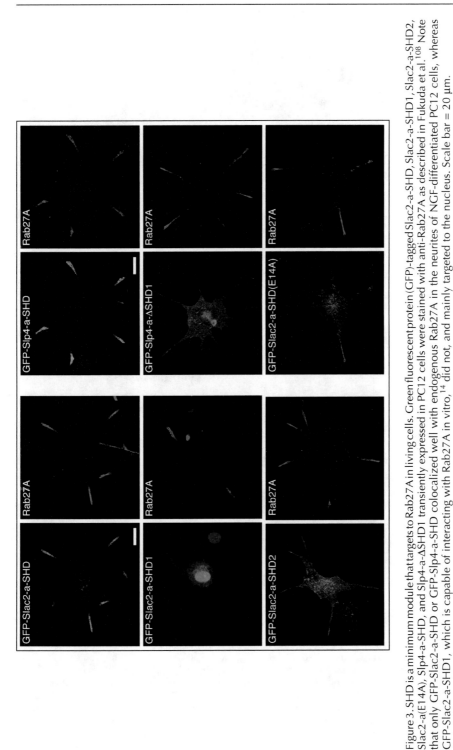

Figure 3. SHD is a minimum module that targets to Rab27A in living cells. Green fluorescent protein (GFP)-tagged Slac2-a-SHD, Slac2-a-SHD1, Slac2-a-SHD2, Slac2-a(E14A), Slp4-a-SHD, and Slp4-a-ΔSHD1 transiently expressed in PC12 cells were stained with anti-Rab27A as described in Fukuda et al.[108] Note that only GFP-Slac2-a-SHD or GFP-Slp4-a-SHD colocalized well with endogenous Rab27A in the neurites of NGF-differentiated PC12 cells, whereas GFP-Slac2-a-SHD1, which is capable of interacting with Rab27A in vitro,[14] did not, and mainly targeted to the nucleus. Scale bar = 20 μm.

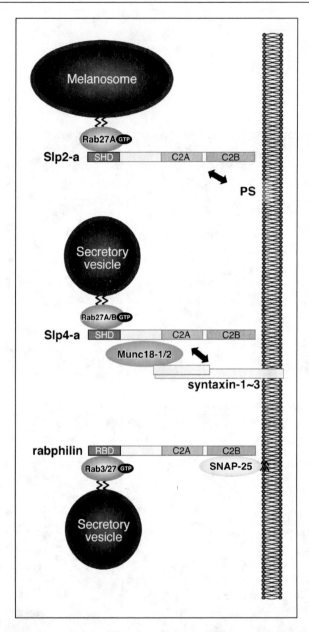

Figure 4. Proposed function of Rab27 effectors in the docking of Rab27A-containing organelles to the plasma membrane. Top) Slp2-a anchors the melanosome to the plasma membrane in melanocytes through simultaneously interacting with Rab27A on the melanosome via the SHD and with PS in the plasma membrane via the C2A domain.[107] Middle) Slp4-a is involved in the dense-core vesicle docking step through simultaneously interacting with Rab27A on the vesicle via the SHD and with Munc18·syntaxin complex or syntaxin via the linker domain.[112] Bottom) Rabphilin promotes docking of the dense-core vesicle to the plasma membrane through simultaneously interacting with Rab3/Rab27 on the vesicle via the RBD and with SNAP-25 at the plasma membrane via the C2B domain[131] (see the text for details).

membrane via the C2A domain, and thereby promotes melanosome anchoring to the plasma membrane in melanocytes (Fig. 4, top). Knockdown of Slp2-a by the specific siRNA (small interfering RNA) results in a reduction in the numbers of peripheral melanosomes in melanocytes (referred to as "peripheral dilution phenotype") and induction of a rounded cell shape, rather than their normal elongated (or dendritic) shape.[107] Although the Slp2-a C2A domain is required for maintenance of the melanocyte morphology, the exact mechanism by which Slp2-a regulates cell morphology remains unknown. Since Slp2-a is also expressed in other cell types besides melanocytes,[101] Slp2-a may be involved in the docking step of Rab27A-containing organelles to the plasma membrane in these cells.

Role of Slp4-a/granuphilin-a in Secretory Vesicle Docking to the Plasma Membrane

Slp4-a was originally identified as granuphilin-a, which is specifically localized on insulin-containing vesicles in pancreatic β-cells, however, recent studies have shown that Slp4-a is also expressed in dense-core vesicles in PC12 cells[108] and amylase-containing vesicles in parotid acinar cells.[109] In contrast to the other Slp members, Slp4-a exhibits several unique biochemical features. First, the SHD of Slp4-a is capable of interacting with Rab3 and Rab8 isoforms in addition to Rab27 isoforms in vitro, whereas others specifically interact with Rab27 isoforms.[13,106,110,111] Second, the SHD of Slp4-a is capable of interacting with Rab27A(T23N), which mimics the GDP-bound form of Rab27A, and the others are not.[110] Third, the linker domain of Slp4-a, which is not conserved among the Slp family members, interacts with Munc18-1, Munc18-1·syntaxin-1a complex, and Munc18-2·syntaxin-2/3 complex.[110,112] Syntaxin-1a has been reported to interact with the SHD in a Rab27A-dependent manner,[113] but this interaction has not been observed in other studies.[110,111] Finally, expression of Slp4-a, but not other Slps, in pancreatic β-cells or PC12 cells strongly attenuates Ca^{2+}-dependent hormone secretion,[104,108,110-113] although Slp4-a expression increases the number of vesicles docked to the plasma membrane.[114] Consistent with these in vitro cell culture findings, Slp4-a knockout mice contain fewer insulin-containing vesicles docked to the plasma membrane in islet β-cells but possess increased insulin secretion activity.[115] It is of great interest that the level of expression of both syntaxin-1a and Munc18-1 is also reduced in Slp4-a knockout mice.[115] All these observations suggest that Slp4-a simultaneously interacts with Rab27A/B on the secretory vesicles via the N-terminal SHD and with a certain syntaxin·Munc18 complexes in the plasma membrane via the linker domain, and thereby promotes the docking of the dense-core vesicle to the plasma membrane in secretory cells (Fig. 4, middle). Consistent with this notion, Rab27A or Munc18-1 itself has been shown to control the docking step of secretory vesicles in certain cell types.[116-118]

Role of Other Slps in Secretory Vesicle Exocytosis

Slp1 (also called JFC1) is characterized as a phosphatidylinositol 3,4,5-trisphosphate (PIP_3)-binding protein[119] and as an Akt substrate.[120] Slp1 has been shown to control secretion of prostate-specific antigen (PSA) by prostate cell lines,[121] although the mechanism by which Slp1 regulates PSA secretion is largely unknown. Nevertheless, it is tempting to speculate that Slp1 simultaneously interacts with Rab27A on the granules via the SHD, and with PIP_3 in the plasma membrane via the C2A domain, and promotes granule docking to the plasma membrane, the same as Slp2-a does.

Almost nothing is known about the expression and localization of Slp3-a and Slp5, two Ca^{2+}-dependent type Slps[98,102] that promote dense-core vesicle exocytosis when expressed in PC12 cells.[110] It is currently unknown whether Slp3-a and Slp5 are involved in the docking step of secretory vesicles.

Role of Rabphilin in Rab27-Dependent Membrane Trafficking

The Doc2/rabphilin family in mice, rats, and humans consists of three Doc2 isoforms (α, β, and γ)[122,123] and one rabphilin (Table 1), and only the rabphilin has been retained during evolution (from *C. elegans* to humans).[17] Doc2α and rabphilin are mainly expressed in neural tissues and certain neuroendocrine cells, whereas Doc2β and Doc2γ are expressed ubiquitously. Involvement of Doc2α/β or rabphilin in regulated exocytosis was intensively investigated by biochemical, structural, and genetic analyses before 2001,[9,124-126] and there have been only a few extensions of research on this family since 2002. One of the most important findings concerning this family is that rabphilin functions as a Rab27 effector, rather than as a Rab3 effector, especially in invertebrates.[127,128] In the final section, I describe the recent discovery of the Rab27 effector function of rabphilin in regulated exocytosis.

Genetic analysis of rabphilin mutant animals[125,126] and biochemical studies[129,130] have indicated that rabphilin is targeted to secretory vesicles independently of the function of Rab3. In early 2002, our group discovered that the Rab-binding domain of rabphilin also interacts in vitro with Rab27, the closest subfamily of Rab3 in the phylogenetic tree.[13,106] Although the endogenous expression level of Rab27A in PC12 cells is lower than that of Rab3A, rabphilin preferentially interacts with Rab27A, because rabphilin binds Rab27A with much higher affinity than Rab3A.[127,128] It should be noted that invertebrate rabphilin specifically binds Rab27, but does not bind Rab3 or Rab8, indicating that rabphilin has functioned as a Rab27 effector during evolution.[127] It has recently been proposed that rabphilin promotes docking of dense-core vesicles to the plasma membrane in PC12 cells through simultaneous interaction with Rab3/27 on the dense-core vesicle via the RBD and with SNAP-25 at the plasma membrane via the C2B domain (Fig. 4, bottom).[131] In contrast to Slp4-a described above, expression of rabphilin significantly increases the depolarization-induced exocytotic fusion events in PC12 cells.[110,131]

Concluding Remarks

The C-type tandem C2 protein family is a large family of putative membrane trafficking proteins found in a variety of species from invertebrates, to vertebrates, and to plants (Fig. 1 and Table 1). Although more than 1000 of papers on the C-type tandem C2 proteins have been published in the literature, this chapter has mainly dealt with recent advances in research on the molecular mechanism of membrane trafficking mediated by the C-type tandem C2 protein family. If the reader is interested in learning more about this family and would like to know of other outstanding studies on the role of the C-type tandem C2 protein family that have been published before 2001, the following reviews provide the details on individual subfamilies of C-type tandem C2 proteins: for the Syt family (see refs. 4-6), for the Doc2/rabphilin family (see refs. 9,132), and for the Slp family (see refs. 10-12). Tremendous advances have been made in research on C-type tandem C2 protein during the past five years, including on the role of Syt I in multiple stages of synaptic vesicle trafficking (Fig. 2), involvement of other Syt isoforms in specific types of Ca^{2+}-regulated exocytosis (e.g., lysosomal exocytosis in fibroblasts, acrosome reaction in sperm, and glutamate release from astrocytes), and novel docking machinery consisting of Rab27 and its effectors (Slp and rabphilin; Fig. 4). By contrast, the tissue distribution, subcellular localization, and function of more than half of the C-type tandem C2 proteins remain unknown. Therefore, one important direction of future research will be the elucidation of the localization and function of individual C-type tandem C2 proteins of unknown function. Another important direction of research will be the determination of functional relationships between the distinct families of C-type tandem C2 proteins during regulated exocytosis, because most secreting cells express members of several families of C-type tandem C2 proteins (e.g., Syt, Slp, and Doc2/rabphilin family) in a single cell type. Intensive studies will be required to achieve full understanding of the roles of the C-type tandem C2 proteins in membrane trafficking at the molecular level.

Acknowledgement

This work was supported in part by the Ministry of Education, Culture, Sports, and Technology of Japan (Grants 15689006, 16044248, 17024065, and 17657067), by the Kato Memorial Bioscience Foundation (to M. Fukuda), by the NOVARTIS Foundation (Japan) for the Promotion of Science, and by the Life Science Foundation of Japan. I thank members of the Fukuda Initiative Research Unit in RIKEN for preparing the manuscripts and for helpful discussions.

References

1. Burgoyne RD, Morgan A. Secretory granule exocytosis. Physiol Rev 2003; 83(2):581-632.
2. Nalefski EA, Falke JJ. The C2 domain calcium-binding motif: Structural and functional diversity. Protein Sci 1996; 5(12):2375-2390.
3. Fukuda M, Mikoshiba K. Synaptotagmin-like protein 1-3: A novel family of C-terminal-type tandem C2 proteins. Biochem Biophys Res Commun 2001; 281(5):1226-1233.
4. Fukuda M. Synaptotagmins, Ca^{2+}- and phospholipid-binding proteins that control Ca^{2+}-regulated membrane trafficking. Recent Res Dev Chem Phys Lipids 2003; 1:15-51.
5. Südhof TC. Synaptotagmins: Why so many? J Biol Chem 2002; 277(10):7629-7632.
6. Chapman ER. Synaptotagmin: A Ca^{2+} sensor that triggers exocytosis? Nat Rev Mol Cell Biol 2002; 3(7):498-508.
7. Shirataki H, Kaibuchi K, Sakoda T et al. Rabphilin-3A, a putative target protein for smg p25A/rab3A p25 small GTP-binding protein related to synaptotagmin. Mol Cell Biol 1993; 13(4):2061-2068.
8. Orita S, Naito A, Sakaguchi G et al. Physical and functional interactions of Doc2 and Munc13 in Ca^{2+}-dependent exocytotic machinery. J Biol Chem 1997; 272(26):16081-16084.
9. Duncan RR, Shipston MJ, Chow RH. Double C2 protein: A review. Biochimie 2000; 82(5):421-426.
10. Fukuda M. Slp and Slac2, novel families of Rab27 effectors that control Rab27-dependent membrane traffic. Recent Res Dev Neurochem 2002; 5:297-309.
11. Cheviet S, Waselle L, Regazzi R. Noc-king out exocrine and endocrine secretion. Trends Cell Biol 2004; 14(10):525-528.
12. Fukuda M. Versatile role of Rab27 in membrane trafficking: Focus on the Rab27 effector families. J Biochem (Tokyo) 2005; 137(1):9-16.
13. Kuroda TS, Fukuda M, Ariga H et al. The Slp homology domain of synaptotagmin-like proteins 1-4 and Slac2 functions as a novel Rab27A binding domain. J Biol Chem 2002; 277(11):9212-9218.
14. Fukuda M. Synaptotagmin-like protein (Slp) homology domain 1 of Slac2-a/melanophilin is a critical determinant of GTP-dependent specific binding to Rab27A. J Biol Chem 2002; 277(42):40118-40124.
15. Kwon OJ, Gainer H, Wray S et al. Identification of a novel protein containing two C2 domains selectively expressed in the rat brain and kidney. FEBS Lett 1996; 378(2):135-139.
16. Fukuda M, Mikoshiba K. The N-terminal cysteine cluster is essential for membrane targeting of B/K protein. Biochem J 2001; 360(2):441-448.
17. Fukuda M. Molecular cloning, expression, and characterization of a novel class of synaptotagmin (Syt XIV) conserved from Drosophila to humans. J Biochem (Tokyo) 2003; 133(5):641-649.
18. Fukuda M, Mikoshiba K. Tac2-N, an atypical C-type tandem C2 protein localized in the nucleus. FEBS Lett 2001; 503(2-3):217-218.
19. Craxton M. Synaptotagmin gene content of the sequenced genomes. BMC Genomics 2004; 5(1):43.
20. Fukuda M. Molecular cloning and characterization of human, rat, and mouse synaptotagmin XV. Biochem Biophys Res Commun 2003; 306(1):64-71.
21. Schulz TA, Creutz CE. The tricalbin C2 domains: Lipid-binding properties of a novel, synaptotagmin-like yeast protein family. Biochemistry 2004; 43(13):3987-3995.
22. Fukuda M, Kanno E, Ogata Y et al. Mechanism of the SDS-resistant synaptotagmin clustering mediated by the cysteine cluster at the interface between the transmembrane and spacer domains. J Biol Chem 2001; 276(43):40319-40325.
23. Fukuda M, Kowalchyk JA, Zhang X et al. Synaptotagmin IX regulates Ca^{2+}-dependent secretion in PC12 cells. J Biol Chem 2002; 277(7):4601-4604.
24. Shin OH, Maximov A, Lim BK et al. Unexpected Ca^{2+}-binding properties of synaptotagmin 9. Proc Natl Acad Sci USA 2004; 101(8):2554-2559.
25. Hui E, Bai J, Wang P et al. Three distinct kinetic groupings of the synaptotagmin family: Candidate sensors for rapid and delayed exocytosis. Proc Natl Acad Sci USA 2005; 102(14):5210-5214.

26. Mackler JM, Drummond JA, Loewen CA et al. The C_2B Ca^{2+}-binding motif of synaptotagmin is required for synaptic transmission in vivo. Nature 2002; 418(6895):340-344.
27. Yoshihara M, Littleton JT. Synaptotagmin I functions as a calcium sensor to synchronize neurotransmitter release. Neuron 2002; 36(5):897-908.
28. Stevens CF, Sullivan JM. The synaptotagmin C2A domain is part of the calcium sensor controlling fast synaptic transmission. Neuron 2003; 39(2):299-308.
29. Nishiki T, Augustine GJ. Dual roles of the C2B domain of synaptotagmin I in synchronizing Ca^{2+}-dependent neurotransmitter release. J Neurosci 2004; 24(39):8542-8550.
30. Fukuda M, Moreira JE, Liu V et al. Role of the conserved WHXL motif in the C terminus of synaptotagmin in synaptic vesicle docking. Proc Natl Acad Sci USA 2000; 97(26):14715-14719.
31. Chieregatti E, Witkin JW, Baldini G. SNAP-25 and synaptotagmin 1 function in Ca^{2+}-dependent reversible docking of granules to the plasma membrane. Traffic 2002; 3(7):496-511.
32. Tsuboi T, Rutter GA. Multiple forms of "kiss-and-run" exocytosis revealed by evanescent wave microscopy. Curr Biol 2003; 13(7):563-567.
33. Bai J, Wang CT, Richards DA et al. Fusion pore dynamics are regulated by synaptotagmin*t-SNARE interactions. Neuron 2004; 41(6):929-942.
34. Poskanzer KE, Marek KW, Sweeney ST et al. Synaptotagmin I is necessary for compensatory synaptic vesicle endocytosis in vivo. Nature 2003; 426(6966):559-563.
35. Nicholson-Tomishima K, Ryan TA. Kinetic efficiency of endocytosis at mammalian CNS synapses requires synaptotagmin I. Proc Natl Acad Sci USA 2004; 101(47):16648-16652.
36. Llinás RR, Sugimori M, Moran KA et al. Vesicular reuptake inhibition by a synaptotagmin I C2B domain antibody at the squid giant synapse. Proc Natl Acad Sci USA 2004; 101(51):17855-17860.
37. Tucker WC, Weber T, Chapman ER. Reconstitution of Ca^{2+}-regulated membrane fusion by synaptotagmin and SNAREs. Science 2004; 304(5669):435-438.
38. Zhang X, Kim-Miller MJ, Fukuda M et al. Ca^{2+}-dependent synaptotagmin binding to SNAP-25 is essential for Ca^{2+}-triggered exocytosis. Neuron 2002; 34(4):599-611.
39. Bai J, Tucker WC, Chapman ER. PIP_2 increases the speed of response of synaptotagmin and steers its membrane-penetration activity toward the plasma membrane. Nat Struct Mol Biol 2004; 11(1):36-44.
40. Rickman C, Archer DA, Meunier FA et al. Synaptotagmin interaction with the syntaxin/SNAP-25 dimer is mediated by an evolutionarily conserved motif and is sensitive to inositol hexakisphosphate. J Biol Chem 2004; 279(13):12574-12579.
41. Wu Y, He Y, Bai J et al. Visualization of synaptotagmin I oligomers assembled onto lipid monolayers. Proc Natl Acad Sci USA 2003; 100(4):2082-2087.
42. Walther K, Diril MK, Jung N et al. Functional dissection of the interactions of stonin 2 with the adaptor complex AP-2 and synaptotagmin. Proc Natl Acad Sci USA 2004; 101(4):964-969.
43. Fernandez I, Arac D, Ubach J et al. Three-dimensional structure of the synaptotagmin 1 C2B-domain: Synaptotagmin 1 as a phospholipid binding machine. Neuron 2001; 32(6):1057-1069.
44. Earles CA, Bai J, Wang P et al. The tandem C2 domains of synaptotagmin contain redundant Ca^{2+} binding sites that cooperate to engage t-SNAREs and trigger exocytosis. J Cell Biol 2001; 154(6):1117-1123.
45. Fukuda M, Mikoshiba K. The function of inositol high polyphosphate binding proteins. Bioessays 1997; 19(7):593-603.
46. Mackler JM, Reist NE. Mutations in the second C2 domain of synaptotagmin disrupt synaptic transmission at Drosophila neuromuscular junctions. J Comp Neurol 2001; 436(1):4-16.
47. Fukuda M, Yamamoto A, Mikoshiba K. Formation of crystalloid endoplasmic reticulum induced by expression of synaptotagmin lacking the conserved WHXL motif in the C terminus: Structural importance of the WHXL motif in the C2B domain. J Biol Chem 2001; 276(44):41112-41119.
48. Han W, Rhee JS, Maximov A et al. C-terminal ECFP fusion impairs synaptotagmin 1 function: Crowding out synaptotagmin 1. J Biol Chem 2005; 280(6):5089-5100.
49. Fukuda M. Vesicle-associated membrane protein-2/synaptobrevin binding to synaptotagmin I promotes O-glycosylation of synaptotagmin I. J Biol Chem 2002; 277(33):30351-30358.
50. Han W, Rhee JS, Maximov A et al. N-glycosylation is essential for vesicular targeting of synaptotagmin 1. Neuron 2004; 41(1):85-99.
51. Kang R, Swayze R, Lise MF et al. Presynaptic trafficking of synaptotagmin I is regulated by protein palmitoylation. J Biol Chem 2004; 279(48):50524-50536.
52. Atiya-Nasagi Y, Cohen H, Medalia O et al. O-glycosylation is essential for intracellular targeting of synaptotagmins I and II in nonneuronal specialized secretory cells. J Cell Sci 2005; 118(7):1363-1372.

53. Voets T, Moser T, Lund PE et al. Intracellular calcium dependence of large dense-core vesicle exocytosis in the absence of synaptotagmin I. Proc Natl Acad Sci USA 2001; 98(20):11680-11685.
54. Kreft M, Kuster V, Grilc S et al. Synaptotagmin I increases the probability of vesicle fusion at low $[Ca^{2+}]$ in pituitary cells. Am J Physiol Cell Physiol 2003; 284(2):C547-554.
55. Shoji-Kasai Y, Yoshida A, Sato K et al. Neurotransmitter release from synaptotagmin-deficient clonal variants of PC12 cells. Science 1992; 256(5065):1821-1823.
56. Fukuda M. RNA interference-mediated silencing of synaptotagmin IX, but not synaptotagmin I, inhibits dense-core vesicle exocytosis in PC12 cells. Biochem J 2004; 380(3):875-879.
57. Craxton M, Goedert M. Synaptotagmin V: A novel synaptotagmin isoform expressed in rat brain. FEBS Lett 1995; 361(2-3):196-200.
58. Hudson AW, Birnbaum MJ. Identification of a nonneuronal isoform of synaptotagmin. Proc Natl Acad Sci USA 1995; 92(13):5895-5899.
59. Iezzi M, Kouri G, Fukuda M et al. Synaptotagmin V and IX isoforms control Ca^{2+}-dependent insulin exocytosis. J Cell Sci 2004; 117(15):3119-3127.
60. Iezzi M, Eliasson L, Fukuda M et al. Adenovirus-mediated silencing of synaptotagmin 9 inhibits Ca^{2+}-dependent insulin secretion in islets. FEBS Lett 2005; 579(23):5241-5246.
61. Haberman Y, Grimberg E, Fukuda M et al. Synaptotagmin IX, a possible linker between the perinuclear endocytic recycling compartment and the microtubules. J Cell Sci 2003; 116(21):4307-4318.
62. Haberman Y, Ziv I, Gorzalczany Y et al. Classical protein kinase C(s) regulates targeting of synaptotagmin IX to the endocytic recycling compartment. J Cell Sci 2005; 118(8):1641-1649.
63. Fukuda M, Kanno E, Mikoshiba K. Conserved N-terminal cysteine motif is essential for homo- and heterodimer formation of synaptotagmins III, V, VI, and X. J Biol Chem 1999; 274(44):31421-31427.
64. Sugita S, Shin OH, Han W et al. Synaptotagmins form a hierarchy of exocytotic Ca^{2+} sensors with distinct Ca^{2+} affinities. EMBO J 2002; 21(3):270-280.
65. Rickman C, Craxton M, Osborne S et al. Comparative analysis of tandem C2 domains from the mammalian synaptotagmin family. Biochem J 2004; 378(2):681-686.
66. Butz S, Fernandez-Chacon R, Schmitz F et al. The subcellular localizations of atypical synaptotagmins III and VI: Synaptotagmin III is enriched in synapses and synaptic plasma membranes but not in synaptic vesicles. J Biol Chem 1999; 274(26):18290-18296.
67. Grimberg E, Peng Z, Hammel I et al. Synaptotagmin III is a critical factor for the formation of the perinuclear endocytic recycling compartment and determination of secretory granules size. J Cell Sci 2003; 116(1):145-154.
68. Saegusa C, Fukuda M, Mikoshiba K. Synaptotagmin V is targeted to dense-core vesicles that undergo calcium-dependent exocytosis in PC12 cells. J Biol Chem 2002; 277(27):24499-24505.
69. Fukuda M, Mikoshiba K. A novel alternatively spliced variant of synaptotagmin VI lacking a trans-membrane domain: Implications for distinct functions of the two isoforms. J Biol Chem 1999; 274(44):31428-31434.
70. Michaut M, De Blas G, Tomes CN et al. Synaptotagmin VI participates in the acrosome reaction of human spermatozoa. Dev Biol 2001; 235(2):521-529.
71. Hutt DM, Baltz JM, Ngsee JK. Synaptotagmin VI and VIII and syntaxin 2 are essential for the mouse sperm acrosome reaction. J Biol Chem 2005; 280(21):20197-20203.
72. Roggero CM, Tomes CN, De Blas GA et al. Protein kinase C-mediated phosphorylation of the two polybasic regions of synaptotagmin VI regulates their function in acrosomal exocytosis. Dev Biol 2005; 285(2):422-435.
73. Babity JM, Armstrong JN, Plumier JC et al. A novel seizureinduced synaptotagmin gene identified by differential display. Proc Natl Acad Sci USA 1997; 94(6):2638-2641.
74. Adolfsen B, Saraswati S, Yoshihara M et al. Synaptotagmins are trafficked to distinct subcellular domains including the postsynaptic compartment. J Cell Biol 2004; 166(2):249-260.
75. Dai H, Shin OH, Machius M et al. Structural basis for the evolutionary inactivation of Ca^{2+} binding to synaptotagmin 4. Nat Struct Mol Biol 2004; 11(9):844-849.
76. Wang CT, Grishanin R, Earles CA et al. Synaptotagmin modulation of fusion pore kinetics in regulated exocytosis of dense-core vesicles. Science 2001; 294(5544):1111-1115.
77. Machado HB, Liu W, Vician LJ et al. Synaptotagmin IV overexpression inhibits depolarization-induced exocytosis in PC12 cells. J Neurosci Res 2004; 76(3):334-341.
78. Fukuda M, Kanno E, Ogata Y et al. Nerve growth factor-dependent sorting of synaptotagmin IV protein to mature dense-core vesicles that undergo calcium-dependent exocytosis in PC12 cells. J Biol Chem 2003; 278(5):3220-3226.
79. Zhang Q, Fukuda M, Van Bockstaele E et al. Synaptotagmin IV regulates glial glutamate release. Proc Natl Acad Sci USA 2004; 101(25):9441-9446.

80. Ferguson GD, Vician L, Herschman HR. Synaptotagmin IV: Biochemistry, genetics, behavior, and possible links to human psychiatric disease. Mol Neurobiol 2001; 23(2-3):173-185.
81. Ibata K, Fukuda M, Hamada T et al. Synaptotagmin IV is present at the Golgi and distal parts of neurites. J Neurochem 2000; 74(2):518-526.
82. Fukuda M, Yamamoto A. Effect of forskolin on synaptotagmin IV protein trafficking in PC12 cells. J Biochem (Tokyo) 2004; 136(2):245-253.
83. Ibata K, Hashikawa T, Tsuboi T et al. Nonpolarized distribution of synaptotagmin IV in neurons: Evidence that synaptotagmin IV is not a synaptic vesicle protein. Neurosci Res 2002; 43(4):401-406.
84. Robinson IM, Ranjan R, Schwarz TL. Synaptotagmins I and IV promote transmitter release independently of Ca^{2+} binding in the C2A domain. Nature 2002; 418(6895):336-340.
85. Fukuda M, Ibata K, Mikoshiba K. A unique spacer domain of synaptotagmin IV is essential for Golgi localization. J Neurochem 2001; 77(3):730-740.
86. Fukuda M, Mikoshiba K. Characterization of KIAA1427 protein as an atypical synaptotagmin (Syt XIII). Biochem J 2001; 354(2):249-257.
87. Huynh DP, Scoles DR, Nguyen D et al. The autosomal recessive juvenile Parkinson disease gene product, parkin, interacts with and ubiquitinates synaptotagmin XI. Hum Mol Genet 2003; 12(20):2587-2597.
88. Sugita S, Han W, Butz S et al. Synaptotagmin VII as a plasma membrane Ca^{2+} sensor in exocytosis. Neuron 2001; 30(2):459-473.
89. Fukuda M, Ogata Y, Saegusa C et al. Alternative splicing isoforms of synaptotagmin VII in the mouse, rat and human. Biochem J 2002; 365(1):173-180.
90. Fukuda M, Mikoshiba K. Mechanism of the calcium-dependent multimerization of synaptotagmin VII mediated by its first and second C2 domains. J Biol Chem 2001; 276(29):27670-27676.
91. Fukuda M, Katayama E, Mikoshiba K. The calcium-binding loops of the tandem C2 domains of synaptotagmin VII cooperatively mediate calcium-dependent oligomerization. J Biol Chem 2002; 277(32):29315-29320.
92. Fukuda M, Kanno E, Satoh M et al. Synaptotagmin VII is targeted to dense-core vesicles and regulates their Ca^{2+}-dependent exocytosis in PC12 cells. J Biol Chem 2004; 279(50):52677-52684.
93. Wang P, Chicka MC, Bhalla A et al. Synaptotagmin VII is targeted to secretory organelles in PC12 cells, where it functions as a high-affinity calcium sensor. Mol Cell Biol 2005; 25(19):8693-8702.
94. Martinez I, Chakrabarti S, Hellevik T et al. Synaptotagmin VII regulates Ca^{2+}-dependent exocytosis of lysosomes in fibroblasts. J Cell Biol 2000; 148(6):1141-1149.
95. Reddy A, Caler EV, Andrews NW. Plasma membrane repair is mediated by Ca^{2+}-regulated exocytosis of lysosomes. Cell 2001; 106(2):157-169.
96. Andrews NW, Chakrabarti S. There's more to life than neurotransmission: The regulation of exocytosis by synaptotagmin VII. Trends Cell Biol 2005; 15(11):626-631.
97. Chakrabarti S, Kobayashi KS, Flavell RA et al. Impaired membrane resealing and autoimmune myositis in synaptotagmin VII-deficient mice. J Cell Biol 2003; 162(4):543-549.
98. Fukuda M. The C2A domain of synaptotagmin-like protein 3 (Slp3) is an atypical calcium-dependent phospholipid-binding machine: Comparison with the C2A domain of synaptotagmin I. Biochem J 2002; 366(2):681-687.
99. Potter GB, Facchinetti F, Beaudoin IIIrd GM et al. Neuronal expression of synaptotagmin-related gene 1 is regulated by thyroid hormone during cerebellar development. J Neurosci 2001; 21(12):4373-4380.
100. Hutt DM, Cardullo RA, Baltz JM et al. Synaptotagmin VIII is localized to the mouse sperm head and may function in acrosomal exocytosis. Biol Reprod 2002; 66(1):50-56.
101. Fukuda M, Saegusa C, Mikoshiba K. Novel splicing isoforms of synaptotagmin-like proteins 2 and 3: Identification of the Slp homology domain. Biochem Biophys Res Commun 2001; 283(2):513-519.
102. Kuroda TS, Fukuda M, Ariga H et al. Synaptotagmin-like protein 5: A novel Rab27A effector with C-terminal tandem C2 domains. Biochem Biophys Res Commun 2002; 293(3):899-906.
103. Serano J, Rubin GM. The Drosophila synaptotagmin-like protein bitesize is required for growth and has mRNA localization sequences within its open reading frame. Proc Natl Acad Sci USA 2003; 100(23):13368-13373.
104. Yi Z, Yokota H, Torii S et al. The Rab27a/granuphilin complex regulates the exocytosis of insulin-containing dense-core granules. Mol Cell Biol 2002; 22(6):1858-1867.
105. Strom M, Hume AN, Tarafder AK et al. A family of Rab27-binding proteins: Melanophilin links Rab27a and myosin Va function in melanosome transport. J Biol Chem 2002; 277(28):25423-25430.

106. Fukuda M. Distinct Rab binding specificity of Rim1, Rim2, rabphilin, and Noc2: Identification of a critical determinant of Rab3A/Rab27A recognition by Rim2. J Biol Chem 2003; 278(17):15373-15380.

107. Kuroda TS, Fukuda M. Rab27A-binding protein Slp2-a is required for peripheral melanosome distribution and elongated cell shape in melanocytes. Nat Cell Biol 2004; 6(12):1195-1203.

108. Fukuda M, Kanno E, Saegusa C et al. Slp4-a/granuphilin-a regulates dense-core vesicle exocytosis in PC12 cells. J Biol Chem 2002; 277(42):39673-39678.

109. Imai A, Yoshie S, Nashida T et al. The small GTPase Rab27B regulates amylase release from rat parotid acinar cells. J Cell Sci 2004; 117(10):1945-1953.

110. Fukuda M. Slp4-a/granuphilin-a inhibits dense-core vesicle exocytosis through interaction with the GDP-bound form of Rab27A in PC12 cells. J Biol Chem 2003; 278(17):15390-15396.

111. Coppola T, Frantz C, Perret-Menoud V et al. Pancreatic β-cell protein granuphilin binds Rab3 and Munc-18 and controls exocytosis. Mol Biol Cell 2002; 13(6):1906-1915.

112. Fukuda M, Imai A, Nashida T et al. Slp4-a/granuphilin-a interacts with syntaxin-2/3 in a Munc18-2-dependent manner. J Biol Chem 2005; 280(47):39175-39184.

113. Torii S, Zhao S, Yi Z et al. Granuphilin modulates the exocytosis of secretory granules through interaction with syntaxin 1a. Mol Cell Biol 2002; 22(15):5518-5526.

114. Torii S, Takeuchi T, Nagamatsu S et al. Rab27 effector granuphilin promotes the plasma membrane targeting of insulin granules via interaction with syntaxin 1a. J Biol Chem 2004; 279(21):22532-22538.

115. Gomi H, Mizutani S, Kasai K et al. Granuphilin molecularly docks insulin granules to the fusion machinery. J Cell Biol 2005; 171(1):99-109.

116. Trambas CM, Griffiths GM. Delivering the kiss of death. Nat Immunol 2003; 4(5):399-403.

117. Kasai K, Ohara-Imaizumi M, Takahashi N et al. Rab27a mediates the tight docking of insulin granules onto the plasma membrane during glucose stimulation. J Clin Invest 2005; 115(2):388-396.

118. Korteweg N, Maia AS, Thompson B et al. The role of Munc18-1 in docking and exocytosis of peptide hormone vesicles in the anterior pituitary. Biol Cell 2005; 97(6):445-455.

119. Catz SD, Johnson JL, Babior BM. The C2A domain of JFC1 binds to 3'-phosphorylated phosphoinositides and directs plasma membrane association in living cells. Proc Natl Acad Sci USA 2002; 99(18):11652-11657.

120. Johnson JL, Pacquelet S, Lane WS et al. Akt regulates the subcellular localization of the Rab27a-binding protein JFC1 by phosphorylation. Traffic 2005; 6(8):667-681.

121. Johnson JL, Ellis BA, Noack D et al. The Rab27a-binding protein, JFC1, regulates androgen-dependent secretion of prostate-specific antigen and prostatic-specific acid phosphatase. Biochem J 2005; 391(3):699-710.

122. Fukuda M, Mikoshiba K. Doc2γ, a third isoform of double C2 protein, lacking calcium-dependent phospholipid binding activity. Biochem Biophys Res Commun 2000; 276(2):626-632.

123. Fukuda M, Saegusa C, Kanno E et al. The C2A domain of double C2 protein γ contains a functional nuclear localization signal. J Biol Chem 2001; 276(27):24441-24444.

124. Sakaguchi G, Manabe T, Kobayashi K et al. Doc2α is an activity-dependent modulator of excitatory synaptic transmission. Eur J Neurosci 1999; 11(12):4262-4268.

125. Schluter OM, Schnell E, Verhage M et al. Rabphilin knock-out mice reveal that rabphilin is not required for rab3 function in regulating neurotransmitter release. J Neurosci 1999; 19(14):5834-5846.

126. Staunton J, Ganetzky B, Nonet ML. Rabphilin potentiates soluble N-ethylmaleimide sensitive factor attachment protein receptor function independently of rab3. J Neurosci 2001; 21(23):9255-9264.

127. Fukuda M, Kanno E, Yamamoto A. Rabphilin and Noc2 are recruited to dense-core vesicles through specific interaction with Rab27A in PC12 cells. J Biol Chem 2004; 279(13):13065-13075.

128. Fukuda M. Rabphilin and Noc2 function as Rab27 effectors that control Ca^{2+}-regulated exocytosis. Recent Res Dev Neurochem 2004; 7:57-69.

129. Chung SH, Joberty G, Gelino EA et al. Comparison of the effects on secretion in chromaffin and PC12 cells of Rab3 family members and mutants: Evidence that inhibitory effects are independent of direct interaction with Rabphilin3. J Biol Chem 1999; 274(25):18113-18120.

130. Joberty G, Stabila PF, Coppola T et al. High affinity Rab3 binding is dispensable for Rabphilin-dependent potentiation of stimulated secretion. J Cell Sci 1999; 112(20):3579-3587.

131. Tsuboi T, Fukuda M. The C2B domain of rabphilin directly interacts with SNAP-25 and regulates the docking step of dense-core vesicle exocytosis in PC12 cells. J Biol Chem 2005; 280(47):39253-39259.

132. Sasaki T, Shirataki H, Nakanishi H et al. Rab3A-rabphilin-3A system in neurotransmitter release. Adv Second Messenger Phosphoprotein Res 1997; 31:279-294.

133. Li C, Ullrich B, Zhang JZ et al. Ca^{2+}-dependent and -independent activities of neural and nonneural synaptotagmins. Nature 1995; 375(6532):594-599.

CHAPTER 5

The Synapsins and the Control of Neuroexocytosis

Pietro Baldelli,* Anna Fassio, Anna Corradi, Flavia Valtorta and Fabio Benfenati

Abstract

The synapsins have been the first synaptic vesicle-associated proteins to be discovered thanks to their prominent ability to be phosphorylated by a variety of protein kinases. At present, the synapsin family in mammals consists of at least 10 isoforms encoded by three distinct genes and composed by a mosaic of conserved and variable domains. The synapsins are highly conserved evolutionarily and synapsin homologues have been described in invertebrates and lower vertebrates. The synapsins are implicated in multiple interactions with synaptic vesicle proteins and phospholipids, actin and protein kinases. Via these interactions, the synapsins play multiple roles in synaptic transmission, including control of synapse formation, regulation of synaptic vesicle trafficking, neurotransmitter release and expression of short-term synaptic plasticity phenomena. This chapter tries to summarize the main functional features of the synapsins that have emerged in the last 20 years, in order to provide a framework for interpreting the complex role played by these phosphoproteins in synaptic physiology.

Introduction

The release of classical neurotransmitters (NTs) occurs at specialized sites of the plasma membrane, named active zones, by exocytotic fusion of small synaptic vesicles (SVs). The uniform loading of SVs with a discrete amount of NT[8] is reflected by the reproducibility in the size of the postsynaptic response elicited by each exocytotic event referred to a NT quantum. At variance with nonneuronal cells, neuroexocytosis is characterized by: (1) an "explosive" rate of NT release, many orders of magnitude faster than that of nonneuronal cells; (2) the ability to operate at various levels of efficiency depending on the microenviromental conditions and the previous "history" of the neuron; and (3) the ability to sustain repetitive high frequency NT release over a long period of time with strong reliability. The molecular features that confer such properties to neuroexocytosis in neurons are: (i) the high colocalization of Ca^{2+} channels with fusion competent SVs which allows an extremely rapid Ca^{2+}-dependent exocytosis, (ii) the existence of a strategically localized reserve pool of SV buffering the depletion of the readily released pool during sustained repetitive release and (iii) the presence of efficient recycling mechanisms active at the presynaptic membrane that prevent the rapid depletion of SVs during a sustained repetitive release. Such recycling mechanisms are contributed by a fast and direct endocytotic pathway operating at the active zones ("kiss & run" mechanism) and by a slower clathrin-mediated endocytosis active at periactive zones.

*Corresponding Author: Pietro Baldell—Department of Experimental Medicine, Section of Human Physiology, University of Genova, Viale Benedetto XV, 3, 16132 Genova, Italy. Emai: pietro.baldelli@unige.it

Molecular Mechanisms of Exocytosis, edited by Romano Regazzi. ©2007 Landes Bioscience and Springer Science+Business Media.

The remarkable properties of neurotransmitter release are generated by the activity of a number of proteins that are localized within the presynaptic terminal and participate in synapse formation, maintenance and function. Among many presynaptic actors which have been identified in the last 20 years, the most abundant phosphoproteins are the synapsins, a highly conserved multigene family of neuron-specific, SV-associated phosphoproteins.

Synapsins exist in all organisms endowed with a nervous system and, in mammals, are encoded by three distinct genes (SYNI, SYNII and SYNIII) located in chromosome X, 3 and 22, respectively.[16,19,24] They are composed of a mosaic of individual and shared domains, the latter of which are highly conserved during evolution (Fig. 1). Synapsins I and II are stably expressed at synapses of mature neurons, where they associate with the cytoplasmic surface of small SVs, whereas the expression of synapsin III is developmentally controlled and not strictly confined to synaptic terminals (Fig. 2). Synapsins are excellent substrates for a large array of protein kinases including protein kinase A, Ca^{2+}/calmodulin-dependent protein kinases (CaMK) I, II and IV, mitogen-activate protein (MAP) kinase and cyclin-dependent kinase-1, that phosphorylate them on distinct serine residues. Synapsins interact in vitro with lipid and protein

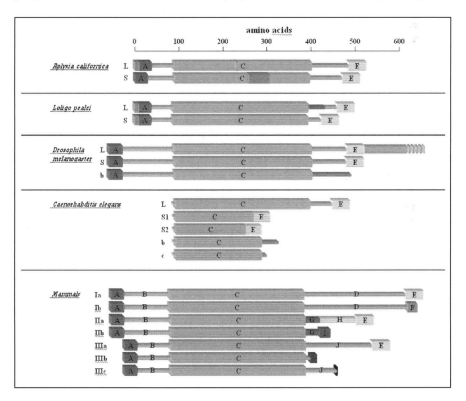

Figure 1. Evolutionary conservation of the synapsins. Synapsins have been cloned from a variety of species, from invertebrates to man. Synapsins are composed of a mosaic of conserved and individual domains that are schematically represented in blocked color form and indicated by letters A-J. The length of the polypeptide chains is shown at the top in number of amino acid residues. Different shades or colors depicted within domains represent different sequences (e.g., within domain C of *Aplysia* synapsin). In the figure, highly conserved domains are shown as thick colored boxes. Domains A, C, and E are defined by significant homology to their mammalian counterparts. While in mammals, synapsins are coded by three distinct genes, in lower vertebrates and invertebrates one single gene gives rise to multiple synapsin isoforms.

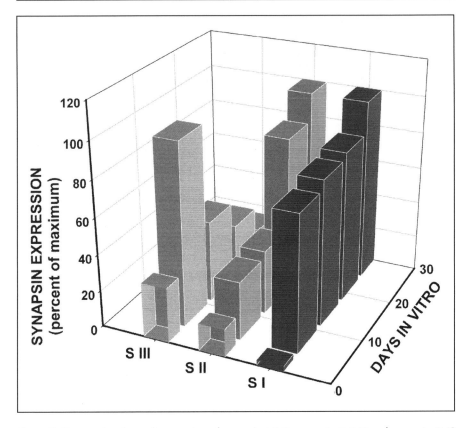

Figure 2. Temporal pattern of expression of synapsin I (S I), synapsin II (S II) and synapsin III (S III) in primary hippocampal neurons as a function of the days in vitro (DIV). Expression at various times is shown in percentage of the maximal level of expression over the analyzed time window. The three synapsin isoforms have a clearly different expression pattern, with synapsin III peaking at 7 DIV and decreasing thereafter and synapsins I and II increasing rapidly (synapsin I) or more slowly (synapsin II) during the time in vitro to reach the maximal level of expression after completion of synaptic maturation (21-28 DIV).

components of SVs, as well as with various cytoskeletal proteins including actin, and control multiple aspects of synapse structure and function, from synaptogenesis and regulation of SV trafficking to modulation of short-term synaptic plasticity. Here, we will describe the functional studies which have outlined the role played by synapsins in regulation of NT release. In the first part, we will focus on those studies that led to the proposed model of a predocking mechanism of action of the synapsins. Then, we will critically summarize the recently growing body of evidence suggesting that synapsins, in addition to their predocking action, directly control the efficiency of synaptic transmission and the rate of NT release by acting at the post-docking level. For further details concerning structure, biochemistry, genetics, cellular and molecular biology and developmental role of the synapsins, the reader is referred to more extensive reviews in references 3,11,16,19,24.

The Synapsins and the Reserve Pool of Synaptic Vesicles

A large body of experimental evidence obtained in reconstituted nerve terminals has proposed that the synapsins reversibly cross-link SVs to each other and to the actin-based cytoskeletal meshwork. This action is believed to be important for the formation and maintenance of a reserve pool of SVs as well as in the fine regulation of the balance between the reserve pool and a pool of SVs ready to undergo exocytosis in an activity- and phosphorylation-dependent manner.

In order to perturb synapsin function at the nerve terminal and define its functional role in neuroexocytosis, two main experimental approaches have been used: (i) microinjection into large presynaptic terminals of invertebrate or lower vertebrate neurons of exogenous synapsin, antibodies to synapsin or peptides derived from evolutionary conserved synapsin sequences; (ii) deletion of one or more of the synapsin genes in mice by gene knockout (KO) technology. Both techniques have advantages and limitations.

Microinjection studies are potentially the best method to acutely study functional changes induced by perturbation of intraterminal synapsin levels. It allows to interfere directly with synapsin function and follow the effects generated by the injected agent in real-time. However, injection of proteins, peptides or antibodies could have nonspecific effects for the relative high concentrations that are often required and for the possibility that the injected agent undergoes a nonphysiological targeting within the neuron. Genetic studies are probably the best approach to provide answers to the ultimate function of a given protein, i.e., the function that cannot be compensated during development by other genes, but again they have some drawbacks. First of all, very often a change in a single protein promotes a series of homeostatic responses in downstream processes that allow neuronal systems to respond to the initial manipulation with secondary changes, making the interpretation of the phenotype difficult. Moreover, the effects of specific gene deletions are often attenuated by the presence of homologue gene products with redundant functions. This is particularly true in the case of the synapsins that are encoded by three genes. Despite these limitations, the knocking out of synapsin genes represents a fundamental technique to study in vivo the action of synapsin on development, synaptogenesis, maintenance and function of synapses in a long-term time scale.

Pioneer experiments testing the effect of exogenous synapsin I in squid giant synapses showed that the injection of dephosphorylated synapsin I decreased the amplitude and rate of rise of postsynaptic potentials, whereas the injection of either phosphorylated synapsin I or heat-inactivated dephosphorylated synapsin I were ineffective. Conversely, injection of CaMKII increased the rate of rise and the amplitude of postsynaptic potentials.[27,28] Analysis of synaptic noise in the same system revealed that dephosphorylated synapsin I reduced the rate of spontaneous and evoked quantal release, whereas the injection of CaMKII increased evoked release without affecting the frequency of spontaneous miniature events.[29] Further data obtained in vertebrate goldfish neurons showed that the presynaptic injection of dephosphorylated synapsin I reduced both spontaneous and evoked synaptic transmission.[17] Internalization of dephosphorylated synapsin, phosphorylated synapsin or activated CaMKII into rat brain synaptosomes using freeze-thaw permeabilization confirmed the results obtained by in vivo injections.[32,33]

These data suggested an initial model in which dephosphorylated synapsin I inhibits synaptic transmission without interfering directly with the release process, but recruiting SVs to the reserve pool and inhibiting SV mobilization to the readily releasable pool, a process that can be reverted upon phosphorylation. Studies on the physical distribution of the protein in response to a depolarising stimulus conducted in frog nerve muscle preparation showed that synapsin I dissociation from the SV membrane is not a prerequisite for fusion and that under high frequency electrical stimulation synapsin I partially dissociates from SVs during exocytosis and reassociates with the SV membrane following endocytosis.[46,47] In agreement with the latter data, phosphorylation of synapsin I in rat brain synaptosomes treated with depolarising agents is associated with a rapid translocation of the protein from the membrane fraction to the synaptosol.[40] These data have been recently confirmed in living hippocampal neurons, in which synapsin was found to disperse in the presynaptic terminal and preterminal axon during depolarization and to recluster at SV sites following return to the resting state.[5] In these studies it was also found that the rates of dispersion and reclustering are indeed controlled by synapsin phosphorylation and dephosphorylation, respectively, and that CaMK-mediated phosphorylation controls SV mobilization at low frequency of stimulation, whereas MAP kinase phosphorylation is recruited at both low and high frequencies of stimulation.[5,6]

Ultrastructural studies were consistent with early functional studies. In living lamprey reticulospinal axons forming en passant synapses, presynaptic injection of an anti-synapsin antibody, specifically recognizing sequences of the synapsin domain E (Fig. 1), caused the disappearance of SVs distal to the synaptic cleft (reserve pool), leaving unaffected the SVs docked at the active zone. The depletion of the reserve pool was in turn associated with a markedly enhanced depression following high, but not low, frequency stimulation.[36] Consistently, the presynaptic injection of a highly conserved peptide fragment of the synapsin domain E into the squid giant synapse greatly reduced the number of SVs far from the active zone and increased the rate and extent of synaptic depression, indicating that domain E, present in both synapsin isoforms expressed in squid (Fig. 1), is essential for the synapsin-mediated maintenance and regulation of the SV reserve pool.[18] Closely similar results were obtained after the injection of a highly conserved peptide derived from the synapsin domain C.[20] Interestingly, both peptides inhibit the synapsin-actin interactions, providing a common mechanism for the physiological and ultrastructural effects of the peptides from domains E and C.

A fundamental contribution to the study of the role of synapsin in NT release derives from genetic experiments in mice in which synapsin genes have been inactivated to generate single, double and triple KO animals.[7,9,15,26,37-39,44,45] All strains of KO mice were viable and fertile. Despite the absence of gross defects in brain morphology or behaviour, synapsin I and synapsin II (but not synapsin III) KO mice as well as double synapsin I/II and triple synapsin I/II/III KO mice exhibited early onset spontaneous and sensory stimuli-evoked (audiogenic) epileptic seizures. Attacks consisted of partial, secondarily generalized "grand mal" attacks followed by post-seizure grooming.[38] Electroencephalogram analysis showed that subconvulsive electrical stimulation in the amygdala was able to induce seizures when applied to synapsin mutant mice.[26] Typically, seizures develop after 2-3 months of age and become more frequent with age. The incidence of seizures is higher in synapsin II than in synapsin I KO mice and is proportional to the number of inactivated synapsin genes. While the synapsin II and I/II KO mice have been reported to have impaired contextual conditioning[41] and triple KO mice exhibited impaired motor coordination and defective spatial learning,[15] a detailed analysis of the behavioural phenotype of the synapsin KO mice is still lacking.

Ultrastructural and physiological abnormalities observed in adult synapsin mutant mice largely confirmed and validated the data obtained by injection studies. Synapsin I, II and I/II KO mice showed a selective decrease in the total number of SVs, as demonstrated by a decrease in the levels of most SV markers (Fig. 3) and by electron microscopy of central synapses.[14,26,38,44] Similarly to what observed with the injection studies,[19,36] the nerve terminal ultrastructure showed a dramatic decrease and disassembly of SVs in the reserve pool, while SVs docked at active zones were only poorly affected.[26,44] In synapsin I KO mice, SV depletion was accompanied by a strong impairment in glutamate release from cortical synaptosomes and by a greater delay in the recovery of synaptic transmission after NT depletion by high frequency stimulation.[26]

The study of SV recycling at individual synaptic boutons using FM dyes showed that the number of exocytosed SVs during brief action potential trains and the total recycling SV pool are significantly reduced in synapsin I KO mice, while the kinetics of endocytosis and SV repriming appear normal.[39] The results were similar to those obtained in a different strain of synapsin KO mice by an independent laboratory,[38] except that (i) in double KO mice the SV depletion was not restricted to the reserve pool, but affected to the same extent the readily releasable pool of SVs and (ii) there was no detectable increase in synaptic depression induced by 30 sec of repetitive stimulation at 10 Hz in synapsin I KO mice. However, depression was increased in synapsin II KO mice and further enhanced in I/II double KO mice, suggesting a participation also of synapsin I in the build-up of depression.

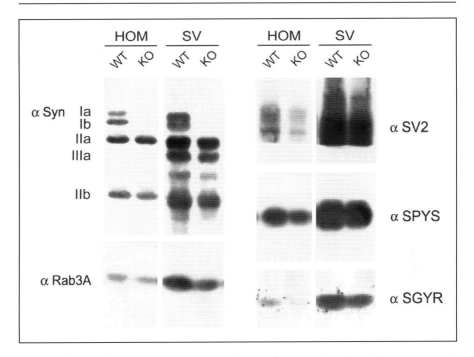

Figure 3. The specific decrease in SV density in central synapses is reflected by a decrease in the expression of the major synaptic vesicle proteins. Homogenate (HOM) and purified SV fractions obtained from wild-type (WT) and synapsin I KO mice were analyzed by immunoblotting for their content in synapsin isoforms, Rab 3A and the integral SV proteins SV2, synaptophysin (SPYS) and synaptogyrin (SGYR). Note the high enrichment of SV markers in purified SVs and the similar and homogeneous decrease in the levels of virtually all SV markers (including the products of the synapsin I and II genes), except for Rab3A levels whose marked decrease suggests an additional function of synapsin in Rab3 targeting to SVs.

The molecular basis of the epileptic phenotype observed in synapsin deficient mice are still far from being elucidated. It has been hypothesized that synaptic depression during repetitive stimulation contributes to seizure development by causing an imbalance between excitatory and inhibitory systems. This imbalance is attributable to the fact that inhibitory GABAergic interneurons experience high frequency firing that may make GABA release particularly sensitive to the relative SV depletion induced by synapsin deletion. Terada and coworkers[45] investigated the impairment of inhibitory transmission in cultured hippocampal synapses from synapsin I KO mice and demonstrated that inhibitory, but not excitatory, synapses become easily fatigued upon repeated application of hypertonic sucrose and recover slowly from depression. Stimulated terminals showed a decrease in the number of SVs in the reserve pool, but not in the readily releasable pool, that was slightly more intense in GABAergic terminals than in glutamatergic ones. However, the young age of the hippocampal neurons used in this study (8 DIV), a stage in which synaptogenesis is in progress and the formed synapses are still immature, suggests that the observed effects could be ascribed, at least in part, to a defect in synaptogenesis rather than to a change in the mature exocytosis machinery.

Taken together, the effects observed in synapsin I, II, and I/II KO mice are in general agreement with the data obtained by injection studies and strongly support the predocking model in which synapsins I and II participate in the formation and maintenance of the reserve pool of SVs (Fig. 4). This pool provides a strategically localized SV reserve, buffering the depletion of the readily released pool when sustained and repetitive release overrides the tonic SV recycling capacity of the terminal through the direct (kiss & stay/kiss & run) or clathrin-mediated endocytosis.[16,19]

Synapsin III, the most recently identified member of the synapsin family, plays a role in synaptic function and NT release that appears completely distinct from that of synapsin I or II. First of all, synapsin III is expressed early during neuronal development and its expression is downregulated in mature neurons,[10] while the product of the other two synapsin genes have an opposite pattern of expression (Fig. 2). Mice lacking synapsin III exhibited a marked delay in neurite outgrowth, no change in SV density, an increase in the size of the recycling pool of SVs and a significant decrease in synaptic depression,[9] in sharp contrast with what has been observed in synapsins I and II KO mice.[26,39] These data indicate a unique nonredundant role for synapsin III in the regulation of NT release.

One of the most intriguing functions of synapsin III is its ability to limit the size of the recycling pool of SVs that allows more SVs to be recruited for NT release during repetitive stimulation in synapsin III KO mice. It is possible that synapsin III, highly expressed in early stages of synaptogenesis, may serve to tether SVs to the cytoskeleton and keep them from recycling during synaptic activity as previously suggested for synapsins I and II.[1,34] However,

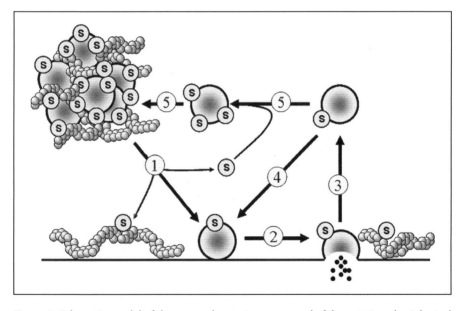

Figure 4. Schematic model of the exo-endocytosis process and of the putative physiological role of the synapsins. Evoked neurotransmitter release is a multi-step process in which SVs, after being released from the reserve pool where they are bound to the actin cytoskeleton (step 1), dock to presynaptic membrane and undergo the sequential steps of priming and Ca^{2+}-triggered fusion (step 2). After fusion, SVs are retrieved through a process of endocytosis (step 3) and either become competent for a new round of exocytosis (step 4) or are recaptured in the reserve pool (step 5). Synapsins (S) modulate this cycle by acting at various levels, namely: (i) synapsins partially dissociate from SVs and actin upon activity-dependent phosphorylation (step 1), making SVs available for exocytosis; (ii) upon dissociation, phosphorylated synapsins diffuse within the nerve terminal and the preterminal regions of the axon (step 1); (iii) by acting at the active zone level, synapsins increase the rate of the post-docking events of priming and/or fusion (step 2), possibly by interacting with the dynamic submembrane actin cytoskeleton meshwork or by removing the inhibitory action of Rab3 on the fusion process; and (iv) upon dephosphorylation, synapsins reassociate with SVs after endocytosis and promote the recruitment of SVs to the reserve pool (step 5).

the marked decrease in the SV population observed in synapsin I and II KO mice indicates that, while synapsins I and II have profound effects on SV clustering and stability,[2,35] synapsin III may be devoid of this activity. Although no physiological evidence for a role of ATP binding to synapsins has been provided thus far, the differential effect of Ca^{2+} on ATP binding to synapsins (Ca^{2+} inhibits ATP binding to synapsin III and stimulates ATP binding to synapsin I)[21] suggests another potential molecular difference between synapsins I and III.

Notwithstanding the absence of an overt or latent epileptic phenotype, synapsin III KO mice also showed an impairment of GABAergic transmission, while excitatory transmission was unaffected. These results leave open the possibility that the function of synapsin III in inhibitory terminals may differ from that at excitatory synapses. A recent study on synapsin I/II/III triple KO mice investigated this possibility in great detail. Excitatory and inhibitory synaptic transmission was differentially altered in these mice: excitatory synapses exhibited normal basal transmission, but decreased number of SV in the reserve pool and marked depression, whereas inhibitory synapses exhibited impaired basal transmission, mild changes in the number of SV and no changes in depression. Although these observations leave completely open the physiological basis of the increased seizure propensity of synapsin I, II, I/II and I/II/III KO mice, but not of synapsin III KO mice, they demonstrate that the synapsins have a critical role in maintaining the balance between excitatory and inhibitory synapses in brain networks.[15]

The Synapsins and Short-Term Synaptic Plasticity

There are only relatively few data concerning the role played by synapsins on short-term synaptic plasticity and their interpretation is still debated. Field EPSPs recorded in hippocampal slices of synapsin I KO mice exhibited increased paired-pulse facilitation (PPF),[37,38] but no effect was observed on post-tetanic potentiation (PTP).[38] On the other hand, synapsin II and I/II KO mice showed no changes in PPF, but a dramatic decrease of PTP.[38] Cultured hippocampal neurons (7-14 DIV) obtained from synapsin III or I/II/III KO mice showed no changes in PPF.[9,15]

In cholinergic synapses of *Aplysia californica*, the functional ablation of synapsin by antibody injection produced a virtual disappearance of PTP, that was substituted by an intense post-tetanic depression. In the same study, basal synaptic transmission was not altered, but PPF was significantly decreased at physiological Ca^{2+} concentrations. However, decreasing the release probability by lowering the Ca^{2+}/Mg^{2+} ratio to remove synaptic depression revealed that PPF was not affected by synapsin neutralization.[22] Finally, presynaptic injection of the peptide fragment of domain E in squid giant synapses dramatically decreased postsynaptic potential in response to a single action potential but did not affect PPF.[18] Thus, most of the available data indicate that PPF is not a primary target of synapsin action in excitatory terminals, although the function of synapsin on short-term plasticity of inhibitory synapses remains completely unexplored. Moreover, PPF has an intrinsic kinetics of tens of milliseconds, a time-range much faster than the time necessary for the Ca^{2+}-dependent mobilization of SVs from the reserve to the readily releasable pool to occur.[51] Thus, the possibility exists that synapsin affects PPF either through a post-docking mechanism (see below) or by altering the baseline level of transmission and indirectly influencing the magnitude of short-term plasticity, as recently suggested.[9,19,22,39,45] Indeed, at most synapses an increase in the initial probability of NT release decreases the magnitude of synaptic enhancement (lower PPF), and, conversely, a decrease in the probability of release results in larger synaptic enhancement or smaller synaptic depression (higher PPF).[51] Although the absence of synapsins seems not to impair PPF, a recent finding indicates that synapsin I is necessary for the increase in PPF promoted by the constitutive activation of the Ras/MAP kinase pathway.[25]

At variance with PPF, synapsins appear to have a definite role in the presynaptic expression of PTP. In fact, both genetically altered mice and invertebrate synapses exhibit a marked impairment in PTP after genetic deletion or neutralization of synapsin I and/or synapsin II.[22,38]

Although synapsin I KO mice showed no detectable changes in PTP, the almost double effect on PTP observed in synapsin I/II double KO mice as compared to the single synapsin II KO mice strongly indicates that synapsin I also plays a role in regulating PTP and that the absence of changes in PTP observed in synapsin I KO mice is attributable to the compensatory effect of synapsin II. PTP is characterized by a time-course in the order of seconds in mammals, a time sufficiently long to involve SV mobilization from the reserve pool. Thus, the action of synapsins on PTP are consistent with the predocking mechanism model described above, in which Ca^{2+} accumulation induced by tetanic stimulation activates a Ca^{2+}-dependent phosphorylation of synapsin releasing SVs from the actin cytoskeleton and increasing their availability for exocytosis.

The basis for the partial disagreement among some of the observed effects of synapsin deletion on short-term plasticity could be several-fold: (i) except for some of the most recent studies,[9,15] a separate analysis of inhibitory and excitatory synapses was not carried out; (ii) the data were obtained using different neuronal preparations, i.e., either primary cultures of hippocampal neurons or acute slices; (iii) primary cultures were used at different stages of maturation and, in most studies, before a complete synaptic maturation had occurred; and (iv) it is experimentally very difficult to measure the baseline level of synaptic transmission, especially in slices. For instance, studies of PPF were performed in brain slices of synapsin KO mice through extracellular stimulation evoking a response that was detected with extracellular electrodes (field EPSPs).[37,38] Under these conditions, the amplitude of the evoked response does not provide a measure of baseline transmission because it reflects the activation of many presynaptic fibres and depends upon several factors (i.e., position of the electrodes, stimulation intensity, slice viability). Thus, the previous studies did not determine whether the observed changes in plasticity were direct effects on facilitation or depression or whether they were secondary to changes in some quantal parameter characterizing the efficiency of synaptic transmission, such as initial release probability and release rate. Single cell patch-clamp recordings represent a more adequate experimental approach, although this technique has to be used with caution. In particular, using neuronal cultures obtained by KO mice, a change in the amplitude of evoked postsynaptic currents can be due to impaired synaptogenesis and/or neurite elongation that dramatically decrease the number of functioning synapses and consequently the number of SV released in response to presynaptic stimulation. Only a detailed quantal analysis of miniature currents and a noise analysis of evoked postsynaptic currents will provide the quantal parameter of neurotransmission necessary to interpret the effects on short-term plasticity.

The Synapsins and Release Probability and Kinetics

According to the general model of synapsins tethering SVs to the actin cytoskeleton at a distance from the active zones and releasing them upon activity through phosphorylation-dependent dissociation, SVs recruited to the readily releasable pool should be depleted of synapsins. Although this general picture is still valid and accounts for most of the physiological data, it has become clear that synapsins also have a function at the membrane stages of release after SVs have docked to the active zones. Several observations support the latter view: (i) SVs in the readily releasable pool are only partially depleted of synapsins, and about 20-35% of the synapsin molecules associated with SVs in the reserve pool remain associated with actively recycling SVs during high frequency stimulation;[47] (ii) while in resting synapses synapsins are preferentially confined to the reserve pool, during synaptic activity synapsins are detected in association with SVs of the readily releasable pool and with uncoated recycled SVs;[4] (iii) synapsins colocalize with actin in the dynamic filamentous cytomatrix present in sites of intense SV recycling.[4]

The possibility that synapsins could play some role in the post-docking stages of neurexocytosis, initially suggested by the uncertain effects on PPF, was recently demonstrated by growing evidence showing that synapsins can directly affect the probability and the rate of NT release. The first functional evidence suggesting a possible post-docking effect of synapsins was obtained by studying SV dynamics with styryl FM dyes. In hippocampal neurons from synapsin I KO mice, the reduction in the total functional recycling SV pool size was found to be associated with a decrease in the total number of SVs which undergo exocytosis during brief trains of action potentials (20 impulses) at individual synaptic boutons.[39] While the former observation was in agreement with the decrease in the reserve pool of SVs,[26,44] the latter result was rather unexpected, since stimuli in this range would be expected to draw solely upon the readily releasable pool of SVs which appears relatively intact in synapsin I KO mice, and suggests a decrease in release probability.

More recently, the presynaptic injection of a peptide corresponding to the highly conserved region of domain E of squid synapsin into the squid giant synapse completely inhibited NT release in the absence of appreciable changes in the number of docked SV.[18] Interestingly, this effect was accompanied by an increase in the rise and decay times of postsynaptic currents. The kinetics of release was also profoundly altered in cholinergic synapses of *Aplysia californica* injected with a specific antibody to snail synapsins.[22] In this study, the rise time of the evoked postsynaptic current was significantly slowed in the absence of any changes in decay time and mean amplitude of postsynaptic response. Closely similar results were obtained with the injection of a conserved peptide derived from the C domain.[20]

A post-docking action of synapsins is likely to be involved also in the decrease of evoked inhibitory postsynaptic currents (eIPSCs) observed in CA3 pyramidal neurons from hippocampal slices (P10-14) of synapsin I KO mice.[45] Mutant mice showed a decrease in the amplitude and an increase in the coefficient of variation of eIPSCs, while the amplitude of miniature IPSCs was not affected, suggesting that synapsin I deficiency reduces the efficiency of inhibitory synaptic transmission by decreasing the number of SV released by a single action potential. The decrease of eIPSCs observed in cultured hippocampal autaptic neurons (7-9 DIV) from synapsin III KO mice[9] could be also attributable to a decreased release probability, although a more detailed electrophysiological analysis is necessary to exclude other possibilities.

Taken together, the data summarized here strongly suggest that the synapsins are also involved in the post-docking steps of release. By directly or indirectly regulating priming and/or fusion reactions, the synapsins may play a role in determining the rate and the amount of docked SVs released in response to the action potential (Fig. 4). Such post-docking action could be accounted for by interactions of the synapsins with the dynamic actin cytoskeleton present at the active and periactive zones and/or with presynaptic proteins involved in the priming/fusion steps. On the one hand, it has recently been shown that the synapsin domain E and domain C peptides have the ability to inhibit the binding of endogenous synapsins to actin,[20] suggesting that an interaction with actin at the active zone may play a role in the post-docking effects of synapsin. On the other hand, synapsins have been recently shown to interact with the SV-associated G protein Rab3A and to modulate Rab3 cycling and GTPase activity within nerve terminals.[13,14] As Rab 3A has been proposed to limit the amount of NT released in response to the Ca^{2+} signal in a late step that follows docking and priming,[12] in principle the post-docking effect of synapsins can be achieved through the removal of the Rab3-mediated inhibitory constraint on quantal release.

The Synapsins, Long-Term Plasticity and Learning

Previous studies have shown that electrically-induced long-term potentiation (LTP) increases the phosphorylation of synapsin I at its CaM kinase II sites immediately after stimulation, with an effect that persists for over 30 min and is fully blocked by the NMDA receptor antagonist D(2)-2-amino-5-phosphonopentanoic acid (APV).[31] Moreover LTP-like potentiation produced by β-adrenergic agonists and protein kinase C activators produces a dose-dependent increase in the phosphorylation of synapsin I at its CaM kinase II sites.[30,43]

Otherwise, a large body of previous data demonstrated that synapsin I KO mice have normal LTP and normal or slightly impaired learning.[37,41,42] However, very recent observations put forward a role for synapsin I in LTP and learning. In fact, deletion of synapsin I blocked the enhancement of LTP, of spatial learning and of contextual fear conditioning associated with a constitutive activation of the H-Ras/ERK signalling pathway.[25] These results suggest that synapsin I could be the main presynaptic effector for certain forms of LTP triggered by the activation of the Ras/MAP kinase pathway. A growing body of evidence suggests that BDNF could be the presynaptic activator of H-Ras/ERK/synapsin I signalling pathway. Indeed, BDNF/trkB signalling can modulate presynaptic function and learning[48-50] and BDNF increases synaptosomal glutamate release through an ERK-dependent phosphorylation of synapsin I.[23]

Conclusions

In this chapter we have summarized and attempted to compose into a unifying frame the numerous physiological observations and hypotheses on synapsin function that have been put forward over the last 15 years in a large array of experimental systems, from reconstituted or isolated nerve terminals to mice bearing deletions in single and multiple synapsin genes. The emerging picture, summarized in Figure 4, is complex, as expected from a complex family of proteins that includes several isoforms with partly redundant functions and distinct developmental and regional patterns of expression and that are targets of multiple signal transduction pathways. Notwithstanding this complexity, the extremely high evolutionary conservation and the overt deficits in synaptic function and neural circuit activity observed in their absence strongly support a central role of the synapsins in the regulation of information transfer among neurons.

References

1. Benfenati F, Valtorta F, Chieregatti E et al. Interaction of free and synaptic vesicle-bound synapsin I with F-actin. Neuron 1992; 8:377-386.
2. Benfenati F, Valtorta F, Rossi MC et al. Interactions of synapsin I with phospholipids: Possible role in synaptic vesicle clustering and in the maintenance of bilayer structures. J Cell Biol 1993; 123:1845-1855.
3. Benfenati F, Onofri F, Giovedi S. Protein-protein interactions and protein modules in the control of neurotransmitter release. Phil Trans R Soc London B 1999; 354:243-257.
4. Bloom O, Evergreen E, Tomilin N et al. Colocalization of synapsin and actin during synaptic vesicle recycling. J Cell Biol 2003; 161:737-747.
5. Chi P, Greengard P, Ryan TA. Synapsin dispersion and reclustering during synaptic activity. Nat Neurosci 2001; 4:1187-1193.
6. Chi P, Greengard P, Ryan TA. Synaptic vesicle mobilization is regulated by distinct synapsin I phosphorylation pathways at different frequencies. Neuron 2003; 38:69-78.
7. Chin LS, Li L, Ferreira A et al. Impairment of axonal development and synaptogenesis in hippocampal neurons of synapsin I-knockout mice. Proc Natl Acad Sci USA 1995; 92:9230-9234.
8. Del Castillo J, Katz B. Quantal components of the end-plate potential. J Physiol 1954; 124:560-73.
9. Feng J, Chi P, Blanpied TA et al. Regulation of neurotransmitter release by synapsin III. J Neurosci 2002; 22:4372-4380.
10. Ferreira A, Kao HT, Feng J et al. Synapsin III: Developmental expression, subcellular localization, and role in axon formation. J Neurosci 2000; 20:3736-3744.
11. Ferreira A, Rapoport M. The synapsins: Beyond the regulation of neurotransmitter release. Cell and Mol Life Sci 2002; 59:589-595.
12. Geppert M, Goda Y, Stevens CF et al. The small GTP-binding protein Rab3A regulates a late step in synaptic vesicle fusion. Nature 1997; 387:810-814.
13. Giovedi S, Vaccaro P, Valtorta F et al. Synapsin is a novel Rab3 effector protein on small synaptic vesicles. I. Identification and characterization of the synapsin I-Rab3 interactions in vitro and in intact nerve terminals. J Biol Chem 2004a; 279:43760-43768.
14. Giovedi S, Darchen F, Valtorta F et al. Synapsin is a novel Rab3 effector protein on small synaptic vesicles. II. Functional effects of the Rab3A-synapsin I interaction. J Biol Chem 2004b; 279:43769-43779.
15. Gitler D, Takagishi Y, Feng J et al. Different presynaptic roles of synapsins at excitatory and inhibitory synapses. J Neurosci 2004; 24:11368-11380.
16. Greengard P, Valtorta F, Czernik AJ et al. Synaptic vesicle phosphoproteins and regulation of synaptic function. Science 1993; 259:780-785.
17. Hackett JT, Cochran SL, Greenfield Jr LJ et al. Synapsin I injected presynaptically into goldfish Mauthner axons reduces quantal synaptic transmission. J Neurophys 1990; 63:701-706.

18. Hilfiker S, Schweizer FE, Kao HT et al. Two sites of action for synapsin domain E in regulating neurotransmitter release. Nat Neurosci 1998; 1:29-35.
19. Hilfiker S, Pieribone VA, Czernik AJ et al. Synapsins as regulators of neurotransmitter release. Phil Trans R Soc London B 1999; 354:269-279.
20. Hilfiker S, Benfenati F, Doussau F et al. Structural domains involved in the regulation of transmitter release by synapsins. J Neurosci 2005; 25:2658-2669.
21. Hosaka M, Sudhof TC. Synapsin III, a novel synapsin with an unusual regulation by Ca2+. J Biol Chem 1998; 273:13371-13374.
22. Humeau Y, Doussau F, Vitiello F et al. Synapsin controls both reserve and releasable synaptic vesicle pools during neuronal activity and short-term plasticity in Aplysia. J Neurosci 2001; 21:4195-4206.
23. Jovanovic JN, Czernik AJ, Fienberg AA et al. Synapsins as a mediators of BDNF-enhanced neurotransmitter release. Nat Neurosci 2000; 3:323-9.
24. Kao HT, Porton B, Hilfiker S et al. Molecular evolution of the synapsin gene family. J Exp Zoology 1999; 285:360-377.
25. Kushner SA, Elgersma Y, Murphy GG et al. Modulation of presynaptic plasticity and learning by the H-ras/extracellular signal-regulated kinase/synapsin I signaling pathway. J Neurosci 2005; 25:9721-9734.
26. Li L, Chin LS, Shupliakov O et al. Impairment of synaptic vesicle clustering and of synaptic transmission, and increased seizure propensity, in synapsin I-knockout mice. Proc Nat Acad Sci USA 1995; 92:9235-9239.
27. Llinas R, McGuinness TL, Leonard CS et al. Intraterminal injection of synapsin I or calcium/calmodulin-dependent protein kinase II alters neurotransmitter release at the squid giant synapse. Proc Natl Acad Sci USA 1985; 82:3035-3039.
28. Llinas R, Gruner JA, Sugimori M et al. Regulation by synapsin I and Ca^{2+}-calmodulin-dependent protein kinase II of the transmitter release in squid giant synapse. J Physiol (London) 1991; 436:257-282.
29. Lin JW, Sugimori M, Llinas RR et al. Effects of synapsin I and calcium/calmodulin-dependent protein kinase II on spontaneous neurotransmitter release in the squid giant synapse. Proc Natl Acad Sci USA 1990; 87:8257-8261.
30. Malenka RC, Madison DV, Nicoll RA. Potentiation of synaptic transmission in the hippocampus by phorbol esters. Nature 1986; 321:175-177.
31. Nayak AS, Moore CI, Browning MD. Ca^{2+}-calmodulin-dependent protein kinase II phosphorylation of the presynaptic protein synapsin I is persistently increased during long-term potentiation. Proc Natl Acad Sci 1996; 93:15451-6.
32. Nichols RA, Sihra TS, Czernik AJ et al. Calcium/calmodulin-dependent protein kinase II increases glutamate and noradrenaline release from synaptosomes. Nature 1990; 343:647-651.
33. Nichols RA, Chilcote TJ, Czernik AJ et al. Synapsin I regulates glutamate release from rat brain synaptosomes. J Neurochem 1992; 58:783-785.
34. Nielander HB, Onofri F, Schaeffer E et al. Phosphorylation-dependent effects of synapsin IIa on actin polymerization and network formation. Eur J Neurosci 1997; 9:2712-2722.
35. Pera I, Stark R, Kappl M et al. Using the atomic force microscope to study the interaction between two solid supported lipid bilayers and the influence of synapsin I. Biophys J 2004; 87:2446-2455.
36. Pieribone VA, Shupliakov O, Brodin L et al. Distinct pools of synaptic vesicles in neurotransmitter release. Nature 1995; 375:493-497.
37. Rosahl TW, Geppert M, Spillane D et al. Short-term synaptic plasticity is altered in mice lacking synapsin I. Cell 1993; 75:661-670.
38. Rosahl TW, Spillane D, Missler M et al. Essential functions of synapsins I and II in synaptic vesicle regulation. Nature 1995; 375:488-493.
39. Ryan TA, Li L, Chin LS et al. SV recycling in synapsin I knockout mice. J Cell Biol 1996; 134:1219-1227.
40. Sihra TS, Wang JK, Gorelick FS et al. Translocation of synapsin I in response to depolarization of isolated nerve terminals. Proc Nat Acad Sci USA 1989; 86:8108-8112.
41. Silva AJ, Rosahl TW, Chapman PF et al. Impaired learning in mice with abnormal short-lived plasticity. Current Biology 1996; 6:1509-1518.
42. Spillane D, Rosahl TW, Sudhof TC et al. Long-term potentiation in mice lacking synapsins. Neuropharmacology 1995; 34:1573-9.
43. Stanton PK, Sarvey JM. Norepinephrine regulates long-term potentiation of both the population spike and dendritic EPSP in hippocampal dentate gyrus. Brain Res Bull 1987; 18:115-119.

44. Takei Y, Harada A, Takeda S et al. Synapsin I deficiency results in the structural change in the presynaptic terminals in the murine nervous system. J Cell Biol 1995; 131:1789-1800.
45. Terada S, Tsujimoto T, Takei Y et al. Impairment of inhibitory synaptic transmission in mice lacking synapsin I. J Cell Biol 1999; 145:1039-1048.
46. Torri-Tarelli F, Villa A, Valtorta F et al. Redistribution of synaptophysin and synapsin I during α-latrotoxin-induced release of neurotransmitter at the neuromuscular junction. J Cell Biol 1990; 110:449-459.
47. Torri-Tarelli F, Bossi M, Fesce R et al. Synapsin I partially dissociates from synaptic vesicles during exocytosis induced by electrical stimulation. Neuron 1992; 9:1143-1153.
48. Tyler WJ, Alonso M, Bramham CR et al. From acquisition to consolidation: On the role of brain-derived neurotrophic factor signaling in hippocampal-dependent learning. Lear Mem 2002; 9:224-37.
49. Xu B, Gottschalk W, Chow A et al. The role of brain-derived neurotrophic factor receptors in the mature hippocampus: Modulation of long-term potentiation through a presynaptic mechanism involving TrkB. J Neurosci 2000; 20:6888-97.
50. Zhang X, Poo MM. Localized synaptic potentiation by BDNF requires local protein synthesis in developing axon. Neuron 2002; 36:675-88.
51. Zucker RS, Regehr WG. Short-term synaptic plasticity. Annu Rev Phys 2002; 64:355-405.

CHAPTER 6

Phospholipase D:
A Multi-Regulated Lipid-Modifying Enzyme Involved in the Late Stages of Exocytosis

Marie-France Bader* and Nicolas Vitale

Abstract

Phospholipase D1, a signaling-activated enzyme that generates phosphatidic acid, has become recognized as a key player in regulated exocytosis in organisms from yeast to mammals. A variety of mechanisms have been proposed to explain how the generation of phosphatidic acid at sites of exocytosis facilitates secretion in mammalian cells, and this remains an active area of investigation. Recent findings suggest that phosphatidic acid may play a biophysical role to generate negative curvature and thus promote fusion of secretory vesicles with the plasma membrane.

Introduction

Phospholipase D (PLD) is a signal transduction-regulated enzyme that catalyzes the hydrolysis of phosphatidylcholine to phosphatidic acid (PA).[1,2] Several evidences support a role for PLD in vesicular trafficking events. The first arose with the discovery that members of the small GTPase ARF family are effective activators of PLD.[3,4] ARF1 is a well-known regulator of membrane budding from the Golgi, and this finding raised the hypothesis that generation of PA recruits coat proteins to budding vesicles.[5,6] The PLD requirement in yeast meiosis was then demonstrated: PLD mediates an unusual form of regulated exocytosis in yeast that results in the formation of membrane organelles critical to spore formation.[7-9] Mammalian PLD has since then been linked to regulated exocytosis in many secretory cell types, including endocrine and neuroendocrine cells as well as neurons.[10-15]

In mammalian cells, two PLD species, PLD1 and PLD2, and their splice variants have been characterized.[16,17] Both have been proposed to play roles in membrane trafficking. PLD1 is most frequently associated with exocytosis, although it has also been reported to function in endocytosis and phagocytosis.[10,11,13-15,18,19] In contrast, PLD2 has been often associated with endocytosis and phagocytosis, but roles for it in secretion have also been proposed.[13,18-23] PLD1 is generally associated to peri-nuclear vesicles, although it can be found on the plasma membrane and most likely cycles between these sites.[10,24,25] PLD2 typically exhibits a location at the plasma membrane, but also most likely cycles through early and recycling endosomes during endocytic events.[17,22,26] However, despite the fact that both enzymes mediate the same biochemical reaction, exhibit common upstream signaling pathways, and are often coincidentally localized in cells, they do not appear to undertake redundant cellular functions. Hence, in

*Corresponding Author: Marie-France Bader—Department of Neurotransmission and Neuroendocrine Secretion, CNRS UMR-7168/LC2, 5 rue Blaise Pascal, 67084 Strasbourg, France. Email: bader@neurochem.u-strasbg.fr

Molecular Mechanisms of Exocytosis, edited by Romano Regazzi. ©2007 Landes Bioscience and Springer Science+Business Media.

many cases overexpression of the wild-type or catalytically-inactive forms of PLD1 affect a signaling or cell trafficking event in a manner that is not reproduced by the overexpression of the corresponding PLD2 isoform, and vice versa.[10,13,20,22] Thus, PLD1 and PLD2 play isoform-specific functions in membrane trafficking.

The role of PLD1 in secretion has been studied in several settings, including the release of hormones from endocrine cells, mast cell degranulation, release of IL-8 from epithelial cells and insulin-stimulated movement of the glucose transporter Glut4 to the cell surface.[10,13-15,27] In this chapter, we will describe a series of experiments that we have performed over the past several years to examine the role of PLD1 in regulated exocytosis in neuroendocrine cells and in neurons.[10,11,28]

Molecular Tools to Probe the Cellular Functions of PLD Enzymes

Pharmacological approaches to assess the cellular functions of PLD have been based primarily on the use of alcohols to divert the production of PA.[1] In the presence of a primary alcohol, PLD can catalyze a transphosphatidylation reaction that exchanges the polar head group of the phospholipid substrate with the given alcohol to form the corresponding phosphatidyl-alcohol.[29] This unique property of PLD has been used to reveal PLD activation and function in many cell biological processes. However, there is a risk of false positive results, since alcohols can mediate nonspecific inhibition, and false negatives, since even at the highest levels that can be tolerated before encountering cellular toxicity, not all PLD-mediated PA production is suppressed.[30]

Molecular tools used to probe more directly for the implication of PLD include overexpression of the wild-type isoforms and expression of the catalytically-inactive point mutants, PLD1(K898R) and PLD2(K758R).[31] The mechanisms through which the inactive PLD mutants function as a dominant negative allele is not entirely clear. For instance PLD1 does not dimerize for function and its activators (small GTPases of the ARF and Rho families, and classical isoforms of protein kinase C) and substrate (phosphatidylcholine) are fairly abundant.[32] However, both PLD1 and PLD2 have to perform their function at specific intracellular locations, and the catalytically-inactive enzymes most likely function through competitive substitution for the endogenous isoforms at those locations. Use has also been made of ARF1 and ARF6 point mutant alleles that specifically lack the ability to activate PLD1 while retaining all other known downstream effectors function.[28,30] Interference RNA is beginning to make its impact on the PLD field and offers a complementary loss-of-function approach that should help to support findings generated through overexpression of the dominant negative alleles and the use of alcohols.[20,22,33,34]

Phospholipase D1 Localizes to Granule Docking Sites in Neuroendocrine Chromaffin and PC12 Cells and Plays a Role in Exocytosis

Adrenal chromaffin cells, along with their tumor cell derivatives, PC12 cells, are neuroendocrine cells that have a prominent place among the models having provided insight into the molecular machinery underlying calcium-regulated exocytosis.[35-39] Taking advantage of the transphosphatidylation reaction catalyzed by PLD in the presence of ethanol, we measured PLD activity in resting and stimulated chromaffin cells as a first attempt to assess the implication of PLD in exocytosis. PLD activation was found to be calcium-dependent and occurred with kinetics very similar to those of the exocytotic response.[40] Subcellular fractionation experiments revealed that this calcium-induced PLD activity was specifically associated with the plasma membrane.[40] The use of isoform specific antibodies confirmed the presence of PLD1b at the plasma membrane in chromaffin and PC12 cells.[10] Interestingly, PLD1 colocalizes with SNAP-25, a component of the SNARE complex known to be involved in the docking and fusion of secretory granules with the plasma membrane.

We used more recently a variety of direct means to test the idea that the activation of PLD represents an important step in the exocytotic process in chromaffin and PC12 cells. We first examined the effect of the overexpression of PLD1, PLD2, and their catalytically-inactive mutants in PC12 cells. Inactive PLD1 dramatically inhibited secretion, whereas PLD1 robustly stimulated it.[10] In contrast, neither the active nor inactive allele of PLD2 had any effect in this system. We also microinjected inactive PLD1 mutant protein into chromaffin cells and monitored secretion by amperometry. This method offers the possibility to resolve the release of the contents of single secretory granules.[41,42] Analysis of individual release events in injected cells revealed an apparent reduction in the initial rate of release, possibly reflecting a defect in membrane fusion and/or pore expansion.[10] These data support the idea that PLD1 might be implicated in a late step of exocytosis close to membrane fusion. It is noteworthy that a similar conclusion was reached by electrophysiological analysis of neurotransmitter release from Aplysia neurons: the inactive PLD1 mutant was found to block neurotransmission by affecting the fusogenic status of the presynaptic release sites.[11] In line with these findings, PLD1 and possibly PLD2, have been since then implicated in the calcium-dependent fusion of vesicles with the plasma membrane in various secretory cell types.[13,15]

The precise role of PLD-produced PA at the site of exocytotic fusion remains an evolving research question. Potential roles for PA can be divided into four primary categories. First, PA can serve as substrate for enzymes like phosphatidic acid phosphohydrolase or phospholipase A2, to generate other signaling lipids such as diacylglycerol or lysophosphatidic acid, respectively. Second, PA can act as a lipid anchor, recruiting proteins required for vesicle priming or fusion events.[43-46] Third, PA can act as a signaling lipid to stimulate enzymes such as phosphatidylinositol-4 phosphate 5 kinase, whose product, phosphatidylinositol 4,5 bisphosphate, also plays critical roles in exocytosis.[47-52] Finally, PA is a cone shaped lipid that has been described as an essential partner for proteins in the basic fusion machinery.[53-57] Hence, a number of fusion reactions, including viral fusion, vesicular trafficking and exocytosis, are reversibly inhibited by lipids of positive curvature, and promoted by lipids of negative curvature. The predicted effect of the local generation of PA due to the activation of PLD1 at the plasma membrane is to promote a negative curvature of the cytoplasmic leaflet (Fig. 1). Thus, PA may trigger or facilitate hemifusion intermediates and may thereby be required for the

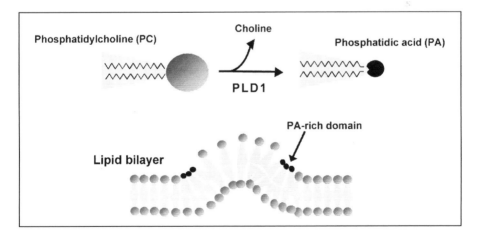

Figure 1. PLD1 hydrolyzes the membrane phospholipid phosphatidylcholine (PC) to generate phosphatidic acid (PA) and choline. PA is a cone shaped lipid. The predicted effect of the local generation of PA at the plasma membrane is to promote a negative curvature of the cytoplasmic leaflet and thereby bending of the lipid bilayer.

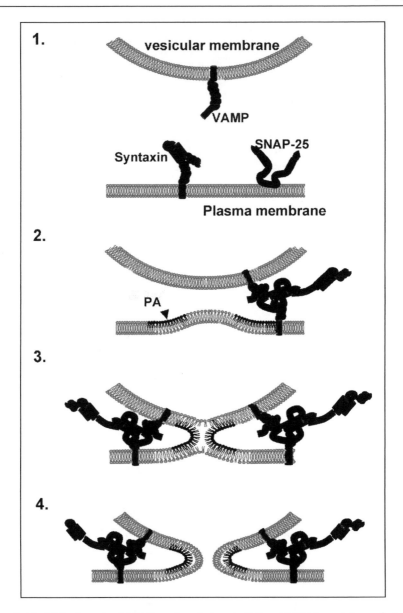

Figure 2. Model for the role of PLD1-produced phosphatidic acid (PA) in membrane fusion. In secretagogue-stimulated cells, vesicle and plasma membranes (1) are brought in close proximity through the formation of SNARE complexes (2). The local elevation of PA generated by the plasma membrane-bound PLD1 at the vesicle docking site promotes membrane bending (2), destabilization of the phospholipid bilayers and formation of the hemifusion intermediates (3) required for the opening of exocytotic fusion pores (4).

formation of the exocytotic fusion pore (Fig. 2). Most of these roles have been demonstrated using in vitro assay systems. Additional experiments are now required to establish convincingly the relative importance of each mechanism in vivo.

Regulation of PLD1 Activity in the Course of Exocytosis

The implication of PA in the late stages of exocytosis such as the fusion process implies a tight control of the PLD1 enzymatic activity at the plasma membrane. PLD1 is a multi-module protein that can be regulated by both ARF and Rho GTPases and by protein kinase C-dependent signaling pathways. ARF GTPases have been linked to regulated secretion in various cell types.[12,58-60] In chromaffin and PC12 cells, ARF6 is associated with the dense core secretory granules.[28,40,61] Evidence that ARF6 plays a role in chromaffin granule exocytosis came first from studies showing that myristoylated peptides corresponding in sequence to the amino-terminal end of ARF6 inhibit noradrenaline secretion when introduced in the cytosol of permeabilized chromaffin cells.[61] These results were then substantiated by the fact that, among various inactive or constitutively active ARF mutants, ARF6 is the sole protein able to modify the secretory response in PC12 cells.[28] Expression of the dominant negative GDP-loaded ARF6 mutant strongly inhibits secretion, whereas the constitutively active GTP-loaded ARF6 significantly stimulates it.[28] Interestingly, expression of a GTP-loaded ARF6(N48I) mutant that specifically lost its ability to activate PLD1 inhibits secretion in PC12 cells, revealing that PLD1 plays a major contribution in the pathway by which ARF6 regulates exocytosis.[28]

It is noteworthy that the activation/inactivation cycle of ARF6 is intimately linked to the exocytotic reaction. For instance, subcellular fractionation experiments reveal that ARF6 bound to secretory granules is in its inactive GDP-bound state whereas active GTP-bound ARF6 is detected only in plasma membrane-containing fractions prepared from secretagogue-stimulated cells.[28] Accordingly, ARNO, a guanine nucleotide exchange factor promoting the activation of ARF6, is present on the plasma membrane in chromaffin and PC12 cells.[28,62] Expression of a catalytically-inactive ARNO mutant in PC12 cells or introduction of anti-ARNO antibodies in the cytoplasm of permeabilized chromaffin cells inhibits secretion to an extent similar to the inactive GDP-bound ARF6 mutant.[62] Thus, ARNO is implicated in the exocytotic pathway, most likely through the regulation of ARF6. Since ARNO is specifically associated with the plasma membrane, ARF6 can be activated only following the recruitment and docking of granules to the plasma membrane.[28,62] Consequently, PLD1 becomes activated and generates PA only at the granule docking sites in secretagogue-stimulated cells (Fig. 3). This spatial and temporal regulation of PLD1 activity is in line with the idea that local elevation of PA is required between granules and the plasma membrane to trigger or facilitate membrane merging at the site of contact.

Exocytosis in neuroendocrine cells is above all a calcium-regulated event, and additional mechanisms are likely to occur at the plasma membrane to link ARF6-dependent PLD1 activity to variations in cytosolic calcium. Ral proteins constitute a family of monomeric GTPases that have the potential to be directly activated by elevation of intracellular calcium levels.[63] Interestingly, RalA has been shown to interact directly with PLD1 and to enhance ARF-stimulated PLD1 activity.[64,65] We recently described a series of experiments indicating that the small GTPase RalA controls exocytosis in chromaffin and PC12 cells.[66] Briefly, expression of a constitutively active RalA mutant enhanced secretagogue-evoked secretion from PC12 cells, whereas expression of a constitutively GDP-bound mutant or silencing of the RalA gene by RNA interference led to a strong inhibition of the secretory response.[66] We also found that RalA is directly implicated in the activation of the ARF6-dependent PLD1 at the plasma membrane. Moreover, using various RalA mutants selectively impaired in their ability to activate downstream effectors, we demonstrated that PLD1 activation is essential for the activation of secretion by GTP-loaded RalA.[66] Together, these results are in line with the idea that RalA stimulates exocytosis by interacting with the ARF6-dependent PLD1 at the plasma membrane. Our data indicate that RalA is activated in response to membrane depolarization, which directly triggers calcium influx through voltage-gated calcium channels. Moreover, RalA is known to contain a calmodulin-binding site and Ca^{2+}-calmodulin binding stimulates directly GTP loading and RalA activation.[67] Thus, RalA may represent an additional checkpoint that integrates calcium signals to PLD1 and its lipid-modifying activity at the exocytotic sites (Fig. 3).

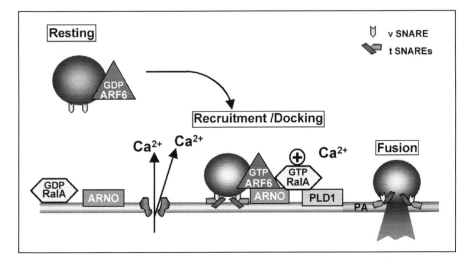

Figure 3. Model for the regulation of PLD1 in the course of exocytosis. In neuroendocrine chromaffin cells, ARF6 is associated with the secretory granule membrane whereas RalA, the ARF6-guanine exchange factor ARNO and PLD1 are found at the plasma membrane. Stimulation with a secretagogue and the consequent elevation in cytosolic calcium triggers the recruitment and SNAREmediated docking of granules to the plasma membrane. Docking allows the transient interaction of ARF6 with ARNO and its GTP loading. Activated ARF6 together with Ca^{2+}-activated RalA synergistically stimulate PLD1 to produce PA at the granule docking sites. This local generation of PA at the site where membranes are pulled together through the formation of the SNARE complexes results in hemifusion and the formation of the exocytotic fusion pore.

ARF and PLD1 in Neurotransmitter Release

Evoked release of neurotransmitter in neurons has many homologies with hormone secretion performed by neuroendocrine cells; in both cell types, secretion arises from an entry of Ca^{2+} that in turn triggers the fusion of a sub-population of primed vesicles docked at the plasma membrane.[68] Thus, the localization and functions of the molecular components participating to the release machinery are often conserved between neurons and neuroendocrine cells. The implication of PLD1 in neuroendocrine cell exocytosis led us to probe the idea that PLD1 fulfills a similar role in neurotransmitter release. PLD1 is present in membrane fractions prepared from rat brain synaptosomes and it colocalizes with synaptophysin, a marker of synaptic vesicles, in cultured cerebellar granule cells, indicating that PLD1 is present in areas specialized in neurotransmitter release.[11] To determine the possible involvement of PLD1 in neurotransmitter release, a catalytically-inactive PLD1 mutated protein was injected into *Aplysia* cholinergic neurons. This mutant was found to be a potent inhibitor of acetylcholine release and, as revealed by analyzing the fluctuations in amplitude of postsynaptic responses, it affected the number of release sites.[11] Thus, by analogy with the function proposed in hormonal secretion, PLD1 may induce lipid modifications in the presynaptic plasma membrane required for fusion of synaptic vesicles and neurotransmitter release.

The mechanisms regulating PLD1 activity in neuronal exocytosis remain to be explored. GTP-bound ARF6 has been found at the plasma membrane in cortical neuron synapses and several arguments support the participation of ARF6 in synaptic vesicle exocytosis.[69] For instance, in *Xenopus* spinal neurons, microinjection of a guanine nucleotide exchange factor for ARF proteins (mSec7-1) increases the frequency of spontaneous release.[70] mSec7-1 was found to enhance the amplitude of evoked responses without affecting the size of the quantum, and it increased the depression rate of synaptic depression induced by repetitive stimulation.[70] From these data, it has been proposed that mSec7-1, most likely by activating ARF6, facilitates the probability of a mature vesicle to fuse in response to Ca^{2+} entry and/or increases the size of the readily releasable pool of synaptic vesicles. The possible functional links between ARF6 and PLD1 in neurons remains, however, to be investigated. It is also interesting to note that the GTPase Rac1 is associated with synaptic vesicles in neurons. Hence, lethal toxin, which inactivates Rac, and the inactive PLD1 mutant, similarly inhibit acetylcholine release from *Aplysia* neurons by reducing the number of functional release sites without affecting the probability of release or the size of the quantum.[11,71] Thus, Rac1 might as well be an activator of the plasma membrane-bound PLD1 in the cascade leading to neurotransmitter release.

Conclusions

The wide variety of systems in which PLD1 has been examined reveals it as a major player in regulated exocytosis in mammalian cells. Many questions remain about the underlying mechanisms through which PLD1 functions and more generally about its physiological roles. With the furthered development of RNAi approaches and the generation of knockout animals, we are closer than ever to be able to answer these important questions. Specific pharmalogical PLD1 and PLD2 inhibitors and activators are critically missing; they could have clinical utility to promote or inhibit regulated exocytosis as appropriate in a wide variety of diseases, such as insulin secretion in type II diabetes and immune mediator release in autoimmune syndromes, respectively.

Reference

1. McDermott M, Wakelam MJ, Morris AJ. Phospholipase D. Biochem Cell Biol 2004; 82:225-253.
2. Jenkins GM, Frohman MA. Phospholipase D: A lipid centric review. Cell Mol Life Sci 2005; 62:2305-2316.
3. Brown HA, Gutowski S, Moomaw CR et al. ADP-ribosylation Factor, a small GTP-dependent regulatory protein, stimulates phospholipase D activity. Cell 1993; 75:1137-1144.
4. Cockcroft S, Thomas GM, Fensome A et al. Phospholipase D: A downstream effector of ARF in granulocytes. Science 1994; 263:523-526.
5. Ktistakis NT, Brown HA, Waters MG et al. Evidence that phospholipase D mediates ADP ribosylation factor- dependent formation of Golgi coated vesicles. J Cell Biol 1996; 134:295-306.
6. Roth MG. Lipid regulators of membrane traffic through the Golgi complex. Trends Cell Biol 1999; 9:174-179.
7. Rose K, Rudge SA, Frohman MA et al. Phospholipase D signaling is essential for meiosis. Proc Natl Acad Sci USA 1995; 92:12151-12155.
8. Rudge SA, Morris AJ, Engebrecht J. Relocalization of Phospholipase D activity mediates membrane formation during meiosis. J Cell Biol 1998; 140:81-90.
9. Rudge SA, Engebrecht J. Regulation and function of PLDs in yeast. Biochim Biophys Acta 1999; 1439:167-174.
10. Vitale N, Caumont AS, Chasserot-Golaz S et al. Phospholipase D1: A key factor for the exocytotic machinery in neuroendocrine cells. EMBO J 2001; 20:2424-2434.
11. Humeau Y, Vitale N, Chasserot-Golaz S et al. A role for phospholipase D1 in neurotransmitter release. Proc Natl Acad Sci USA 2001; 98:15300-15305.
12. Cockcroft S, Way G, O'Luanaigh N et al. Signalling role for ARF and phospholipase D in mast cell exocytosis stimulated by crosslinking of the high affinity FcepsilonR1 receptor. Mol Immunol 2002; 38:1277-1282.
13. Choi WS, Kim YM, Combs C et al. Phospholipases D1 and D2 regulate different phases of exocytosis in mast cells. J Immunol 2002; 168:5682-5689.
14. Hughes WE, Elgundi Z, Huang P et al. Phospholipase D1 regulates secretagogue-stimulated insulin release in pancreatic β-cells. J Biol Chem 2004; 279:27534-27541.
15. Huang P, Altshuller YM, Hou JC et al. Insulin-stimulated plasma membrane fusion of Glut4 glucose transporter-containing vesicles is regulated by phospholiopase D1. Mol Biol Cell 2005; 16:2614-2623.
16. Hammond SM, Altshuller YM, Sung TC et al. Human ADP-ribosylation factor-activated phosphatidylcholine-specific Phospholipase D Defines a New and Highly Conserved Gene Family. J Biol Chem 1995; 270:29640-29643.

17. Colley WC, Sung TC, Roll R et al. Phospholipase D2, a distinct phospholipase D isoform with novel regulatory properties that provokes cytoskeletal reorganization. Curr Biol 1997; 7:191-201.
18. Shen Y, Xu L, Foster D. Role for phospholipase D in receptor-mediated endocytosis. Mol Cell Biol 2001; 21:595-602.
19. Iyer SS, Barton JA, Bourgoin S et al. Phospholipases D1 and D2 coordinately regulate macrophage phagocytosis. J Immunol 2004; 173:2615-2623.
20. Bhattacharya M, Babwah AV, Godin C et al. Ral and phospholipase D2-dependent pathway for constitutive metabotropic glutamate receptor endocytosis. J Neurosci 2004; 24:8752-8761.
21. Koch T, Brandenburg LO, Schulz S et al. ADP-ribosylation factor-dependent phospholipase D2 activation is required for agonist-induced μ-opioid receptor endocytosis. J Biol Chem 2003; 278:9979-9985.
22. Du G, Huang P, Lian BT et al. Phospholipase D2 localizes to the plasma membrane and regulates angiotensin II receptor endocytosis. Mol Biol Cell 2004; 15:1024-1030.
23. Denmat-Ouisse LA, Phebidias C, Honkavaara P et al. Regulation of constitutive protein transit by phospholipase D in HT29-cl19A cells. J Biol Chem 2001; 276:48840-48846.
24. Lucocq J, Manifava M, Bi K et al. Immunolocalisation of phospholipase D1 on tubular vesicular membranes of endocytic and secretory origin. Eur J Cell Biol 2001; 80:508-520.
25. Du G, Huang P, Vitale N et al. Regulation of Phospholipase D subcellular cycling through coordination of multiple membrane association motifs. J Cell Biol 2003; 62:305-315.
26. Sarri E, Pardo R, Fensome-Green A et al. Endogenous phospholipase D2 localizes to the plasma membrane of RBL-2H3 mast cells and can be distinguished from ADP ribosylation factor-stimulated phospholipase D1 activity by its specific sensitivity to oleic acid. Biochem J 2003; 369:319-329.
27. Wang L, Cummings R, Usatyuk P et al. Involvement of phospholipases D1 and D2 in sphingosine 1-phosphate-induced ERK (extracellular-signal-regulated kinase) activation and interleukin-8 secretion in human bronchial epithelial cells. Biochem J 2002; 367:751-760.
28. Vitale N, Chasserot-Golaz S, Bailly Y et al. Calcium-regulated exocytosis of dense core vesicles requires the activation of ARF6 by ARNO at the plasma membrane. J Cell Biol 2002; 159:79-89.
29. Kobayashi M, Kanfer JN. Phosphatidylethanol formation via transphosphatidylation by rat brain synaptosomal phospholipase D. J Neurochem 1987; 48:1597-1603.
30. Skippen A, Jones DH, Morgan CP et al. Mechanism of ADP ribosylation factor-stimulated phosphatidylinositol 4,5-bisphosphate synthesis in HL60 cells. J Biol Chem 2002; 277:5823-5831.
31. Sung TC, Roper RL, Zhang Y et al. Mutagenesis of phospholipase D defines a superfamily including a trans- Golgi viral protein required for poxvirus pathogenicity. EMBO J 1997; 16:4519-4530.
32. Hammond SM, Jenco JM, Nakashima S et al. Characterization of two alternately spliced forms of Phospholipase D1. Activation of the purified enzymes by PIP2, ARF, and Rho family monomeric G-proteins and protein kinase C-α. J Biol Chem 1997; 272:3860-3868.
33. Fang Y, Park IH, Wu AL et al. PLD1 regulates mTOR signaling and mediates Cdc42 activation of S6K1. Curr Biol 2003; 13:2037-2044.
34. Waselle L, Gerona RR, Vitale N et al. Role of phosphoinositide signaling in the control of insulin exocytosis. Mol Endocrinol 2005; 19:3097-3106.
35. Livett BG. Chromaffin cells: Roles for vesicle proteins and Ca2+ in hormone secretion and exocytosis. Trends Pharmacol Sci 1993; 14:345-348.
36. Burgoyne RD, Morgan A. Analysis of regulated exocytosis in adrenal chromaffin cells: Insights into NSF/SNAP/SNARE function. Bioessays 1998; 20:328-335.
37. Rettig J, Neher E. Emerging roles of presynaptic proteins in Ca++-triggered exocytosis. Science 2002; 298:781-785.
38. Bader MF, Holz RW, Kumakura K et al. Exocytosis: The chromaffin cell as a model system. Ann NY Acad Sci 2002; 971:178-183.
39. Bader MF, Doussau F, Chasserot-Golaz S et al. Coupling actin and membrane dynamics during calcium-regulated exocytosis: A role for Rho and ARF GTPases. Biochim Biophys Acta 2004; 1742:37-49.
40. Caumont AS, Galas MC, Vitale N et al. Regulated exocytosis in chromaffin cells. Translocation of ARF6 stimulates a plasma membrane-associated phospholipase D. J Biol Chem 1998; 273:1373-1379.
41. Chow RH, von Ruden L, Neher E. Delay in vesicle fusion revealed by electrochemical monitoring of single secretory events in adrenal chromaffin cells. Nature 1992; 356:60-63.
42. Wightman RM, Jankowski JA, Kennedy RT et al. Temporally resolved catecholamine spikes correspond to single vesicle release from individual chromaffin cells. Proc Nat Acad Sci USA 1991; 88:10754-10758.
43. Rizzo MA, Shome K, Watkins SC et al. The recruitment of Raf-1 to membranes is mediated by direct interaction with phosphatidic acid and is independent of association with Ras. J Biol Chem 2000; 275:23911-23918.

44. Manifava M, Thuring JW, Lim ZY et al. Differential binding of traffic-related proteins to phosphatidic acid- or phosphatidylinositol (4,5)- bisphosphate-coupled affinity reagents. J Biol Chem 2001; 276:8987-8994.
45. Wagner ML, Tamm LK. Reconstituted syntaxin1a/SNAP25 interacts with negatively charged lipids as measured by lateral diffusion in planar supported bilayers. Biophys J 2001; 81:266-275.
46. Nakanishi H, de los Santos P, Neiman AM. Positive and negative regulation of a SNARE protein by control of intracellular localization. Mol Biol Cell 2004; 15:1802-1815.
47. Jenkins GH, Fisette PL, Anderson RA. Type I phosphatidylinositol 4-phosphate 5-kinase isoforms are specifically stimulated by phosphatidic acid. J Biol Chem 1994; 269:11547-11554.
48. Honda A, Nogami M, Yokozeki T et al. Phosphatidylinositol 4-phosphate 5-kinase alpha is a downstream effector of the small G protein ARF6 in membrane ruffle formation. Cell 1999; 99:521-532.
49. Hay JC, Fisette PL, Jenkins GH et al. ATP-dependent inositide phosphorylation required for Ca^{2+}-activated secretion. Nature 1995; 374:173-177.
50. Di Paolo G, Moskowitz HS, Gipson K et al. Impaired PtdIns(4,5)P2 synthesis in nerve terminals produces defects in synaptic vesicle trafficking. Nature 2004; 431:415-422.
51. Cremona O, De Camilli P. Phosphoinositides in membrane traffic at the synapse. J Cell Sci 2001; 114:1041-1052.
52. Cockcroft S, De Matteis A. Inositol lipids as spatial regulators of membrane traffic. J Membr Biol 2001; 180:187-194.
53. Blackwood RA, Smolen JE, Transue A et al. Phospholipase D activity facilitates Ca2+-induced aggregation and fusion of complex liposomes. Am J Physiol 1997; 272:C1279-1285.
54. Siegel DP. The modified stalk mechanism of lamellar/inverted phase transitions and its implications for membrane fusion. Biophys J 1999; 76:291-313.
55. Zimmerberg J, Chernomordik LV. Membrane fusion. Adv Drug Deliv Rev 1999; 38:197-205.
56. Kozlovsky Y, Chernomordik LV, Kozlov MM. Lipid intermediates in membrane fusion: Formation, structure, and decay of hemifusion diaphragm. Biophys J 2002; 83:2634-2651.
57. Kooijman EE, Chupin V, de Kruijff B et al. Modulation of membrane curvature by phosphatidic acid and lysophosphatidic acid. Traffic 2003; 4:162-174.
58. Yang CZ, Mueckler M. ADP-ribosylation factor 6 (ARF6) defines two insulin-regulated secretory pathways in adipocytes. J Biol Chem 1999; 274:25297-25300.
59. Lawrence JT, Birnbaum MJ. ADP-ribosylation factor 6 regulates insulin secretion through plasma membrane phosphatidylinositol 4,5-bisphosphate. Proc Natl Acad Sci USA 2003; 100:13320-13325.
60. Matsukawa J, Nakayama K, Nagao T et al. Role of ADP-ribosylation factor 6 (ARF6) in gastric acid secretion. J Biol Chem 2003; 278:36470-36475.
61. Galas MC, Helms JB, Vitale N et al. Regulated exocytosis in chromaffin cells. A potential role for a secretory granule-associated ARF6 protein. J Biol Chem 1997; 272:2788-2793.
62. Caumont AS, Vitale N, Gensse M et al. Identification of a plasma membrane-associated guanine nucleotide exchange factor for ARF6 in chromaffin cells. Possible role in the regulated exocytotic pathway. J Biol Chem 2000; 275:15637-15644.
63. Wolthuis RM, Franke B, van Triest M et al. Activation of the small GTPase Ral in platelets. Mol Cell Biol 1998;18:2486-2491.
64. Luo JQ, Liu X, Frankel P et al. Functional association between Arf and RalA in active phospholipase D complex. Proc Natl Acad Sci USA 1998; 95:3632-3637.
65. Kim JH, Lee SD, Han JM et al. Activation of phospholipase D1 by direct interaction with ADP-ribosylation factor 1 and RalA. FEBS Lett 1998; 430:231-235.
66. Vitale N, Mawet J, Camonis J et al. The small GTPase RalA controls exocytosis of large dense core secretory granules by interacting with ARF6-dependent phospholipase D1. J Biol Chem 2005; 280:29921-29928.
67. Clough RR, Sidhu RS, Bhullar RP. Calmodulin binds RalA and RalB and is required for the thrombin-induced activation of Ral in human platelets. J Biol Chem 2002; 277:28972-28980.
68. Murthy VN, De Camilli P. Cell biology of the presynaptic terminal. Annu Rev Neurosci 2003; 26:701-728.
69. Krauss M, Kinuta M, Wenk MR et al. ARF6 stimulates clathrin/AP-2 recruitment to synaptic membranes by activating phosphatidylinositol phosphate kinase type Igamma. J Cell Biol 2003; 162:113-124.
70. Ashery U, Koch H, Scheuss V et al. A presynaptic role for the ADP ribosylation factor (ARF)-specific GDP/GTP exchange factor msec7-1. Proc Natl Acad Sci USA 1999; 96:1094-1099.
71. Humeau Y, Popoff MR, Kojima H et al. Rac GTPase plays an essential role in exocytosis by controlling the fusion competence of release sites. J Neurosci 2002; 22:7968-7981.

CHAPTER 7

Lipid Rafts as Regulators of SNARE Activity and Exocytosis

Christine Salaün and Luke H. Chamberlain*

Abstract

Lipid rafts, cholesterol and sphingolipid rich microdomains of the plasma membrane, have been implicated in the regulation of several intracellular pathways. Interestingly, components of the SNARE membrane fusion machinery associate with raft domains, and recent work suggests that this interaction may play an important role in regulated exocytosis. Here, we review the relationship between rafts and SNAREs, and discuss how rafts might participate in regulated exocytosis.

Introduction

Exocytosis describes the fusion of intracellular vesicles with the plasma membrane (PM).[1] Exocytic events are required for many aspects of cellular life, including protein and lipid trafficking to the PM and the secretion of a range of essential molecules from cells, such as neurotransmitters and hormones. A diverse range of intracellular membrane compartments undergo exocytosis, including synaptic vesicles, recycling endosomes, caveolae and lysosomes. Certain viruses also fuse with the PM of cells to facilitate the release of the viral genetic material into the cytoplasm of the target cell. Cellular and viral fusion are catalysed by specific proteins, SNAREs and fusion proteins respectively, and these fusion events are thought to occur through grossly similar mechanisms.[1]

During the course of membrane fusion, a narrow aqueous connection (the early fusion pore) forms between the extracellular and the intralumenal spaces. This is followed by a flux of lipids from one membrane to the other and the enlargement of a full fusion pore. These observations led to two distinct models for fusion, mainly differing by the nature (proteic or lipidic) of the early fusion pore.[1] An important observation is that the opening of the early fusion pore occurs before any lipid mixing, leading to the hypothesis that this early pore was composed of proteins (similar to a channel or a gap-junction).[2] This proteic fusion model describes a macromolecular pore complex spanning the two apposed bilayers, which would be formed by the transmembrane helices of fusion molecules (such as the SNARE protein Syntaxin,[3] the viral glycoprotein HA,[2] viral fusion peptides,[2] or the assembly of the Vo-ATPase during homotypic vacuolar fusion in yeast)[4] (see Fig. 1A).

Fusion can however be accomplished between pure lipid bilayers in vitro (in the absence of any proteins) with kinetics similar to that observed in vivo.[5] Moreover, in vivo fusion is sensitive to the addition of exogenous lipids, suggesting an important role for lipids in fusion

*Corresponding Author: Luke H. Chamberlain—Henry Wellcome Laboratory of Cell Biology, Division of Biochemistry and Molecular Biology, Davidson Building, IBLS, University of Glasgow, Glasgow G12 8QQ, U.K. Email: l.chamberlain@bio.gla.ac.uk

Molecular Mechanisms of Exocytosis, edited by Romano Regazzi. ©2007 Landes Bioscience and Springer Science+Business Media.

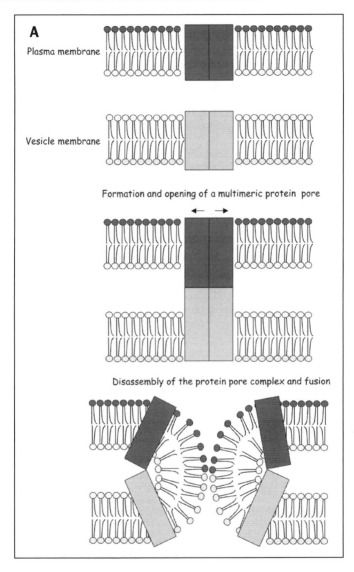

Figure 1. A) Exocytosis via a proteic fusion pore. A pore complex spanning the two bilayers forms; this consists of an oligomeric assembly of transmembrane helices of fusion proteins (represented by rectangles). Pore expansion and fusion correlates with disassembly of the complex. Figure continued on next page.

(see below). This led to the hypothesis that fusion was conducted by a rearrangement of the lipids composing the two apposed bilayers.[6,7] According to this hypothesis, the fusion reaction transits through a lipid structure called a stalk that expands into a hemifusion intermediate, where the two proximal monolayers are fused. The hemifusion structure then breaks into an early pore that expands into a stable lipidic fusion pore. The delayed mixing of lipids observed (occurring only after the opening of the early fusion pore) is explained by a 'restriction' of the lipid flux around the hemifusion structure (Fig. 1B).

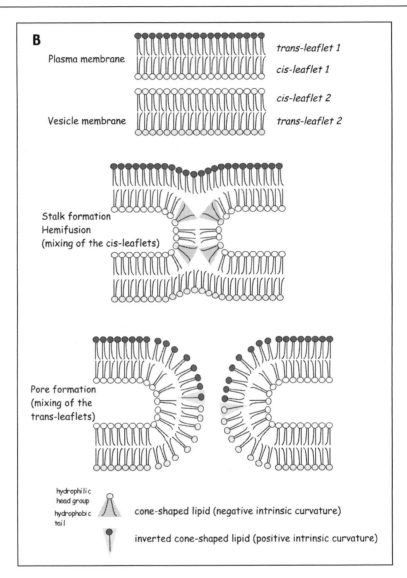

Figure 1, continued. B) Lipidic model of fusion. Mixing of the cis leaflets leads to the formation of a negatively curved structure called a stalk. This hemifusion state is stabilised by cone-shaped lipids present in the cis monolayers. Merging of the trans monolayers then leads to pore formation. The fusion pore has a positive curvature and is stabilised by inverted cone-shaped lipids in the trans monolayers. It is important to note that this lipidic model of fusion can nevertheless be catalysed by fusion protein interactions (although the proteins themselves do not constitute the pore). For simplicity, proteins have not been shown in this illustration, although a hypothetical model of SNARE protein interactions in hemifusion are shown in Figure 4.

The nature (proteic or lipidic) of the early fusion pore is still a matter of hot debate. However recent studies in viral[8-12] and cellular fusion[13-16] have suggested that these pathways do transit through a lipid hemifusion intermediate. A stalk structure between fusing membrane bilayers has also been characterised in vitro.[17] The important role played by lipids in the process of exocytosis is therefore becoming increasingly evident.

Lipids are characterised by the particular structure they adopt in contact with water. Generally, depending on the diameter of lipid head groups and acyl chains, lipids can be classified as cylinders, cones, or inverted cones (Fig. 1B). Lipids that have an inverted cone shape have an intrinsic positive curvature (such as lysophospholipids) and spontaneously form micelles, whereas cone-shaped lipids have an intrinsic negative curvature (such as oleic acid or diacylglycerol) and form inverted micelles. The net curvature of the stalk is negative and its formation is affected by the addition of each type of lipids: it is favoured by the addition, to the cis leaflets, of lipids having a negative curvature and inhibited by lipids displaying a positive curvature (Fig. 1B). In contrast, the net curvature of the pore tends to be positive, and pore formation is therefore favoured by lipids having an intrinsic positive curvature and inhibited by the opposite type of lipids, when added to the trans leaflets (Fig. 1B). The sensitivity of fusion events to the addition of lipids has been demonstrated in vivo.[7] As lipids directly influence fusion, it is therefore likely that the type of lipids surrounding the proteins defining the site of exocytosis (SNAREs or viral proteins) are crucial for its completion.[18]

Lipid Rafts, Caveolae and Detergent-Resistant Membranes

Mammalian cells contain greater than 500 different types of lipid. The distinct chemical and physical nature of these lipids leads to preferences in lipid associations; for example, lipid a may prefer to position itself beside lipid b rather than lipid c. This idea is readily observed in simple model membrane systems where specific lipids cocluster into domains segregated from other lipids.[19] This separation of lipids in vitro has often been used to support the notion that lipids are also spatially organised and restricted in cellular membranes. Of particular interest is the idea that cholesterol may form so-called 'raft' domains with sphingolipids and/or saturated phospholipids not only in model membranes but also in cell membranes.[19-21] In model membranes, cholesterol and sphingolipids (and/or saturated glycerophospholipids) pack tightly together to form a 'liquid ordered' phase that is less fluid than the liquid disordered phase formed by unsaturated glycerophospholipids.

The plasma membrane is not a uniform protein-lipid mosaic, but can contain numerous invaginations/pits, ruffles and extensions, all of which may be enriched in specific proteins and lipids. In 1953, Palade first described the appearance of multilamellar vesicles on the surface of endothelial cells;[22] these structures were later termed 'caveolae' (little caves) by Yamada.[23] Caveolae have a specialised lipid composition: electron microscopy (EM) studies have revealed that these structures are enriched in sphingolipids and cholesterol.[24,25] Caveola formation is dependent upon the caveolin protein family, proteins that interact with cholesterol, and homo- and hetero-oligomerise to form a caveolar coat. The membrane topology of caveolins and their interaction with cholesterol is likely to be important in deforming the plasma membrane to produce the typical flask-like caveolar morphology[21,26,27] (Fig. 2A). The specialised lipid and protein composition of caveolae make these domains resistant to solubilisation by nonionic detergents such as Triton X-100,[28] as visualised using EM.

Interestingly, cholesterol/sphingolipid-rich domains can also be isolated as detergent-resistant membranes (DRMs) from cell membranes lacking both caveolin expression and morphologically identifiable caveolae.[29] These lipid 'raft' domains have been proposed to function in a number of cellular pathways. For example, investigators studying the targeting of membrane proteins to the apical cell surface in polarised epithelial cells reported that apically-destined proteins acquired detergent-resistance during maturation in the Golgi. These apical proteins could be isolated in DRMs also enriched in sphingolipids and cholesterol.[30] Thus, it was proposed that apical membrane proteins associated with detergent-resistant, cholesterol/sphingolipid rich 'lipid raft' domains in polarised epithelial cells, and that this raft association facilitated apical targeting. Since these early studies, caveolae and rafts have been proposed to

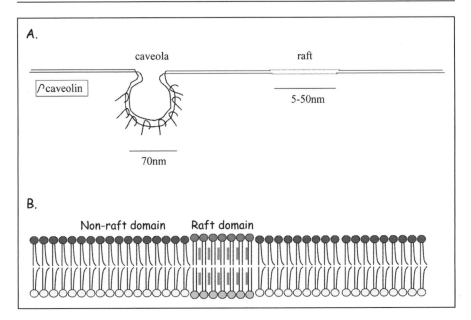

Figure 2. A) Schematic diagram of a caveola and a raft. A major biochemical difference between rafts and caveolae is the structural protein of caveolae, caveolin, which is absent from rafts. Caveolin is thought to form a hairpin loop within the membrane which leads to deformation of the membrane resulting in the typical flask-like morphology of caveolae, as illustrated. The longer acyl chains and high degree of saturation of lipids in raft domains increases the thickness of the membrane and this is also highlighted in the figure. B) Raft domains and caveolae are enriched in saturated phospholipids and cholesterol. The straight acyl chains of the saturated lipids in raft domains facilitates the tight packing of these domains into a liquid-ordered phase. In contrast, non-raft domains, enriched in unsaturated, kinked phospholipids, are thought to exist mainly in a more loosely packed liquid-disordered phase.

function in numerous membrane traffic and intracellular signalling pathways.[31,32] Indeed, the purification of caveolae and rafts as DRMs has allowed a detailed molecular characterisation of the function of these domains to emerge.

Because investigators have used the criterion of detergent-resistance to study both caveolar and raft composition and function, there is often confusion surrounding the designation of these two types of membrane domain; (Fig. 2) shows a schematic of caveolae and lipid raft domains. Whereas rafts are probably present in most (if not all) mammalian cell types, as well as plant and yeast cells, the abundance and indeed presence of caveolae is cell type specific. As caveola formation is dependent upon caveolin, rafts and caveolae are most commonly classified biochemically by the absence or presence of caveolin respectively. At the ultrastructural level the typical flask-like morphology of caveolae aids in their recognition (in conjunction with caveolin labelling), however, caveolae may also display a flattened morphology that is more difficult to distinguish.[33] The size of caveolae and rafts can also differ dramatically: caveolae have an average diameter of 70nm, whereas current estimates suggest that rafts may be as small as 5 nm.[34,35]

Although there is no doubt that caveolae exist, the presence of rafts in cell membranes has been, and remains, a contentious issue.[36] In contrast to caveolae, rafts display no obvious morphology by electron microscopy (although 'raft' proteins can appear clustered by immunogold labelling).[37] Several studies have provided strong evidence supporting the existence of cholesterol-rich rafts in living cells however, these studies often do not provide consensus data on the size or stability of rafts.[34,35,38-41] Nevertheless, it should be noted that a number of studies using different techniques have failed to detect rafts in cell membranes. Arguably the most contentious issue in the raft field is the use of detergents to purify and study rafts. Do detergents accurately report protein association with rafts or do they create the domains we claim to study? The vast majority of studies looking at rafts use detergents (typically Triton X-100) to purify these domains. These studies are at their best when they combine biochemical analysis of detergent-resistant membranes with protein (co)distribution in living or fixed cells. Despite evidence that detergents can facilitate domain formation in model membranes,[42,43] it is important to stress that preexisting plasma membrane domains are resistant to extraction by Triton X-100,[28] shown by EM analysis of caveolae. In addition, many proteins that associate with DRMs have been shown by other criteria to exhibit characteristics expected for raft associated molecules, such as cholesterol-dependent clustering in cell membranes and colocalisation with other raft moleculess.

SNAREs and Rafts

Intracellular membrane fusion reactions in the secretory and endosomal pathways are mediated by a family of proteins called SNAREs.[1] Interestingly, a number of SNARE proteins that function in exocytosis pathways have been suggested to accumulate in raft domains. The first SNARE proteins to be localised to DRMs were syntaxin 3 and TI-VAMP in polarised Madin Darby Canine Kidney cells.[44] It was suggested that raft association was important for apical sorting of these proteins (the basolateral SNARE syntaxin 4 was largely excluded from DRM fractions), rather than having a direct role in SNARE function. Subsequently, two groups reported on the domain localisation of Syntaxin 1A and SNAP-25, plasma membrane SNAREs that function in regulated exocytosis pathways in neuronal and (neuro)endocrine cells. We showed that a proportion of syntaxin 1A and SNAP-25 (20%) were associated with DRMs isolated from the neuroendocrine cell line PC12 cells.[45] Although Jahn and colleagues could not copurify the SNAREs with DRMs, they did report their association with cholesterol-rich domains visualised by immunofluorescence analysis of isolated plasma membrane sheets.[46] Since these initial studies, a number of other laboratories have reported the association of plasma membrane SNARE proteins with DRMs from a variety of cell types.[47-52] SNARE association with cholesterol-rich raft domains has also been confirmed by nondetergent methods.[47] The widespread association of SNARE proteins with DRMs suggests that SNARE function in vivo may be regulated by association with lipid raft domains.

SNARE Targeting to Raft Domains

SNAP25 and its ubiquitous homologue SNAP23 are anchored to membranes by palmitate groups attached to a central cysteine-rich domain.[53,54] Palmitate, a saturated fatty acid, is predicted to be readily accommodated in a tightly-packed raft environment and indeed, palmitoylation has been shown to target many proteins to rafts.[55] The minimal membrane binding/palmitoylation domain has been mapped to amino acids 85-120 of SNAP25, which includes the central cysteine-rich domain.[56] This region of SNAP-25 supports raft targeting of a reporter molecule,[57] indicating that amino acids 85-120 of SNAP25 contain all the necessary information for raft association. Furthermore, chemical deacylation of SNAP25 decreases raft association[57] without removing the protein from membranes.[58] These results suggest that: (i) palmitoylation of SNAP25 drives its association with raft domains; and (ii) protein-protein interactions are probably not required for SNAP-25 to associate with rafts.

Comparison of the central cysteine-rich domains of SNAP25 and SNAP23 reveals an additional cysteine residue in SNAP23 (SNAP23 has 5 cysteines, SNAP25 has 4) (Fig. 3A). Interestingly, mutational analyses showed that the 5-cysteine motif of SNAP23 supports a greater level of raft association than the 4-cysteine motif present in SNAP25: 20% of SNAP25 and greater that 50% of SNAP23 are associated with DRMs in PC12 cells.[57] These results imply that the extent of palmitoylation of SNAP25/23 likely regulates the level of raft association

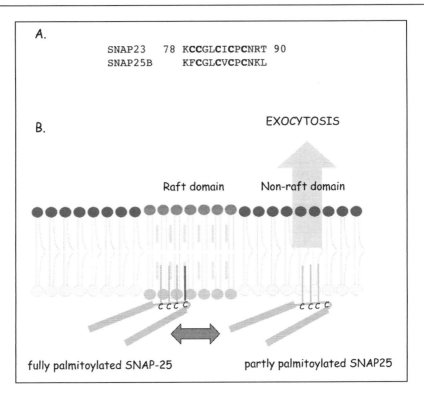

Figure 3. A) Comparison of the cysteine-rich domains of SNAP25 and SNAP23. Note the extra cysteine residue in SNAP23. The cysteine residues in these domains are sites for palmitoylation, and hence SNAP23 has an additional palmitoylation site. B) Model of how dynamic palmitoylation of SNAP25/23 may regulate raft association and exocytosis. The extent of palmitoylation of SNAP25/23 might be regulated by palmitoyl transferases and thioesterases. Increased palmitoylation would enhance the raft partitioning of these proteins, decreasing the extent of exocytosis.

of the proteins. The potential dynamic regulation of SNAP25/23 palmitoylation offers an intriguing method of regulating raft association of these proteins (Fig. 3B).

In contrast to SNAP25/23, it is not clear how syntaxin associates with rafts although there are a number of possibilities to explain this: (i) The transmembrane domain (TMD) of syntaxin may preferentially partition into raft domains. As raft domains are thought to increase the thickness of the lipid bilayer,[59] the long 23-amino acid TMD of Syntaxin 1A may be more readily accommodated in a raft domain. Alternatively, if cysteine residues present within the TMD of syntaxin 1A underwent palmitoylation then this may mediate raft association of syntaxin. (ii) Syntaxin association with rafts may be mediated through its interaction with a binding partner (such as SNAP25). (iii) Syntaxin might interact with a specific raft-localised lipid such as phosphatidylinositol (4, 5) bisphosphate (PIP2), which colocalises at the plasma membrane with syntaxin 1A[60] and has been reported to be enriched in DRMs.[61]

It is of interest to note that Syntaxin 1A does not associate with cholesterol/sphingolipid-rich ordered domains in model membranes,[62,63] suggesting that intrinsic sequences within its TMD may not direct association with more ordered lipid domains. However, repeating these experiments under modified conditions would be likely to yield important information on the potential mechanism of raft association of syntaxin in vivo. For example, purification of syntaxin from insect cells may facilitate the potential palmitoylation of the protein (as opposed to protein expression in *Escherichia coli* which does not allow protein palmitoylation), allowing the importance of syntaxin palmitoylation for raft association to be assessed. Furthermore, repeating the analysis in the presence of palmitoylated SNAP25 or lipids such as PIP2 may also provide important information on the mechanism of syntaxin raft targeting.

Compartmentalisation of SNARE Effectors

Whilst SNARE proteins can clearly function as the 'minimal' membrane fusion machinery in vitro,[64] several other proteins play essential roles in exocytosis in vivo; not surprisingly, many of these additional proteins bind to either isolated SNAREs or to the assembled SNARE complex. SNAP and NSF are essential for the ATP-stimulated disassembly of the SNARE complex following membrane fusion.[65] Munc18a binds Syntaxin 1A forming a complex that prevents SNARE complex assembly.[66] Complexins bind to assembled SNARE complexes,[67] important for SNARE complex stabilisation, oligomerisation and/or coupling to the Ca^{2+}-sensing machinery.[68-70] Finally, synaptotagmins function as Ca^{2+} sensors, and through their interactions with phospholipids and Syntaxin 1A/SNAP25 likely couple increases in intracellular Ca^{2+} concentration to exocytosis.[71]

Despite numerous studies reporting the association of SNAREs with DRMs, few SNARE-binding proteins have been reported to also occupy these domains. By employing chemical crosslinking, we failed to detect a syntaxin 1A:munc18a complex in DRM fractions.[45] A similar finding was also reported in RBL mast cells.[48] Similarly, we did not detect proteins such as SNAP, NSF, complexin or munc13 in DRMs.[45] The only SNAREbinding proteins reported to occupy DRM domains are P/Q-type voltage-gated calcium channels[50] and the vesicle protein and putative calcium sensor synaptotagmin.[52] The proposed cooccupation of raft domains by SNAREs and calcium channels is intriguing as it would (if the proteins were localised to the same raft) allow spatial coupling of the calcium signal and membrane fusion machinery. Interestingly, L-type channels, which couple to exocytosis in cells such as chromaffin and pancreatic beta cells, were excluded from DRMs,[50] suggesting that if rafts do spatially couple Ca^{2+} channels and SNARE proteins that this is cell type-specific. The relevance of synaptotagmin association with DRMs is not clear, but if this interaction occurred on vesicle membranes it may facilitate the close spatial coupling of this putative calcium sensor with the vesicle SNARE, VAMP2, which was also reported to be present in DRMs.[47]

The general absence of SNARE binding proteins may provide clues to the role of rafts in exocytosis. For example, munc18a and munc13 have been suggested to be required for vesicle docking[72] and priming,[73] respectively; the absence of these proteins in DRMs questions whether raft domains are capable of supporting docking/priming reactions. Similarly, as munc18a interaction with syntaxin 1A has also been suggested to regulate the kinetics of vesicle fusion,[74] the absence of this protein from rafts would also be consistent with a model in which raft and nonraft domains supported kinetically distinct forms of exocytosis.

Rafts and Exocytosis

The most widely used tool to disrupt lipid rafts is methyl-beta-cyclodextrin, a cholesterol-extracting agent which removes cholesterol from the outer leaflet of the plasma membrane. Cyclodextrin treatment in most instances has been reported to inhibit exocytosis,[45,46] although insulin secretion from pancreatic beta cells was enhanced following cyclodextrin treatment.[51] These results have been widely interpreted as evidence that cholesterol-rich raft domains act as platforms, or at the least are required, for exocytosis. However, it is essential that results from this type of study are treated with caution: Although cholesterol is enriched in raft domains, it is also present in nonraft membranes. Indeed, one study reported that cyclodextrin preferentially extracts cholesterol from non-raft membranes.[75] Furthermore, cyclodextrin is ineffective at removing a number of bona fide raft proteins from DRM fractions. For example, in adipocytes, cyclodextrin treatment did not allow effective solubilisation of any caveolin (our unpublished observation), arguing strongly that (despite removal of 40% of cellular cholesterol) cyclodextrin may not completely disrupt raft/caveola domains. Another obvious problem with the use of cyclodextrin and other methods of disrupting rafts is that they have pleiotropic effects, affecting many cellular process which may have downstream effects on the pathway being monitored. For example, cholesterol depletion has been suggested to inhibit the formation of clathrin-coated endocytic vesicles;[76] as endo- and exocytosis are tightly coupled, inhibition of endocytosis may indirectly affect measurements of exocytosis.

In light of these problems, it is essential that more specific and less deleterious methods are used to study the role of rafts in regulating SNARE function and exocytosis. Bader and colleagues followed the spatial distribution of exocytic sites at the plasma membrane of adrenal chromaffin cells by monitoring the appearance of dopamine-β-hydroxylase (DBH), a lumenal vesicle protein.[77] When these chromaffin granules fuse with the plasma membrane, DBH becomes exposed to the cell exterior and can be visualised with fluorescently-labelled antibodies. Following cell stimulation, rafts (visualised by cholera toxin binding to the raft-specific lipid, GM1 ganglioside) were suggested to aggregate. This raft aggregation was dependent upon the translocation of the Ca^{2+}-binding protein annexin 2 from the cell interior to the plasma membrane, suggesting that annexin 2 facilitates the cross-linking of raft domains. Intriguingly, DBH was shown to colocalise with cholera toxin-labelled raft domains following cell stimulation. In addition, synaptotagmin I (a vesicle protein) was translocated from detergent soluble to DRM fractions upon stimulation, taken to indicate the fusion of vesicles with plasma membrane raft domains. It will now be important to examine whether SNAREs also colocalise with cholera toxin-labelled GM1 and DBH. The colocalisation of chromaffin granule proteins with raft markers at the plasma membrane also merits further discussion. Given that the diameter of chromaffin granules is in the region of 280 nm, full fusion of the granule into the membrane would incorporate a surface area of approximately 0.25 μm^2.[78] Incorporation of 0.25μm^2 of granule membrane into a raft domain would be predicted to cause raft dispersion (or raft enlargement) and it is not clear what effect this would have on the rafts visualised using fluorescent cholera toxin. It will also be interesting to investigate further the association of the granule protein synaptotagmin with DRMs following cell stimulation: Does this imply fusion at raft domains, or that Ca^{2+} binding to synaptotagmin specifically increases the affinity of this protein for raft domains present in the plasma membrane or vesicle membrane?

We examined, more specifically, the requirement for raft association of SNAREs for exocytosis.[79] In this case, cysteine mutants of SNAP25 and SNAP23 with altered affinities for raft domains were employed. Toxin resistant forms of SNAP25/23 were used to rescue exocytosis in PC12 cells expressing the catalytic subunit of botulinum neurotoxin E (BoNT/E), which cleaves and inactivates endogenous SNAP25. The results from this study highlighted an inverse correlation between the extent of raft association of SNAREs and the extent of exocytosis. Importantly, these differences were normalised following raft disruption, arguing that the different effects of the mutant SNAP25/23 proteins on exocytosis were a consequence of their altered distribution in cholesterol-rich raft domains. At first sight these results appear inconsistent with a model in which exocytosis occurs in raft domains. However, as exocytosis was not directly visualised in this study, it may be too simplistic to infer that rafts do not support exocytosis; for example, exocytic events with different release kinetics may occur in raft/nonraft domains that would be missed in an assay system measuring only gross content release. Another possibility is that SNARE association with rafts is finely tuned for specific exocytic pathways; increasing the raft association of SNAP25 in PC12 cells may perturb this fine balance and thus have a negative effect on exocytosis. Similarly, mutating SNAP23 to give it a similar level of raft association as SNAP25 may allow it to function better in exocytosis pathways in which SNAP25 normally functions. It will be interesting to examine whether SNAP23 mutants with a decreased raft affinity enhance exocytosis in cells that employ SNAP23 as a SNARE rather than SNAP25 (such as mast cells and adipocytes). Similarly, the development of SNAP25 mutants with a decreased raft affinity will provide essential information on the role of raft association of SNAREs in exocytosis.

Possible Roles for Lipid Rafts in Exocytosis

In the following models, we have attempted to incorporate ideas consistent with our observation that increased DRM association of SNAP25/23 inhibits exocytosis and also with the work of Bader and colleagues who measured a close association of granule proteins with raft domains following stimulation of exocytosis.

Regulating the Kinetics of Exocytosis

As discussed by us previously,[18] raft and nonraft pools of SNARE proteins may support different types of exocytosis. It has become clear over recent years that membrane fusion in mammalian cells can occur by a number of kinetically distinct mechanisms. The textbook view of membrane fusion was that bilayer mixing led to fusion pore expansion, where the vesicle became part of the plasma membrane and was then recycled via a clathrin-dependent pathway.[80] It is now thought that both synaptic vesicle exocytosis and exocytosis of secretory granules can also occur by a more transient form of exocytosis called 'kiss and run'.[81-87] In this pathway a transient fusion pore is formed between the vesicle and plasma membrane, allowing the release of a limited amount of vesicle content before the pore reseals and the vesicle becomes intact once again. Following kiss and run exocytosis the vesicle can either be released from the plasma membrane or remain in close contact to the initial fusion site.

Several recent studies have identified kiss and run exocytosis as a significant mode of secretion of granule contents in neuroendocrine/endocrine cells,[85-87] which may be the major pathway of exocytosis under certain conditions.[87] Kiss and run exocytosis offers several advantages over complete merger of the secretory granule with the plasma membrane, not least that it may allow selective release of small molecules such as neurotransmitters, catecholamines and ions, whilst larger molecules such as peptides are retained. Kiss and run exocytosis could also increase the effective life-time of a secretory granule.

Dynamin performs a central role in vesicle scission from the plasma membrane in pathways such as clathrin-dependent endocytosis and internalisation of caveolae.[80] Furthermore, this protein functions in the closure of fusion pores during kiss and run exocytosis.[85-87] In addition to dynamin, a number of proteins have been reported to modify the kinetics of exocytosis or determine the choice between kiss and run and full fusion. For example, synaptotagmin isoforms I and IV were suggested to regulate the choice between kiss and run and full fusion, with synaptotagmin IV overexpression increasing the number of kiss and run events.[88,89] Fusion pore dynamics have also been shown to be affected by phorbol ester treatment[90] (which activates endogenous protein kinase C), overexpression of the SNARE regulators CSP[91] and complexin,[92] and expression of a munc18a mutant with reduced binding affinity for syntaxin 1A.[74]

As yet there is no data available on the possible role of membrane domains in the regulation of kiss and run and full fusion events. However, as several proteins have been shown to influence membrane fusion dynamics, the selective enrichment or exclusion of exocytic proteins in raft domains may make these structures specific sites for either full fusion or kiss and run exocytosis. For example, munc18a was shown by us and others to be absent from detergent-resistant raft domains both in PC12 cells[45] and mast cells.[48] As the interaction of this protein with syntaxin regulates the dynamics of membrane fusion, the absence of this protein from rafts may make these domains more suited to support a specific type of exocytic event. Another interesting possibility is that kiss and run and full fusion events are related to Ca^{2+} levels, as suggested by a study of synaptic vesicle exocytosis, which showed an increased frequency of kiss and run events as the concentration of extracellular Ca^{2+} was raised.[83] This idea is particularly intriguing as P/Q channels are associated with DRMs in neurons, whereas L-type channels are detergent soluble.[50] In neurons, therefore, raft-associated SNAREs may be closely coupled to P/Q-type Ca^{2+} channels and hence exposed to a higher Ca^{2+} concentration than SNAREs outside of rafts; this may result in an increased occurrence of kiss and run events in raft domains. The situation in adrenal chromaffin cells would be different though: exocytosis in these cells is regulated by L-type channels,[93,94] which are detergent-soluble.[50] Thus, in these cells nonraft sites may experience an elevated Ca^{2+} level.

It is also important to stress that although specific proteins can undoubtedly alter the kinetics of exocytosis, the exact lipid composition of the fusing membranes is also likely to play an important role in supporting either full fusion or kiss and run exocytosis. In this regard the specific lipid geometry in raft/nonraft sites may regulate fusion pore formation and expansion (see *Introduction*). For example, glycosphingolipids in the outer leaflet of raft domains with an inverted cone-shaped geometry might facilitate fusion pore formation and expansion, ensuring that exocytic events in raft domains primarily involve full fusion.[18] In this regard, it is interesting to note that in vitro fusion stimulated by the addition of palmitoylated SNAP-23 peptides was enhanced when liposomes contained cholesterol and sphingolipids.[95]

As we have discussed the possible role of raft/nonraft proteins and lipids in supporting either full fusion or kiss and run exocytosis, it is important to reconsider what DBH appearance at the PM reports about exocytosis:[77] It is likely that this assay system measures predominantly full fusion events, without detecting more transient kiss and run exo/endocytosis. Thus, although this assay is likely to provide data on full fusion exocytic events, it will be important to develop other more specific methods to allow monitoring of kiss and run exocytosis at defined sites of the PM. Such an assay system will provide further essential data on the compartmentalisation of exocytic events within raft and nonraft domains.

Restriction of Lipid Diffusion during Membrane Fusion

The simplest interpretation of our analysis of SNAP25/23 cysteine mutants is that an increased level of palmitoylation of these proteins enhances raft association and decreases the level of protein in nonraft domains of the membrane. An alternative interpretation is that differential palmitoylation of SNAP25/23 affects how the protein interacts with rafts: fully palmitoylated SNAP25/23 may reside in the highly-ordered 'core' of the raft domain, whereas partially palmitoylated protein may associate with the edges of raft domains (and this weaker interaction may allow solubilisation of the protein by Triton X-100). In this case, based on the results of our functional studies of cysteine mutants of SNAP25/23, we predict that only the SNAREs localised to the edges of rafts would be able to support fusion. This model is also consistent with the work presented by Bader's group,[77] as DBH would show a partial colocalisation with raft markers following fusion.

The attractiveness of this model is that lipid rafts could function to prevent lipid diffusion during membrane fusion (see *Introduction*). Despite work supporting the idea that transmembrane domains of SNARE proteins may form a fusion pore,[3] recent evidence has provided strong support for SNAREmediated fusion occurring through a lipidic fusion pore, via a hemifusion intermediate.[13-15] In restricted hemifusion, lipids are prevented from entering the hemifusion contact site, a property that is essential to allow mixing of trans lipid monolayers and fusion pore formation. Restricted hemifusion requires a physical barrier to prevent lipid diffusion into and out of the cis membrane contact site. In the case of Influenza HA-mediated fusion, this barrier could be provided by the TMDs of HA multimers.[9] Although associations between the TMDs of syntaxin and VAMP could be important to prevent lipid diffusion, it is also possible that lipid domains could function as diffusion barriers. This idea is particularly attractive if rafts are envisaged to surround the membrane fusion site, with fusion-competent SNAREs localised to the edges of rafts; thus, the more ordered lipid raft structure could prevent lipid diffusion from occurring when membrane fusion occurs in more fluid regions of the membrane (Fig. 4). A caveat to this model is that exocytosis persists in digitonin-permeabilised cells, in which rafts are disrupted.[79]

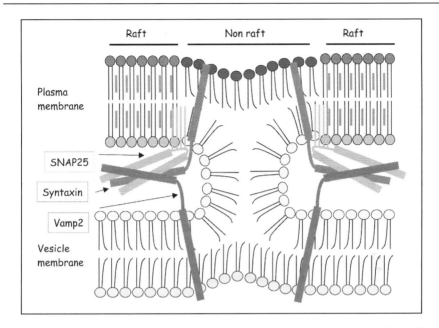

Figure 4. Fusion at the edges of raft domains. Association of SNAP25 with the edges of lipid rafts would mark these sites for exocytosis. Lipid-lipid immiscibility of nonraft and raft domains might be important in this model to prevent lipid diffusion between the fusing membranes, thus facilitating subsequent pore formation.

Perspective and Future Direction

The exocytosis of intracellular vesicles loaded with physiologically important molecules such as neurotransmitters and hormones is tightly regulated by a complex array of protein-protein interactions. Membrane lipids (as well as playing a critical role in the membrane fusion event) are likely to conduct these protein interactions, ensuring that they occur on cue and at specific sites within the cell. The association of SNARE proteins with raft/nonraft domains of the PM presents an intriguing mechanism to spatially coordinate membrane fusion proteins and fusion sites, perhaps generating physically distinct pools of 'active' and 'inactive' SNAREs. Furthermore, the potential regulation of SNAP25/23 raft association by palmitoylation offers a compelling mechanism to modulate the extent of raft association of these SNAREs, and hence the level of exocytosis. Although this review has focussed on regulated exocytosis, many other exocytic events occur within cells. Constitutive fusion of TGN-derived carrier vesicles with the plasma membrane is essential for the continuous delivery of newly synthesised lipids and proteins (such as ion channels and receptors) to the cell surface. In addition, endosome-derived vesicles fuse with the plasma membrane to allow recycling of cell surface receptors. Furthermore, polarised cells can transport macromolecules across the cell in a process termed transcytosis;[96] this pathway can involve the budding of clathrin-coated vesicles or caveolae from one side of the plasma membrane (e.g., apical) and their fusion with the other side of the plasma membrane (e.g., basolateral). Thus within a cell, there is likely to be simultaneous trafficking and fusion of a diverse set of exocytic vesicles with the plasma membrane. How do rafts regulate these many distinct exocytic events? Do rafts/nonraft domains offer the potential to physically separate these diverse exocytic events at the PM? Perhaps the biggest challenge now is to develop techniques that allow rafts to be 'viewed' in live cells; visualising the dynamic association of exocytic proteins and secretory vesicles with raft/nonraft domains will provide essential data on how membrane compartmentalisation regulates the exocytic process.

Acknowledgements
Work in the authors' laboratory is funded by the Wellcome Trust, the Diabetes Research and Wellness Foundation and the Biotechnology and Biological Sciences Research Council.

References
1. Jahn R, Südhof TC. Membrane fusion and exocytosis. Annu Rev Biochem 1999; 68:863-911.
2. Lindau M, Almers W. Structure and function of fusion pores in exocytosis and ectoplasmic membrane fusion. Curr Opin Cell Biol 1995; 7:509-517.
3. Han X, Wang CT, Bai J et al. Transmembrane segments of syntaxin line the fusion pore of Ca^{2+}-triggered exocytosis. Science 2004; 304:289-92.
4. Peters C, Bayer MJ, Buhler S et al. Trans-complex formation by proteolipid channels in the terminal phase of membrane fusion. Nature 2001; 409:581-8.
5. Lee J, Lentz BR. Secretory and viral fusion may share mechanistic events with fusion between curved lipid bilayers. Proc Natl Acad Sci USA 1998; 95:9274-9279.
6. Siegel DP. Energetics and intermediates in membrane fusion: Comparison of stalk and inverted micellar intermediate mechanisms. Biophys J 1993; 65:2124-2140.
7. Chernomordik L. Nonbilayer lipids and biological fusion intermediates. Chem Phys Lipids 1996; 81:203-213.
8. Kemble GW, Danieli T, White JM. Lipid-anchored influenza hemagglutinin promotes hemifusion, not complete fusion. Cell 1994; 76(2):383-91.
9. Chernomordik LV, Frolov VA, Leikina et al. The pathway of membrane fusion catalyzed by influenza hemagglutinin: Restriction of lipids, hemifusion, and lipidic fusion pore formation. J Cell Biol 1998; 140:1369-82.
10. Gaudin Y. Rabies virus-induced membrane fusion pathway. J Cell Biol 2000; 150:601-12.
11. Melikyan GB, Barnard RJ, Abrahamyan LG et al. Imaging individual retroviral fusion events: From hemifusion to pore formation and growth. Proc Natl Acad Sci USA 2005; 102(24):8728-33.
12. Markosyan RM, Cohen FS, Melikyan GB. Time-resolved imaging of HIV-1 Env-mediated lipid and content mixing between a single virion and cell membrane. Mol Biol Cell 2005, (in press).
13. Xu Y, Zhang F, Su Z et al. Hemifusion in SNAREmediated membrane fusion. Nat Struct Mol Biol 2005; 12:417-422.
14. Lu X, Zhang F, McNew JA. Membrane fusion induced by neuronal SNAREs transits through hemifusion. J Biol Chem 2005; 280:30538-30541.
15. Giraudo CG, Hu C, You D et al. SNAREs can promote complete fusion and hemifusion as alternative outcomes. J Cell Biol 2005; 170:249-260.
16. Reese C, Heise F, Mayer A. Trans-SNARE pairing can precede a hemifusion intermediate in intracellular membrane fusion. Nature 2005; 436:410-414.
17. Yang L, Huang HW. Observation of a membrane fusion intermediate structure. Science 2002; 297:1877-18797.
18. Salaün C, James DE, Chamberlain LH. Lipid rafts and the regulation of exocytosis. Traffic 2004; 5:255-264.
19. Edidin M. The state of lipid rafts: From model membranes to cells. Ann Rev Biophys Biomol Struct 2003; 32:257-283.
20. Simons K, van Meer G. Lipid sorting in epithelial cells. Biochemistry 1988; 27:6197-6202.
21. Simons K, Ikonen E. Functional rafts in cell membranes. Nature 1997; 387:569-572.
22. Palade GE. Fine structure of blood capillaries. J Appl Physiol 1953; 24:1424.
23. Yamada E. The fine structure of the gall bladder epithelium of the mouse. J Biochem Biophys Cytol 1955; 1:445-458.
24. Parton RG. Ultrastructural localization of gangliosides; GM1 is concentrated in caveolae. JHistochem Cytochem 1994; 42:155-166.
25. Simionescu N, Lupu F, Simionescu M. Rings of membrane sterols surround the openings of vesicles and fenestrae, in capillary endothelium. J Cell Biol 1983; 97:1592-1600.
26. Rothberg KG, Heuser JE, Donzell WC et al. Caveolin, a protein component of caveolae membrane coats. Cell 1992; 68:673-682.
27. Sargiacomo M, Scherer PE, Tang ZL et al. Oligomeric structure of caveolin: Implications for caveolae membrane organization. Proc Natl Acad Sci USA 1995; 92:9407-9411.
28. Moldovan NI, Heltianu C, Simionescu N et al. Ultrastructural evidence of differential solubility in Triton X-100 of endothelial vesicles and plasma membrane. Exp Cell Res 1995; 219:309-313.
29. Fra AM, Williamson E, Simons et al. Detergent-insoluble glycolipid microdomains in lymphocytes in the absence of caveolae. J Biol Chem 1994; 269:30745-30748.

30. Brown DA, Rose JK. Sorting of GPI-anchored proteins to glycolipid-enriched membrane subdomains during transport to the apical cell surface. Cell 1992; 68:533-544.
31. Simons K, Toomre D. Lipid rafts and signal transduction. Nat Rev Mol Cell Biol 2000; 1:31-41.
32. Ikonen E. Roles of lipid rafts in membrane transport. Curr Opin Cell Biol 2001; 13:470-477.
33. Anderson RGW. The caveolae membrane system. Ann Rev Biochem 1998; 67:199-225.
34. Sharma P, Varma R, Sarasij RC et al. Nanoscale organisation of multiple GPI-anchored proteins in living cell membranes. Cell 2004; 116:577-589.
35. Plowman SJ, Muncke C, Parton RG et al. H-ras, K-ras, and inner plasma membrane raft proteins operate in nanoclusters with differential dependence on the actin cytoskeleton. Proc Natl Acad Sci USA 2005, (in press).
36. Munro S. Lipid rafts: Elusive or illusive? Cell 2003; 115:377-388.
37. Prior IA, Muncke C, Parton RG et al. Direct visualisation of ras proteins in spatially distinct cell surface microdomains. J Cell Biol 2003; 160:165-170.
38. Friedrichson T, Kurzchalia TV. Microdomains of GPI-anchored proteins in living cells revealed by crosslinking. Nature 1998; 394:802-805.
39. Varma R, Mayor S. GPI-anchored proteins are organized in submicron domains at the cell surface. Nature 1998; 394:798-801.
40. Pralle A, Keller P, Florin EL et al. Sphingolipid-cholesterol rafts diffuse as small entities in the plasma membrane of mammalian cells. J Cell Biol 2000; 148:997-1007.
41. Gaus K, Gratton E, Kable EPW et al. Visualising lipid structure and raft domains in living cells with two-photon microscopy. Proc Natl Acad Sci USA 2003; 100:15554-15559.
42. Heerklotz H. Triton promotes domain formation in lipid raft mixtures. Biophys J 2002; 83:2693-2701.
43. Heerklotz H, Szadkowska H, Anderson T et al. The sensitivity of lipid domains to small perturbations demonstrated by the effect of Triton. J Mol Biol 2003; 329:793-799.
44. Lafont F, Verkade P, Galli T et al. Raft association of SNAP receptors acting in apical trafficking in Madin-Darby canine kidney cells. Proc Natl Acad Sci USA 1999; 96:3734-3738.
45. Chamberlain LH, Burgoyne RD, Gould GW. SNARE proteins are highly enriched in lipid rafts in PC12 cells: Implications for the spatial control of exocytosis. Proc Natl Acad Sci USA 2001; 98:5619-5624.
46. Lang T, Bruns D, Wenzel D et al. SNAREs are concentrated in cholesterol-dependent clusters that define docking and fusion sites for exocytosis. EMBO J 2001; 20:2202-2213.
47. Chamberlain LH, Gould GW. The vesicle- and target-SNARE proteins that mediate Glut4 vesicle fusion are localised in detergent-insoluble lipid rafts present on distinct intracellular membranes. J Biol Chem 2002; 277:49750-49754.
48. Pombo I, Rivera J, Blank U. Munc18-2/syntaxin3 complexes are spatially separated from syntaxin3-containing SNARE complexes. FEBS Letts 2003; 550:144-148.
49. Foster LJ, de Hoog CL, Mann M. Unbiased quantitative proteomics of lipid rafts reveals high specificity for signaling factors. Proc Natl Acad Sci USA 2003; 100:5813-5818.
50. Taverna E, Saba E, Rowe J et al. Role of lipid microdomains in P/Q-type calcium channel (Cav2.1) clustering and function in presynaptic membranes. J Biol Chem 2004; 279:5127-5134.
51. Xia F, Gao X, Kwan E et al. Disruption of pancreatic beta cell lipid rafts modifies Kv2.1 channel gating and insulin exocytosis. J Biol Chem 2004; 279:24685-24691.
52. Gil C, Soler-Jover A, Blasi J et al. Synaptic proteins and SNARE complexes are localised in lipid rafts from rat brain synaptosomes. Biochem Biophys Res Comm 2005; 329:117-124.
53. Veit M, Sollner TH, Rothman JE. Multiple palmitoylation of synaptotagmin and the t-SNARE SNAP-25. FEBS Letts 1996; 385:119-123.
54. Vogel K, Roche PA. SNAP-23 and SNAP-25 are palmitoylated in vivo. Biochem Biophys Res Comm 1999; 258:407-410.
55. Melkonian KA, Ostermeyer AG, Chen JZ et al. Role of lipid modifications in targeting proteins to detergent-resistant membrane rafts. Many raft proteins are acylated, while few are prenylated. J Biol Chem 1999; 274:3910-3907.
56. Gonzalo S, Greentree WK, Linder ME. SNAP-25 is targeted to the plasma membrane through a novel membrane-binding domain. J Biol Chem 1999; 274:21313-21318.
57. Salaün C, Gould GW, Chamberlain LH. The SNARE proteins SNAP-25 and SNAP-23 display different affinities for lipid rafts in PC12 cells: Regulation by distinct cysteine-rich domains. J Biol Chem 2005; 280:1236-1240.
58. Gonzalo S, Linder ME. SNAP-25 palmitoylation and plasma membrane targeting require a functional secretory pathway. Mol Biol Cell 1998; 9:585-597.
59. Sprong H, van der Sluijs P, van Meer G. How proteins move lipids and lipids move proteins. Nat Rev Mol Cell Biol 2001; 2:504-513.

60. Aoyagi K, Sugaya T, Umeda M et al. The activation of exocytotic sites by the formation of phosphatidylinositol 4,5-bisphosphate microdomains at syntaxin clusters. J Biol Chem 2005; 280:17346-17352.
61. Hope HR, Pike LJ. Phosphoinositides and phosphoinositide-utilizing enzymes in detergent-insoluble lipid domains. Mol Biol Cell 1996; 7:843-851.
62. Saslowsky DE, Lawrence JC, Henderson RM et al. Syntaxin is efficiently excluded from sphingomyelin-enriched domains in supported lipid bilayers containing cholesterol. J Membr Biol 2003; 194:153-164.
63. Bacia K, Schuette CG, Kahya N et al. SNAREs prefer liquid-disordered over "raft" (liquid-ordered) domains when reconstituted in giant unilameller vesicles. J Biol Chem 2004; 279:37951-37955.
64. Weber T, Zemelman BV, McNew JA et al. SNAREpins: Minimal machinery for membrane fusion. Cell 1998; 92:759-772.
65. Littleton JT, Barnard RJ, Titus SA et al. SNARE complex disassembly by NSF follows synaptic vesicle fusion. Proc Natl Acad Sci USA 2001; 98:12233-12238.
66. Pevsner J, Hsu SC, Braun JE et al. Specificity and regulation of a synaptic vesicle docking complex. Neuron 1994; 13:353-361.
67. McMahon HT, Missler M, Li C et al. Complexins: Cytosolic proteins that regulate SNAP receptor function. Cell 1995; 83:111-119.
68. Reim K, Mansour M, Varoqueaux F et al. Complexins regulate a late step in Ca^{2+}-dependent neurotransmitter release. Cell 2001; 104:71-81.
69. Tokumaru H, Umayahara K, Pellegrini LL et al. SNARE complex oligomerization by synaphin/complexin is essential for synaptic vesicle exocytosis. Cell 2001; 104:421-432.
70. Chen X, Tomchick DR, Kovrigin E et al. Three-dimensional structure of the complexin/SNARE complex. Neuron 2002; 33:397-409.
71. Chapman ER. Synaptotagmin: A Ca^{2+} sensor that triggers exocytosis? Nat Rev Mol Cell Biol 2002; 3:498-508.
72. Voets T, Toonen RF, Brian EC et al. Munc18-1 promotes large dense-core vesicle docking. Neuron 2001; 30:581-591.
73. Augustin I, Rosenmund C, Sudhof TC et al. Munc13-1 is essential for fusion competence of glutamatergic synaptic vesicles. Nature 1999; 400:457-461.
74. Fisher RJ, Pevsner J, Burgoyne RD. Control of fusion pore dynamics during exocytosis by munc18. Science 2001; 291:875-878.
75. Ilangumaran S, Hoessli DC. Effects of cholesterol depletion by cyclodextrin on the sphingolipid microdomains of the plasma membrane. Biochem J 1998; 335:433-440.
76. Rodal SK, Skretting G, Garred O. Extraction of cholesterol with methyl-beta-cyclodextrin perturbs formation of clathrin-coated endocytic vesicles. Mol Biol Cell 1999; 10:961-974.
77. Chasserot-Golaz S, Vitale N, Umbrecht-Jenck E et al. Annexin 2 promotes the formation of lipid microdomains required for calcium-regulated exocytosis of dense-core vesicles. Mol Biol Cell 2005; 16:1108-1119.
78. Wick PF, Trenkle JM, Holz RW. Punctate appearance of dopamine-beta-hydroxylase on the chromaffin cell surface reflects the fusion of individual chromaffin granules upon exocytosis. Neuroscience 1997; 80:847-860.
79. Salaün C, Gould GW, Chamberlain LH. Lipid raft association of SNARE proteins regulates exocytosis in PC12 cells. J Biol Chem 2005; 280:19449-19453.
80. Mousavia SA, Malerod L, Berg T et al. Clathrin-dependent endocytosis. Biochem J 2004; 377:1-16.
81. Artalejo CR, Henley JR, McNiven MA et al. Rapid endocytosis coupled to exocytosis in adrenal chromaffin cells involves Ca^{2+}, GTP and dynamin, but not clathrin. Proc Natl Acad Sci USA 1995; 92:8328-8332.
82. Palfrey HC, Artalejo CR. Vesicle recycling revisited: Rapid endocytosis may be the first step. Neuroscience 1998; 83:969-989.
83. Ales E, Tabares L, Poyato JM et al. High calcium concentrations shift the mode of exocytosis to the kiss and run mechanism. Nat Cell Biol 1999; 1:40-44.
84. Stevens CF, Williams JH. "Kiss and run" exocytosis at hippocampal synapses. Proc Natl Acad Sci USA 2000; 97:12828-12833.
85. Graham ME, O'Callaghan DW, McMahon HT et al. Dynamin-dependent and dynamin-independent processes contribute to the regulation of single vesicle release kinetics and quantal size. Proc Natl Acad Sci USA 2002; 99:7124-7129.
86. Holroyd P, Lang T, Wenzel D et al. Imaging direct, dynamin-dependent recapture of fusing secretory granules on plasma membrane lawns from PC12 cells. Proc Natl Acad Sci USA 2002; 99:16806-16811.

87. Tsuboi T, McMahon HT, Rutter GA. Mechanisms of dense core vesicle recapture following "kiss and run" ("cavicapture") exocytosis in insulin-secreting cells. J Biol Chem 2004; 279:47115-47124.
88. Wang CT, Grishanin R, Earles CA et al. Synaptotagmin modulation of fusion pore kinetics in regulated exocytosis of dense-core vesicles. Science 2001; 294:1111-1115.
89. Wang CT, Lu JC, Bai J et al. Different domains of synaptotagmin control the choice between kiss-and-run and full fusion. Nature 2003; 424:943-947.
90. Graham ME, Fisher RJ, Burgoyne RD. Measurement of exocytosis by amperometry in adrenal chromaffin cells: Effects of clostridial neurotoxins and activation of protein kinase C on fusion pore kinetics. Biochimie 2000; 82:469-479.
91. Graham ME, Burgoyne RD. Comparison of Cysteine-string protein (Csp) and mutant alpha-SNAP overexpression reveals a role for Csp in late steps of membrane fusion in dense-core granule exocytosis in adrenal chromaffin cells. J Neurosci 2000; 20:1281-1289.
92. Archer DA, Graham ME, Burgoyne RD. Complexin regulates the closure of the fusion pore during regulated exocytosis. J Biol Chem 2002; 277:18249-18252.
93. Owen PJ, Marriott DB, Boarder MR. Evidence for a dihydropyridine-sensitive and conotoxin-insensitive release of noradrenaline and uptake of calcium in adrenal chromaffin cells. Br J Pharmacol 1989; 97:133-138.
94. Lopez MG, Villarroya M, Lara B et al. Q- and L-type calcium channels dominate the control of secretion in bovine chromaffin cells. FEBS Letts 1994; 349:331-337.
95. Pallavi B, Nagaraj R. Palmitoylated peptides from the cysteine-rich domain of SNAP-23 cause membrane fusion depending on peptide length, position of cysteines, and extent of palmitoylation. J Biol Chem 2003; 278:12737-12744.
96. Tuma PL, Hubbard AL. Transcytosis: Crossing cellular barriers. Physiol Rev 2003; 83:871-932.

CHAPTER 8

Mast Cells as a Model of Nonneuroendocrine Exocytosis

Cristiana Brochetta and Ulrich Blank*

Abstract

Mast cells are granulated effectors of hematopoietic origin localized to tissues. They participate in innate and acquired immunity by their capacity to release upon stimulation a whole set of inflammatory mediators from sources prestored in lysosome-related secretory granules. In contrast to neuronal and neuroendocrine cells, which release individual specialized secretory granules at the plasma membrane, mast cells are set-up to liberate granular content by compound exocytosis involving both granule-granule fusion and granule-plasma membrane fusion. Such degranulation enables release of virtually all vesicular content in one single stimulatory event supporting a massive inflammatory reaction. Although some fundamental differences exist with neuronal and neuroendocrine cells, recent years have shown that mast cells use the same molecular machinery of membrane fusion for compound exocytosis. Differences rather exist in the regulation of this process, which involves interplay between components of the SNARE fusion machinery and positive and negative regulatory effectors. In recent years, some of the basic principles of fusion during mast cell exocytosis have been characterized and a clearer picture is emerging. Nonetheless, many of the key molecules and intracellular signaling principles that relay the stimulation of cell surface receptors to the secretory apparatus remain to be discovered.

Introduction

The release of inflammatory mediators from mast cell (MC) secretory granules (SG) by exocytosis is fundamental in the initiation of numerous inflammatory processes in tissues.[1-3] These can either be beneficial for the host in immune defenses against bacteria, parasites or viruses or have a detrimental role such as for example in allergies. Although exocytosis is a constitutive cellular process, MC, like certain other cell types (neuronal, neuroendocrine cells, cells from exocrine tissues, other hematopoietic cells), have developed a specialized way of regulated exocytosis. They store large amounts of proinflammatory substances in SGs that accumulate in the cytoplasm. SGs are prevented from fusion with the plasma membrane (PM) unless the cell becomes stimulated. Following activation MCs have the potential to almost entirely release their granular content by compound exocytosis ensuring a maximal biological effect.[4,5] This process is therefore often also referred to as anaphylactic degranulation.[6] A potent stimulus for degranulation is the aggregation of IgE bound to surface high affinity IgE receptors (FcεRI) with specific antigen or allergen.[7,8] However, MCs express numerous other

*Corresponding Author: Ulrich Blank, INSERM U699, Faculté de Médecine X. Bichat, 16 rue Henri Huchard, 75870 Paris Cedex 18, France, Email: ublank@bichat.inserm.fr

Molecular Mechanisms of Exocytosis, edited by Romano Regazzi. ©2007 Landes Bioscience and Springer Science+Business Media.

cell surface receptors that allow them to respond in a highly flexible manner to a multitude of environmental stimuli many of which can also induce degranulation.[1-3]

In recent years considerable advances have been made in the knowledge of the molecular mechanisms involved in MC degranulation. Besides the understanding of the early signaling events initiated by the aggregation of FcεRI progress includes also knowledge about the membrane fusion machinery implicated in the late steps of secretion.[8,9] It has become clear that membrane fusion during MC degranulation relies on the evolutionary conserved fusion machinery of membrane trafficking. These include members of the SNARE (Soluble N-ethyl-maleimide-sensitive factor Attachment protein REceptor) family of proteins that cooperate with a variety of accessory proteins to regulate the secretory apparatus. In this chapter we attempt to integrate the current information on the mechanisms of MC exocytosis as an example for the secretory machinery present in hematopoietic cells as well as our current ideas about its regulation.

Morphological Characteristics of MC Granules

MCs have initially been described by Paul Ehrlich in 1878 by their capacity to show metachromasy (color change) when stained with aniline-derived dyes such as toluidine blue.[10] We now know that this phenomenon is due to the proteoglycan matrix stored in their SGs.[11] As these proteoglycans differ in MC from various tissues two major MC subpopulations have been defined.[12] Thus, connective tissue MCs (CTMC) present for example in skin or peritoneal cavity contain heparin, while mucosal MCs (MMCs) present for example at mucosal surfaces in the lung or in the gut contain chondroitinsulfate.[2,12]

MCs are small cells between 10 to 15 μm and contain in their cytoplasm up to 1000 SG occupying up to 40% of the cell volume. As compared to endo- or exocrine cells, which have specialized SG, those from MCs or other hematopoietic cells are secretory lysosomes.[13,14] These organelles contain a variety of lysosomal hydrolases and are situated at the junction between the endo- and exocytic pathways. Electron microscopic analysis of ultrathin cryosections of primary bone marrow derived MCs (BMMCs) have revealed granular heterogeneity and allowed to describe type I, II and III SG differentiated on the basis of a specific marker and the access to endocytotic tracers.[15] Type I SG are multivesicular and could represent conventional lysosomes able to release their content only under nonphysiological situations. Type II and III SG could represent the compartment regulated by a physiological stimulus. Type II granules show an electron-dense core surrounded by membrane vesicles and are accessible to endocytotic tracers. Type III granules are essentially formed by electron-dense material and are not accessible by endocytotic tracers within a given time frame. Type III granules are found to a large part in highly differentiated MCs. The biogenetic relationship between these three types of SG remains to be elucidated. Nonetheless, it has become clear that the granular compartment and proper loading with inflammatory mediators depends crucially on the expression of proteoglycans as genetic targeting of the heparin biosynthesis pathway or the serglycin core in mice revealed severe defects in SG maturation.[16-18] Furthermore, genetic diseases associated with membrane trafficking such as the Chediak-Higashi syndrome (equivalent of the beige mutation in mice) with a defect in the Lyst gene causes the appearance of giant SG in MC.[19,20]

MC granules contain a large variety of inflammatory mediators that are secreted when the cell becomes activated. The most well known MC and basophil-specific is histamine accounting for many of the symptoms associated with immediate hypersensitivity reactions.[21] However the role of MC proteases such as tryptase and chymase is also increasingly recognized.[22,23] Some of these mediators and their biologic properties are listed in Table 1.

Particular Features of MC Exocytosis

Classical exocytosis observed in neuronal, neuroendocrine or exocrine cells involves generally the fusion of single SG with the PM in discrete events. In neurons, exocytosis is coupled to the rapid endocytosis and regeneration of the SG in order to allow multiple releases in a

Table 1. *Mediators stored in secretory granules of mast cells and some of their biologic activities*

Mediator	Biologic Activities
Histamine	alters vascular permeability, promotes smooth muscle contraction, alteration of immune cell function, mucus production, neuronal activation
Heparin and chondroitin sulfate	anti-coagulant activity, promote angiogenesis, cofactor for tryptase
Tryptase	Tissue remodelling (degradation of collagen); cell proliferation by activation of protease act. Receptors, chemotaxis
Chymase	tissue remodelling (degradation of fibronectin, activation of MMPs); degradation of thrombin,
TNF	proinflammatory cytokine, neutrophil chemotaxis
VEGF	Angiogenesis

minimum amount of time.[24] In MCs virtually all granular content of a cell is extruded in one single stimulatory event, which is therefore also referred to as anaphylactic degranulation. Early ultra-structural[4] and electrophysiological studies[5] have revealed that this massive mobilization of granules is made possible by compound exocytosis (Fig. 1), which besides the fusion of granules at the PM, also involves fusion between granules from deeper inside layers thereby forming degranulation channels.[25] From there the proteoglycan matrix is expelled and bound mediators are liberated by ion exchange due to the alteration in pH and matrix swelling. The expelled proteoglycan core can usually be seen in the tissue surrounding the MC (Fig. 2). Compound exocytosis is also observed in other hematopoietic cells involved in the inflammatory response including eosinophils, neutrophils and platelets.[26-28] Such massive release of

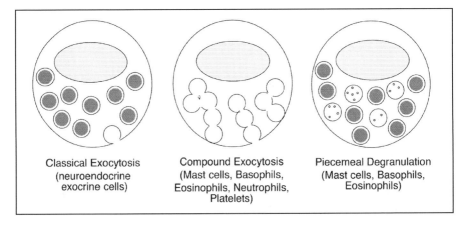

Classical Exocytosis
(neuroendocrine
exocrine cells)

Compound Exocytosis
(Mast cells, Basophils,
Eosinophils, Neutrophils,
Platelets)

Piecemeal Degranulation
(Mast cells, Basophils,
Eosinophils)

Figure 1. The basic modes of exocytosis in regulated-secretion competent cells are illustrated. In neuroendocrine and exocrine cells one usually observes fusion of a single specialized SG at the PM in a given time frame. In MCs and some other cells of hematopoietic origin compound exocytosis can be observed, which involves SG-SG and SG-PM fusion. In MCs, basophils and eosinophils another mode of exocytosis called piecemeal degranulation can be observed, which involves gradual emptying of vesicular content without any observable fusion event.

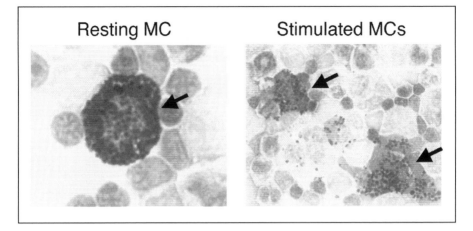

Figure 2. Resting and stimulated mouse peritoneal MCs (arrows) stained with Toluidine Blue. In resting cells one can see a cell whose cytoplama is filled with many cytoplasmic granules that show metachromatic staining. In stimulated cells one can observe extruded proteoglycan cores that still show metachromatic staining.

mediators by compound exocytosis favors an immediate and maximal biologic effect and induces a strong inflammatory reaction. MCs have also be shown to exhibit piecemeal degranulation (PMD)(Fig. 1), a term coined for observed progressive losses of granule particulate contents without any evidences for fusion events.[29,30] PMD exists also in basophils[30,31] and eosinophils[32,33] and occurs in a variety of physiological and pathophysiological situations.[34,35] The mechanisms for PMD are not fully understood. Early ultrastructural analysis revealed the budding of small cytoplasmic granules that move towards the PM leading to a gradual emptying of the granule from where they originate.[33,36] However, it is possible that the progressive emptying of granular content could also result from incomplete or transient fusion events often referred to as kiss and run fusion.[37] Recently, such transient fusion events have been shown to occur at a frequency of approximately 2:1 over full fusion in live imaged MCs.[38]

MCs can participate in multiple cycles of activation for mediator release. However, in comparison to the very fast synaptic vesicle cycle (~1 sec) in neuronal cells, the regeneration of a fully reconstituted and loaded granular compartment can take up to 72 h.[39]

The Basic Secretory Machinery in MCs

Membrane Fusion Proteins in MCs

MC exocytosis relies on the conserved mechanism of membrane fusion implicating SNARE proteins.[8] SNAREs lie on opposing cellular membranes and can form a stable multimeric complex that catalyzes fusion.[40-42] A typical SNARE complex at the PM includes a vesicular SNARE (v-SNARE) such as vesicle associated membrane protein (VAMP) that pairs with two target SNAREs (t-SNAREs) such as synaptosome-associated protein of 23 (ubiquitous) or 25 (neuronal) kDa (SNAP-23/25) and a syntaxin molecule.[41]

MCs express a wide array of SNAREs. These include the t-SNAREs SNAP-23 as well as syntaxins 2, 3 and 4. Expressed VAMP family proteins members include VAMP-2, 3, 7 and 8. Introduction of antibodies directed to SNAP-23[43] inhibited stimulated exocytosis. During exocytosis PM SNAP-23 relocated along degranulation channels that form in the interior of the cell. Overexpression of syntaxin 4,[44] but not syntaxin 2 or 3 also inhibited exocytosis, while overexpression of SNAP-23, but not of a derived VAMP-binding mutant, enhanced MC exocytosis.[45] For VAMP proteins evidences for functional implication remain indirect and are based on the observation that several of them relocate during degranulation. It was found that ectopically expressed GFP-VAMP-2 gradually fuses with the PM following stimulation,[46] while VAMP-8 apparently coalesces forming larger granules.[44] Similarly, VAMP-7-CFP and Syntaxin-3-CFP relocate from granular structures to the PM.[47] A precise definition of the different SNARE proteins implicated in degranulation will nevertheless require additional studies. Yet, it is possible that several different SNARE complexes may form and mediate fusion in agreement with the compound exocytosis mechanism and the above described heterogeneity of MC granules.

Accessory Proteins in MC Membrane Fusion

SNAREs in MCs function with a variety of accessory proteins. One crucial regulator is the ATPase NSF, which disassembles SNARE complexes.[48] Expression of an ATPase-deficient NSF mutant dramatically inhibited MC exocytosis and resulted in accumulation of fusion-inefficient SNARE complexes highlighting the importance of SNARE disassembly or priming in degranulation.[49] MCs also express SM (Sec1/Munc18) family members that bind syntaxin proteins in an isoform-specific manner and are known to play fundamental roles in exocytosis.[50,51] While mice deficient in the neuronal isoform Munc18-1 exhibited a complete deficiency in synaptic vesicle exocytosis,[52] mice deficient in the ubiquitous isoform Munc18-3 showed enhanced externalization of the glucose transporter Glut 4 in derived adipocytes at low concentrations of insulin. In these cells externalization was uncoupled from the regulation by PI3 Kinase suggesting a signaling connection.[53] By contrast, in skeletal muscle and pancreatic islets a decreased transport and insulin sensitivity was found in heterozygous mice revealing some fundamental differences.[54] This is, however, in agreement with described positive and negative regulatory roles of SM proteins.[50,51] While the essential positive role of SM proteins remains unclear, the inhibitory role is thought to result from its binding to syntaxin thereby fixing the latter in a closed conformation unable to engage in SNARE complex formation.[55] In agreement, in Munc18-3 deficient adipocytes despite the decreased expression of its syntaxin 4 partner, the amount of "free" syntaxin 4 was increased allowing enhanced fusion.[53] MC express Munc18-2 and Munc18-3 that form complexes with syntaxin 2, 3 and syntaxin 4, respectively.[56] Overexpression of Munc18-2, but not Munc18-3 inhibits degranulation. This indicates a role for the former, but does not exclude a role for Munc18-3. More definitive studies will await the use of MCs deficient in these proteins. MCs may also express the neuronal isoform Munc18-1 as shown by PCR,[57] however, this has to be confirmed at the protein level.

Small GTPases of the Rab3 family have also been recognized as important regulators of membrane fusion in many cell types. In synaptic vesicle exocytosis they are not in itself essential, but function as a gatekeeper for the normal regulation of calcium-triggered exocytosis.[58,59] In MC, Rab3A, Rab3B and Rab3D are expressed when examined by PCR analysis.[60,61] Protein expression has been confirmed for Rab3A and Rab3D and the latter was shown to translocate from SGs to the PM following stimulation.[61,62] Overexpression of Rab3D or a constitutively active GTP bound mutant, but not of Rab3A, revealed an inhibitory effect in IgE-stimulated RBL MCs.[61] In another study overexpression of Rab3A was also able to inhibit exocytosis.[63] The role of Rab3D has been challenged, because peritoneal MC from Rab3D deficient mice exhibited normal exocytosis in patch clamp studies and did not reveal any increase in granule size as shown in zymogen granules in the pancreas.[64] However, in these studies the role of existing compensatory mechanisms and the use of more physiological stimuli have not been addressed.

Another GTPase that could play a role in MC secretion is Rab27, which exists in two isoforms (Rab27A and Rab27B), known to interact with a variety of Rab27-binding proteins (i.e., the synaptotagmin-like protein (Slp) family with tandem C2 Calcium-binding motifs, the Slac2 family, and Munc13-4) all functioning in membrane trafficking.[65] Interaction with Slac2 family members couples Rab27 function to the actin-based motor myosin Va in melanocytes and neuroendocrine cells.[66-69] This mode of action does not apply to Rab27-deficient cytotoxic T cells (CTLs), which nevertheless are unable to release granular content[70,71] due to a docking defect of SGs.[72] A potential effector of Rab27 in these cells could be Munc13-4, a more ubiquitously expressed isoform of the neuronal priming factor Munc13-1.[73,74] Similar to Rab27, mutations in Munc13-4 led to secretion defects in CTLs.[72] Munc13-4 acts downstream of Rab27 as SGs from deficient CTLs can still dock at the PM, however, are unable to release their content. In RBL MCs, both Rab27 and Munc13-4 are expressed at the protein level and both localize to SGs.[75,76] Exogenous expression of a dominant active Rab27A reduces antigen-induced histamine release by about 30%.[76] No defects, where seen in Rab27A-deficient BMMC.[56] However, it is possible that its function could be compensated by the related Rab27B isoform, which is the major functional isoform in platelets.[77,78] Munc13-4 overexpression in MCs enhanced secretion.[75] These results suggest both proteins could function as regulators of exocytosis in MCs.

A recently described SNARE interacting effector is the cytosolic protein complexin II expressed in brain but also in MCs.[79] Complexin II preferentially binds to assembled SNARE complexes.[80] Upon stimulation of MC it was found to translocate to the PM in a calcium-independent manner where it acts to increase the calcium sensitivity of the fusion machinery.[79] A further accessory protein that regulates the fusion competent state is secretory carrier membrane protein 2 (SCAMP2). SCAMP2 in MCs localizes to SGs and vesicles, but a small fraction is also found at the PM where it colocalizes with Syntaxin 4 and SNAP-23. Expression of an oligopeptide (E-peptide, within the cytoplasmic segment linking the second and third membrane domain of this tetraspanin protein) potently inhibits exocytosis in streptolysin O-permeabilized MCs.[81] SCAMP2 may act at a late step that couples Arf6-stimulated phospholipase D (PLD) activity to formation of fusion pores. This could be in agreement with the proposed implication of PLD isozymes (PLD1 and PLD2) as regulators of MC exocytosis.[82,83] PLD may act upstream of PKC[84] to control multiple pathways including cytoskeletal changes[85] and fusion.[86] Regulatory proteins also include calcium sensors, protein kinases and phosphatases or cytoskeletal proteins linking the machinery to early receptor stimulated events. These are described in more detail below in relationship to their signaling function.

Taken together, MCs express a whole series of fusion accessory proteins. Their function is to prevent the secretory apparatus from unnecessary fusion, render the basic fusion machinery more efficient and allow its connection to activating cellular signals.

Signal Transduction for Exocytosis in MCs

Early Receptor-Mediated Events

The signaling requirements for degranulation have been mostly worked out for IgE-mediated triggering through high affinity IgE receptors (FcεRI) that initiates transmembrane signaling by coupling to nonreceptor tyrosine kinases.[7] This multisubunit ($\alpha\beta\gamma_2$) receptor bears Immunoreceptor Tyrosine-based Activation Motifs (ITAM) in β and γ chains that upon aggregation of α chain-bound IgE with specific antigen become phosphorylated by the src-related protein tyrosine kinase Lyn. This engenders activation of Syk tyrosine kinase through ITAM binding thereby launching an amplification cascade involving multiple signaling adaptors, tyrosine kinases, PI3 Kinases and PLCγ. This cascade leads to the activation of PKC and calcium influx, both events have been demonstrated to represent crucial checkpoints for degranulation.[8]

Targets of Calcium

Although it has been realized more than 50 years ago that calcium is an essential intracellular messenger for secretion in neurons,[87] and somewhat later in MC[88] the molecular targets have remained obscure for many years. Recently, the synaptotagmin (Syt) family of calcium sensors have emerged as prime candidates. Syts are membrane-anchored proteins that bind calcium via conserved tandem calcium and phospholipid-binding C2 domains (C2A and C2B).[89,90] Calcium binding promotes oligomerization[91,92] and membrane phospholipid binding.[93-95] It also promotes binding to a subset of SNAREs (syntaxin, SNAP-25).[96,97] Together these events are thought to regulate calcium-dependent SNARE assembly.[98]

MCs express several Syt family members (Syt I, II, III, and IX),[99-103] however, the physiologic calcium sensor for exocytosis is still unknown. Overexpression of neuronal Syt I enhanced ionomycin-stimulated secretion[104] and evidence for its expression in tissue MCs has been presented.[105] However, the physiologic role in receptor-stimulated cells of this low affinity calcium sensor remains to be validated. Syt II deficiency did not sensibly affect histamine release, but rather blocked release of cathepsin D suggesting that it acts to prevent fusion of genuine lysosomes.[99] Syt III was found to function as a critical regulator of the perinuclear endocytic recycling compartment (ERC) and could possibly regulate SG size,[101] while Syt IX was shown to function in protein export from the ERC to the cell surface.[102] These data suggest that nonneuronal Syts interfere with distinct steps of membrane trafficking in MCs along the endo- and exocytic pathway. Besides classical Syts, other proteins with calcium-binding C2 domains such as Syt-like proteins (Slp 1-5), could potentially also play a role.[65] However, so far they have not been characterized in MCs.

Another calcium sensor is calmodulin (CaM)[106,107] interacting with numerous proteins that could play a role in secretion.[107] Data in MCs using inhibitors ranged from no effect to an inhibitory effect on secretion and are difficult to interpret with respect to the function of this sensor.[108-112] In MCs, potential targets include myosin light chain kinase (MLCK),[113] CaM kinase II, which phosphorylates nonmuscle myosin[114] syntaxin 3[115] and PLD.[116] CaM binds also the vesicular SNARE protein VAMP-2,[117] which collapses into the PM in degranulating MCs.[46] CaM is thought to activate VAMP-2 by liberating it from its lipid-bound state upon binding.[117] In agreement, microinjection of a VAMP-2-derived peptide that blocks CaM-binding into chromaffin cells resulted in inhibition of exocytosis.[118] However, such experiments have not been performed in MCs.

Besides these professional sensors calcium interacts also directly with a variety of signaling molecules that coordinate cell signaling and membrane fusion. A prominent example is PKC described below.

The Role of Kinases and Phosphatases

In addition to its extensively characterized role in the early steps of receptor activation, protein phosphorylation-dephosphorylation also represent an important regulatory mechanism in the late steps of exocytosis in MCs. Several proteins of the membrane fusion machinery have been shown to be phosphorylated by a variety of Ser-Thr kinases, including PKC, CaM kinase II, PKA, Casein kinase II. Phosphorylation could have a variety of functions in the fusion process.[119] This could include mobilisation of fusion proteins from the reserve pool as shown for synapsins in neurons,[120] regulation of the activity state of SNARE proteins[121,122] or a direct interference in the fusion process.[119,123]

In MCs, there is ample evidence that activation of PKC represents an essential signal for secretion. Both a calcium-dependent isoform, PKCβ, and a calcium-independent isoform, PKCδ, have been implicated using depletion-reconstitution studies.[124] The importance of PKCβ has been confirmed in MCs deficient for this isozyme,[125] while PKCδ-deficient MCs actually showed enhanced exocytosis.[126] However, these apparent differences may be explained by indirect effects by PKCδ on negative regulatory mechanisms. Regardless it is clear that PKC regulates late steps in exocytosis. Molecular targets include cytoskeletal proteins such as myosin light and heavy chains necessary for the reorganization of the actino-myosin cortex during secretion.[127,128] PKC can also phosphorylate SNAP-25 and syntaxin 4[129,130] although the in vivo relevance remains to be established. In MCs about 10% of SNAP-23 becomes transiently phosphorylated during degranulation on Ser[95] and Ser[120] within its cysteine-rich linker region likely by PKC.[131] Overexpression of phosphorylation mutants inhibited IgE-stimulated exocytosis suggesting that SNAP-23 phosphorylation modulates degranulation although the precise molecular mechanism remains to be determined. In eosinophils[27] and chromaffin cells[132] patch clamp studies have demonstrated that fusion pore expansion is sensible to PKC inhibitors and/ or can be enhanced by the addition of PMA suggesting a role in the very late steps of exocytosis, but it remained unknown whether this is due to the mentioned phosphorylation of SNAREs, Syts,[123] or other physiologic substrates. PKC also phosphorylates Munc18-1 within its second domain thereby inhibiting its capacity to bind syntaxin.[133,134] Similarly, phosphorylation of Munc18-3 by PKC reduced its affinity to bind syntaxin 2 and 4 in platelets.[135] Such phosphorylation could thus regulate the amount of available fusion-competent syntaxin in the cell. MCs express Munc18-2 and Munc18-3 and preliminary evidence from our laboratory indicates that Munc18-2 also becomes phosphorylated in these cells (unpublished).

Besides PKC, SNARE Kinase (SNAK) could also be a regulator of exocytosis in MC.[136] SNAK phosphorylates SNAP-23 at Ser-Thr residues in MCs thereby increasing the stability of newly synthesized protein. SNAK-dependent phosphorylation could thus indirectly promote SNARE complex formation by increasing the pool of available SNAP-23. A different kinase activity present in Rab3D-containing immunoprecipitates was responsible for specific phosphorylation of Syntaxin 4 in vitro thereby decreasing its SNAP-23 binding capacity.[121] This activity was downregulated after stimulation in a calcium-dependent manner. As already mentioned, other kinases including for example CaM Kinase II, PKA, Casein Kinase II, MLCK have been shown to phosphorylate targets of the secretory machinery and could also play role in MC degranulation. For example, secretion is susceptible to inhibitors of PKA, which may relate to the capacity to phosphorylate the lipid metabolising enyme PLD.[137]

Protein phosphorylation is regulated by a dynamical balance between the action of kinases and phosphatases. Given the evidences for the role of phosphorylation in controlling exocytosis, phosphatases likely also regulate fusion[138,139] including in MCs.[8] Inhibitors of phosphatase PP1 and PP2a reduced secretion even when bypassing early-receptor-mediated signals.[140] Following stimulation of MCs PP2A gets recruited to the PM in a manner that is correlated with the kinetics of secretion. Subsequent studies in RBL MCs demonstrated that both PP1 and PP2A could transiently associate with cortical myosin II suggesting a role in the cytoskeletal rearrangements of the actino-myosin cortex.[141] A crucial target for phosphatase activity could be the PKC-dependent phosphorylation sites in myosin light and heavy chains. Another phosphatase implicated in membrane fusion could also include megakaryocyte cytosolic protein tyrosine phosphatase 2 (MEG2) localized to SG, whose overexpression results in the formation of large granules.[142]

Role of the Cytoskeleton in Secretion

It has been known for many years that secretion in MC is accompanied by an extensive cytoskeletal reorganization characterized by the dissolution of the subcortical actino-myosin complex and formation of F-actin ruffles.[143,144] Similarly, microtubules stretch out into the lamellipodia formed during the stimulation process.[145] Furthermore, SG are embedded in a filamentous network of cytoskeletal structures with hook-like structures.[146] The role of cytoskeletal elements in the secretory process could be manifold. They could provide an appropriate scaffold for signaling proteins and the attachment of cellular compartments and proteins of the fusion machinery. The actino-myosin contractile system or tubulin-based motors could be important for the mobilisation and extrusion of SG content.[147] On the other hand, the actino-myosin terminal web could also represent a substantial barrier for SGs to reach the PM. In agreement, stimulation is accompanied by its dissolution and actin-depolymerizing drugs actually enhance secretion.[56,143] However, in the presence of certain inhibitors, secretion could be uncoupled from cortical F-actin disassembly suggesting that the latter is not absolutely required for secretion to occur.[112] Similarly, treatment with actin-depolymerizing drugs or ectopic expression of a constitutively active mutant of RhoA, Rac1 or CDC42 promoting actin remodeling enhances secretion, while dominant-negative forms inhibited secretion.[148-150]

However, other functional consequences of these treatments, notably enhancement of early cellular signaling pathways have been described, suggesting the existence of divergent signaling pathways for actin remodeling and secretion.[149,151,152] In agreement with a role in early signaling events, bypassing receptor stimulated signaling with PMA/calcium ionophore rescues inhibited levels of secretion.[152] In the same line MC with mutations in the Wiskott-Aldrich syndrome protein WASP or the WASP-interacting protein WIP also show markedly diminished degranulation and parallel affection of early signaling events.[153,154]

The tubulin network also plays an important role in secretion. Inhibitor studies show that microtubule destabilizing or stabilizing drugs strongly affect secretion in MCs.[56,155-157] In stimulated MCs, SG identified by Munc18-2-staining were excluded from F-actin-containing ruffles but appeared to be aligned along newly formed microtubular tracks suggesting microtubular transport mechanisms.[56] In agreement, live imaging of SG in RBL cells revealed bidirectional movement in both resting and activated cells adjacent to microtubules and the rate and extent of stimulated exocytosis was inhibited by the MT depolymerizing drug Colchicine.[158] A recent study showed that the microtubule-dependent translocation of SG labeled by CD63-GFP to the PM occurred in a calcium-independent manner, while the actual fusion was calcium-dependent.[157] Thus, it is possible that similar to other hematopoietic cells with a more locally restricted release from secretory lysosomes,[159] the microtubular system also plays an important role in the secretory process of MC. These data, however, contrast with studies, which do not reveal significant movement of SG during exocytosis in RBL cells.[38,160] Similarly, rat peritoneal MCs with a highly differentiated SG compartment rather show an inward movement of PM components such as SNAP-23 in agreement with the compound mode of exocytosis.[43]

Taken together, at present it is still difficult to propose a cohesive model on the implication of cytoskeletal elements in the late steps of secretion. This is largely due to the complexity of the molecular pathways involved, which do not only impinge on the late steps of exocytosis but on a whole variety of cellular processes. Nevertheless, accumulated data suggest that cytoskeletal functions are an integral part of the cellular fusion machinery in MCs

Evidence for Lipid Raft Domains in MC Exocytosis

Lipid rafts are dynamic assemblies of cholesterol and sphingolipids forming a separate liquid-ordered phase in lipid membranes that are enriched in discrete subsets of proteins.[161] At the PM lipid rafts regulate signal transduction by providing concentrated platforms for signaling proteins.[162] Lipid rafts are also found in biosynthetic and endocytic pathways.[163,164] Evidence exists that regulated secretion also utilizes raft-dependent interactions to achieve cell-type specific sorting of secretory proteins.[165] Treatment with cholesterol-depleting agents considerably affected regulated secretion in several cellular systems.[166-168] Moreover, both vesicular and PM fusion proteins integrate into lipid rafts although the nature of this compartment is still somewhat under dispute.[166,169,170] SGs preferentially dock and fuse at these identified cholesterol-dependent clusters with high preference.[165] Thus, similar to the platforms at the PM for transmembrane signaling such domains may also form functional entities for fusion. In MCs, when examined by sucrose gradient fractionation, SNARE proteins were either found excluded (syntaxin 2), equally distributed between raft and nonraft fractions (syntaxin 4, VAMP-8, VAMP-2), or selectively enriched in rafts (syntaxin 3, SNAP-23).[171] The accessory protein Munc18-2 and 18-3 were found in nonraft fractions, however, small amounts of Munc18-2 also consistently distributed into rafts. Cognate SNARE complexes of syntaxin 3 with SNAP-23 or VAMP-8 were enriched in rafts, while those of syntaxin 3 with Munc18-2 were excluded. These results suggest that regulatory and cognate fusion complexes are spatially separated in MCs and likely also in other cells.

Table 2. Fusion and fusion accessory proteins in mast cells

Proteins t-SNARES	Expression mRNA	Expression protein	Localisation	Evidence for Role in MC Exocytosis	Refs.
SNAP-23	+	+	PM	α-SNAP23 blocks fusion in act. permeabilized cells, SNAP-23 overexpr. enhances exocytosis; relocation of SNAP-23 from PM to degranulation channels in act. MCs	43,45
Syntaxin 2	+	+	?	no effect of overexpression	44
Syntaxin 3	+	+	SG and PM	no effect of overexpression; relocation from SGs to PM in act. MCs	44,47
Syntaxin 4	+	+	PM	overexpression inhibits exocytosis in act. MCs	44
v-SNAREs					
VAMP-2	+	+	SG	relocation from SG to PM in act. MCs	46
VAMP-3	+	+	vesicular	?	44
VAMP-7	+	+	vesicular	relocation of exogenously expressed VAMP-7 from vesicles to PM in act. MCs	47
VAMP-8	+	+	vesicular and SG	coalescence of vesicles and SGs in act. MCs	44
Access. proteins					
Rab3A	+	+	cytoplasmic	overexpression has either no effect or inhibits exocytosis in act. MCs	61,63
Rab3B	+	-	?	?	60
Rab3D	+	+	SG (partially)	overexpression inhibits exocytosis, relocation from SGs to PM, Rab3D deficiency does not affect exocytosis in patch clamp act. MCs	61,62,64
Rab 27	ND	+	SG	overexpression inhibits exocytosis in act. MCs; relocation to PM in act. MCs; no effect in Rab27A-def. BMMC	56,76
Munc18-1	+	-	?	?	57
Munc18-2	+	+	SG	overexpression inhibits exocytosis in act. MCs, relocation of into lamellipodia	56
Munc18-3	+	+	PM	no effect of overexpresssion on exocytosis	56
Munc 13-4	+	+	SGs	overexpression enhances exocytosis in act. MCs	75
SCAMP2	ND	+	SG, PMr	introduction of oligopeptide from cytoplasmic domain inhibits exocytsis in permeab. MCs	81

Continued on next page

Table 2. Continued

Proteins t-SNARES	Expression mRNA	Expression Protein	Localisation	Evidence for Role in MC Exocytosis	Refs.
Complexin II	+	+	Cyt	relocates from cytoplasm to PM in act. MCs, siRNA inhibits exocytosis	79
Synaptotagmin I	+	+	SG (exogen.)	overexpression enhances calcium ionophore act. exocytosis	104,105
Synaptotagmin II	+	+	Lysosome	siRNA augments fusion of genuine lysosomes with SGs	99
Synaptotagmin III	+	+	Endos./SG	siRNA blocks delivery to the ERC and augments SG size	101
Synaptotagmin IX	+	+	ERC	siRNA slows down protein export from the ERC to the cell surface	102

Conclusion

MCs are tissue immune cells that upon stimulation with an appropriate trigger massively release inflammatory mediators by compound exocytosis. This process is fundamental in adaptive and innate immunity but also plays a detrimental role in several inflammatory diseases. Evidence has been presented that MCs use a specific and highly regulated secretory apparatus for compound exocytosis. Besides cognate membrane fusion SNARE proteins a number of accessory regulatory proteins have been characterized. These proteins and the evidence for functional implication are summarized in Table 2. They further connect to highly sophisticate signaling machinery initiated by the activation of receptors at their surface. Together, this molecular machinery coordinately functions to protect the organism from unwanted release of potentially dangerous substances but enable release upon a physiologic activation signal at the surface. In disease, besides inappropriate triggering MCs often also show a hyperactivated phenotype translating inappropriate regulation of this machinery. Thus, future research efforts will continue to focus on the molecular understanding of this complex and highly regulated molecular secretory machinery.

Acknowledgements

The work of U. Blank is supported by INSERM and a grant from the Fondation pour la Recherche Médicale (FRM). C. Brochetta s and U. Blanks research is also supported by a Marie Curie Early Stage Research Training Fellowship of the European Community's Sixth Framework Programme under contract number 504926.

References

1. Marshall JS. Mast-cell responses to pathogens. Nat Rev Immunol 2004; 4(10):787-799.
2. Galli SJ, Kalesnikoff J, Grimbaldeston MA et al. Mast cells as "tunable" effector and immunoregulatory cells: Recent advances. Annu Rev Immunol 2005; 23:749-786.
3. Galli SJ, Nakae S, Tsai M. Mast cells in the development of adaptive immune responses. Nat Immunol 2005; 6(2):135-142.
4. Röhlich P, Anderson P, Uvnäs B. Electron microscope observation on compound 48/80-induced degranulation in mast cells. J Cell Biol 1971; 51:465-483.
5. Alvarez de Toledo G, Fernandez J. Compound versus multigranular exocytosis in peritoneal cells. J Gen Physiol 1990; 95:397-402.
6. Dvorak AM, Massey W, Warner J et al. IgE-mediated anaphylactic degranulation of isolated human skin mast cells. Blood 1991; 77(3):569-578.
7. Kinet JP. The high-affinity IgE receptor (Fc epsilon RI): From physiology to pathology. Annu Rev Immunol 1999; 17:931-972.

8. Blank U, Rivera J. The ins and outs of IgE-dependent mast-cell exocytosis. Trends Immunol 2004; 25(5):266-273.
9. Blank U, Cyprien B, Martin-Verdeaux S et al. SNAREs and associated regulators in the control of exocytosis in the RBL-2H3 mast cell line. Mol Immunol 2002; 38(16-18):1341-1345.
10. Ehrlich P. Beiträge zur Theorie und Praxis histologischer Färbung. Doktorarbeit, Deutschland: Universität Leipzig, 1878.
11. Yurt RW, Leid Jr RW, Austen KF. Native heparin from rat peritoneal mast cells. J Biol Chem 1977; 252(2):518-521.
12. Metcalfe DD, Baram D, Mekori YA. Mast cells. Physiol Rev 1997; 77(4):1033-1079.
13. Griffiths G. Secretory lysosomes - A special mechanism of regulated secretion in haemopoietic cells. Trends Cell Biology 1996; 6:329-332.
14. Blott EJ, Griffiths GM. Secretory lysosomes. Nat Rev Mol Cell Biol 2002; 3(2):122-131.
15. Raposo G, Tenza D, Mecheri S et al. Accumulation of major histocompatibility complex class II molecules in mast cell secretory granules and their release upon degranulation. Mol Biol Cell 1997; 8(12):2631-2645.
16. Humphries DE, Wong GW, Friend DS et al. Heparin is essential for the storage of specific granule proteases in mast cells. Nature 1999; 400(6746):769-772.
17. Forsberg E, Pejler G, Ringvall M et al. Abnormal mast cells in mice deficient in a heparin-synthesizing enzyme. Nature 1999; 400:773-776.
18. Abrink M, Grujic M, Pejler G. Serglycin is essential for maturation of mast cell secretory granule. J Biol Chem 2004; 279(39):40897-40905.
19. Perou CM, Moore KJ, Nagle DL et al. Identification of the murine beige gene by YAC complementation and positional cloning. Nat Genet 1996; 13(3):303-308.
20. Barbosa MD, Nguyen QA, Tchernev VT et al. Identification of the homologous beige and Chediak-Higashi syndrome genes. Nature 1996; 382(6588):262-265.
21. MacGlashan Jr D. Histamine: A mediator of inflammation. J Allergy Clin Immunol 2003; 112(4 Suppl):S53-59.
22. Huang C, Sali A, Stevens RL. Regulation and function of mast cell proteases in inflammation. J Clin Immunol 1998; 18(3):169-183.
23. Tchougounova E, Pejler G, Abrink M. The chymase, mouse mast cell protease 4, constitutes the major chymotrypsin-like activity in peritoneum and ear tissue. A role for mouse mast cell protease 4 in thrombin regulation and fibronectin turnover. J Exp Med 2003; 198(3):423-431.
24. Südhof T. The synaptic vesicule cycle: A cascade of protein-protein interactions. Nature 1995; 375:645-653.
25. Dvorak A. Ultrastructural analysis of human mast cells and basophils. Chem Immunol 1995; 61:1-33.
26. Morgenstern E. The formation of compound granules from different types of secretory granules in human platelets (dense granules and alpha granules). A cryofixation/-substitution study using serial section. Eur J Cell Biol 1995; 68:183-190.
27. Scepek S, Coorssen JR, Lindau M. Fusion pore expansion in horse eosinophils is modulated by Ca2+ and protein kinase C via distinct mechanisms. EMBO J 1998; 17(15):4340-4345.
28. Lollike K, Lindau M, Calafat J et al. Compound exocytosis of granules in human neutrophils. J Leukoc Biol 2002; 71(6):973-980.
29. Crivellato E, Nico B, Mallardi F et al. Piecemeal degranulation as a general secretory mechanism? Anat Rec A Discov Mol Cell Evol Biol 2003; 274(1):778-784.
30. Dvorak HF, Dvorak AM. Basophilic leucocytes: Structure, function and role in disease. Clin Haematol 1975; 4(3):651-683.
31. Dvorak AM, MacGlashan Jr DW, Morgan ES et al. Vesicular transport of histamine in stimulated human basophils. Blood 1996; 88(11):4090-4101.
32. Erjefalt JS, Andersson M, Greiff L et al. Cytolysis and piecemeal degranulation as distinct modes of activation of airway mucosal eosinophils. J Allergy Clin Immunol 1998; 102(2):286-294.
33. Melo RC, Perez SA, Spencer LA et al. Intragranular vesiculotubular compartments are involved in piecemeal degranulation by activated human eosinophils. Traffic 2005; 6(10):866-879.
34. Dvorak A, Tepper R, Weller P et al. Piecemeal degranulation of mast cells in the inflammatory eyelid lesions of interleukin-4 transgenic mice. Evidence of mast cell histamine release in vivo by diamine oxidase-gold enzyme affinity ultrastructural cytochemistry. Blood 1994; 83:3600-3612.
35. Dvorak AM. Ultrastructural features of human basophil and mast cell secretory function. In: Marone GL, Galli LM, SJ, eds. Mast cells and basophils. London: Academic Press, 2000:63-88.
36. Dvorak AM, Hammond ME, Morgan E et al. Evidence for a vesicular transport mechanism in guinea pig basophilic leukocytes. Lab Invest 1980; 42(2):263-276.
37. Schneider SW. Kiss and run mechanism in exocytosis. J Membr Biol 2001; 181(2):67-76.

38. Williams RM, Webb WW. Single granule pH cycling in antigen-induced mast cell secretion [In Process Citation]. J Cell Sci 2000; 113(Pt 21):3839-3850.
39. Galli S, Dvorak A, Dvorak H. Basophils and mast cells: Morphologic insights into their biology, secretory patterns, and function. Prog Allergy 1984; 34:1-141.
40. Söllner T, Whiteheart SW, Brunner M et al. SNAP receptors implicated in vesicle targeting and fusion [see comments]. Nature 1993; 362(6418):318-324.
41. Sutton RB, Fasshauer D, Jahn R et al. Crystal structure of a SNARE complex involved in synaptic exocytosis at 2.4 A resolution. Nature 1998; 395(6700):347-353.
42. Schoch S, Deak F, Konigstorfer A et al. SNARE function analyzed in synaptobrevin/VAMP knockout mice. Science 2001; 294(5544):1117-1122.
43. Guo Z, Turner C, Castle D. Relocation of the t-SNARE SNAP-23 from lamellipodia-like cell surface projections regulates compound exocytosis in mast cells. Cell 1998; 94(4):537-548.
44. Paumet F, Le Mao J, Martin S et al. Soluble NSF attachment protein receptors (SNAREs) in RBL-2H3 mast cells: Functional role of syntaxin 4 in exocytosis and identification of a vesicle-associated membrane protein 8-containing secretory compartment. J Immunol 2000; 164(11):5850-5857.
45. Vaidyanathan VV, Puri N, Roche PA. The last exon of SNAP-23 regulates granule exocytosis from mast cells. J Biol Chem 2001; 276(27):25101-25106.
46. Miesenbock G, De Angelis DA, Rothman JE. Visualizing secretion and synaptic transmission with pH-sensitive green fluorescent proteins. Nature 1998; 394(6689):192-195.
47. Hibi T, Hirashima N, Nakanishi M. Rat basophilic leukemia cells express syntaxin-3 and VAMP-7 in granule membranes. Biochem Biophys Res Commun 2000; 271(1):36-41.
48. Hanson PI, Roth R, Morisaki H et al. Structure and conformational changes in NSF and its membrane receptor complexes visualized by quick-freeze/deep-etch electron microscopy. Cell 1997; 90(3):523-535.
49. Puri N, Kruhlak MJ, Whiteheart SW et al. Mast cell degranulation requires N-ethylmaleimide-sensitive factor-mediated SNARE disassembly. J Immunol 2003; 171(10):5345-5352.
50. Jahn R. Sec1/Munc18 proteins: Mediators of membrane fusion moving to center stage. Neuron 2000; 27(2):201-204.
51. Rizo J, Sudhof TC. Snares and Munc18 in synaptic vesicle fusion. Nat Rev Neurosci 2002; 3(8):641-653.
52. Verhage M, Maia AS, Plomp JJ et al. Synaptic assembly of the brain in the absence of neurotransmitter secretion. Science 2000; 287(5454):864-869.
53. Kanda H, Tamori Y, Shinoda H et al. Adipocytes from Munc18c-null mice show increased sensitivity to insulin-stimulated GLUT4 externalization. J Clin Invest 2005; 115(2):291-301.
54. Oh E, Spurlin BA, Pessin JE et al. Munc18c heterozygous knockout mice display increased susceptibility for severe glucose intolerance. Diabetes 2005; 54(3):638-647.
55. Misura KMS, Scheller RH, Weis WI. Three-dimensional structure of the neuronal-Sec1-syntaxin 1a complex. Nature 2000; 404(6776):355-362.
56. Martin-Verdeaux S, Pombo I, Iannascoli B et al. Analysis of Munc18-2 compartmentation in mast cells reveals a role for microtubules in granule exocytosis. J Cell Sci 2003; 116:325-334.
57. Nigam R, Sepulveda J, Tuvim M et al. Expression and transcriptional regulation of Munc18 isoforms in mast cells. Biochim Biophys Acta 2005; 1728(1-2):77-83.
58. Schluter OM, Khvotchev M, Jahn R et al. Localization versus function of Rab3 proteins. Evidence for a common regulatory role in controlling fusion. J Biol Chem 2002; 277(43):40919-40929.
59. Schluter OM, Schmitz F, Jahn R et al. A complete genetic analysis of neuronal Rab3 function. J Neurosci 2004; 24(29):6629-6637.
60. Oberhauser A, Balan V, Fernandez-Badilla C et al. RT-PCR cloning of Rab3 isoforms expressed in peritoneal mast cells. FEBS Letters 1994; 339:171-174.
61. Roa M, Paumet F, Lemao J et al. Involvement of the ras-like GTPase rab3d in RBL-2H3 mast cell exocytosis following stimulation via high affinity IgE receptors (Fc epsilon RI). J Immunol 1997; 159(6):2815-2823.
62. Tuvim MJ, Adachi R, Chocano JF et al. Rab3D, a small GTPase, is localized on mast cell secretory granules and translocates to the plasma membrane upon exocytosis. American Journal of Respiratory Cell and Molecular Biology 1999; 20(1):79-89.
63. Smith J, Thompson N, Armstrong J et al. Rat Basophilic Leukaemia (RBL) cells overexpressing rab3a have a reversible block in antigen-stimulated exocytosis. Biochemical Journal 1997; 323:321-328.
64. Riedel D, Antonin W, Fernandez-Chacon R et al. Rab3D is not required for exocrine exocytosis but for maintenance of normally sized secretory granules. Mol Cell Biol 2002; 22(18):6487-6497.

65. Fukuda M. Versatile role of Rab27 in membrane trafficking: Focus on the Rab27 effector families. J Biochem (Tokyo) 2005; 137(1):9-16.
66. Wu XS, Rao K, Zhang H et al. Identification of an organelle receptor for myosin-Va. Nat Cell Biol 2002; 4(4):271-278.
67. Fukuda M, Kuroda TS, Mikoshiba K. Slac2-a/melanophilin, the missing link between Rab27 and myosin Va: Implications of a tripartite protein complex for melanosome transport. J Biol Chem 2002; 277(14):12432-12436.
68. El-Amraoui A, Schonn JS, Kussel-Andermann P et al. MyRIP, a novel Rab effector, enables myosin VIIa recruitment to retinal melanosomes. EMBO Rep 2002; 3(5):463-470.
69. Desnos C, Schonn JS, Huet S et al. Rab27A and its effector MyRIP link secretory granules to F-actin and control their motion towards release sites. J Cell Biol 2003; 163(3):559-570.
70. Stinchcombe JC, Barral DC, Mules EH et al. Rab27a is required for regulated secretion in cytotoxic T lymphocytes. J Cell Biol 2001; 152(4):825-834.
71. Haddad EK, Wu X, Hammer IIIrd JA et al. Defective granule exocytosis in Rab27a-deficient lymphocytes from Ashen mice. J Cell Biol 2001; 152(4):835-842.
72. Feldmann J, Callebaut I, Raposo G et al. Munc13-4 is essential for cytolytic granules fusion and is mutated in a form of familial hemophagocytic lymphohistiocytosis (FHL3). Cell 2003; 115(4):461-473.
73. Augustin I, Rosenmund C, Sudhof TC et al. Munc13-1 is essential for fusion competence of glutamatergic synaptic vesicles. Nature 1999; 400(6743):457-461.
74. Ashery U, Varoqueaux F, Voets T et al. Munc13-1 acts as a priming factor for large dense-core vesicles in bovine chromaffin cells. EMBO J 2000; 19(14):3586-3596.
75. Neeft M, Wieffer M, de Jong AS et al. Munc13-4 is an effector of rab27a and controls secretion of lysosomes in hematopoietic cells. Mol Biol Cell 2005; 16(2):731-741.
76. Goishi K, Mizuno K, Nakanishi H et al. Involvement of Rab27 in antigen-induced histamine release from rat basophilic leukemia 2H3 cells. Biochem Biophys Res Commun 2004; 324(1):294-301.
77. Chen D, Guo J, Miki T et al. Molecular cloning and characterization of rab27a and rab27b, novel human rab proteins shared by melanocytes and platelets. Biochem Mol Med 1997; 60(1):27-37.
78. Shirakawa R, Higashi T, Tabuchi A et al. Munc13-4 is a GTP-Rab27-binding protein regulating dense core granule secretion in platelets. J Biol Chem 2004; 279(11):10730-10737.
79. Tadokoro S, Nakanishi M, Hirashima N. Complexin II facilitates exocytotic release in mast cells by enhancing Ca2+ sensitivity of the fusion process. J Cell Sci 2005; 118(Pt 10):2239-2246.
80. Marz KE, Hanson PI. Sealed with a twist: Complexin and the synaptic SNARE complex. Trends Neurosci 2002; 25(8):381-383.
81. Guo Z, Liu L, Cafiso D et al. Perturbation of a very late step of regulated exocytosis by a secretory carrier membrane protein (SCAMP2)-derived peptide. J Biol Chem 2002; 277(38):35357-35363.
82. Brown FD, Thompson N, Saqib KM et al. Phospholipase D1 localises to secretory granules and lysosomes and is plasma-membrane translocated on cellular stimulation. Curr Biol 1998; 8(14):835-838.
83. Choi WS, Kim YM, Combs C et al. Phospholipases D1 and D2 regulate different phases of exocytosis in mast cells. J Immunol 2002; 168(11):5682-5689.
84. Peng Z, Beaven MA. An essential role for phospholipase D in the activation of protein kinase C and degranulation in mast cells. J Immunol 2005; 174(9):5201-5208.
85. O'Luanaigh N, Pardo R, Fensome A et al. Continual production of phosphatidic acid by phospholipase D is essential for antigen-stimulated membrane ruffling in cultured mast cells. Mol Biol Cell 2002; 13(10):3730-3746.
86. Ichikawa S, Walde P. Phospholipase D-mediated aggregation, fusion, and precipitation of phospholipid vesicles. Langmuir 2004; 20(3):941-949.
87. Katz B. The release of neural transmitter substances. Springfield, Illinois: Thomas, 1969.
88. Foreman JC, Mongar JL, Gomperts BD. Calcium ionophores and movement of calcium ions following the physiological stimulus to a secretory process. Nature 1973; 245(5423):249-251.
89. Li C, Ullrich B, Zhang JZ et al. Ca(2+)-dependent and -independent activities of neural and nonneural synaptotagmins. Nature 1995; 375(6532):594-599.
90. Sugita S, Shin OH, Han W et al. Synaptotagmins form a hierarchy of exocytotic Ca(2+) sensors with distinct Ca(2+) affinities. EMBO J 2002; 21(3):270-280.
91. Chapman ER, An S, Edwardson JM et al. A novel function for the second C2 domain of synaptotagmin. Ca2+-triggered dimerization. J Biol Chem 1996; 271(10):5844-5849.
92. Desai RC, Vyas B, Earles CA et al. The C2B domain of synaptotagmin is a Ca(2+)-sensing module essential for exocytosis. J Cell Biol 2000; 150(5):1125-1136.

93. Perin MS, Fried VA, Mignery GA et al. Phospholipid binding by a synaptic vesicle protein homologous to the regulatory region of protein kinase C. Nature 1990; 345(6272):260-263.
94. Davletov BA, Sudhof TC. A single C2 domain from synaptotagmin I is sufficient for high affinity Ca2+/phospholipid binding. J Biol Chem 1993; 268(35):26386-26390.
95. Rizo J, Sudhof TC. C2-domains, structure and function of a universal Ca2+-binding domain. J Biol Chem 1998; 273(26):15879-15882.
96. Chapman ER, Hanson PI, An S et al. Ca2+ regulates the interaction between synaptotagmin and syntaxin 1. J Biol Chem 1995; 270(40):23667-23671.
97. Schiavo G, Stenbeck G, Rothman JE et al. Binding of the synaptic vesicle v-SNARE, synaptotagmin, to the plasma membrane t-SNARE, SNAP-25, can explain docked vesicles at neurotoxin-treated synapses [see comments]. Proc Natl Acad Sci USA 1997; 94(3):997-1001.
98. Chapman ER. Synaptotagmin: A Ca(2+) sensor that triggers exocytosis? Nat Rev Mol Cell Biol 2002; 3(7):498-508.
99. Baram D, Adachi R, Medalia O et al. Synaptotagmin II negatively regulates Ca2+-triggered exocytosis of lysosomes in mast cells. J Exp Med 1999; 189(10):1649-1658.
100. Baram D, Mekori YA, Sagi-Eisenberg R. Synaptotagmin regulates mast cell functions. Immunol Rev 2001; 179:25-34.
101. Grimberg E, Peng Z, Hammel I et al. Synaptotagmin III is a critical factor for the formation of the perinuclear endocytic recycling compartment and determination of secretory granules size. J Cell Sci 2003; 116(Pt 1):145-154.
102. Haberman Y, Grimberg E, Fukuda M et al. Synaptotagmin IX, a possible linker between the perinuclear endocytic recycling compartment and the microtubules. J Cell Sci 2003; 116(Pt 21):4307-4318.
103. Haberman Y, Ziv I, Gorzalczany Y et al. Classical protein kinase C(s) regulates targeting of synaptotagmin IX to the endocytic recycling compartment. J Cell Sci 2005; 118(Pt 8):1641-1649.
104. Baram D, Linial M, Mekori YA et al. Ca2+-dependent exocytosis in mast cells is stimulated by the Ca2+ sensor, synaptotagmin I. J Immunol 1998; 161(10):5120-5123.
105. Kimura N, Shiraishi S, Mizunashi K et al. Synaptotagmin I expression in mast cells of normal human tissues, systemic mast cell disease, and a human mast cell leukemia cell line. J Histochem Cytochem 2001; 49(3):341-346.
106. Ashby MC, Tepikin AV. Polarized calcium and calmodulin signaling in secretory epithelia. Physiol Rev 2002; 82(3):701-734.
107. Chin D, Means AR. Calmodulin: A prototypical calcium sensor. Trends Cell Biol 2000; 10(8):322-328.
108. Peachell PT, Pearce FL. Effect of calmodulin inhibitors on histamine secretion from mast cells. Agents Actions 1985; 16(1-2):43-44.
109. Oishi H, Sasaki T, Nagano F et al. Localization of the Rab3 small G protein regulators in nerve terminals and their involvement in Ca2+-dependent exocytosis. J Biol Chem 1998; 273(51):34580-34585.
110. Gigl G, Hartweg D, Sanchez-Delgado E et al. Calmodulin antagonism: A pharmacological approach for the inhibition of mediator release from mast cells. Cell Calcium 1987; 8(5):327-344.
111. Chakravarty N. The roles of calmodulin and protein kinase C in histamine secretion from mast cells. Agents Actions 1992; 36(3-4):183-191.
112. Sullivan R, Burnham M, Torok K et al. Calmodulin regulates the disassembly of cortical F-actin in mast cells but is not required for secretion. Cell Calcium 2000; 28(1):33-46.
113. Choi OH, Adelstein RS, Beaven MA. Secretion from rat basophilic RBL-2H3 cells is associated with diphosphorylation of myosin light chains by myosin light chain kinase as well as phosphorylation by protein kinase C. J Biol Chem 1994; 269(1):536-541.
114. Buxton DB, Adelstein RS. Calcium-dependent threonine phosphorylation of nonmuscle myosin in stimulated RBL-2H3 mast cells. J Biol Chem 2000; 275(44):34772-34779.
115. Risinger C, Bennett MK. Differential phosphorylation of syntaxin and synaptosome-associated protein of 25 kDa (SNAP-25) isoforms. Journal of Neurochemistry 1999; 72(2):614-624.
116. Min DS, Cho NJ, Yoon SH et al. Phospholipase C, protein kinase C, Ca(2+)/calmodulin-dependent protein kinase II, and tyrosine phosphorylation are involved in carbachol-induced phospholipase D activation in Chinese hamster ovary cells expressing muscarinic acetylcholine receptor of Caenorhabditis elegans. J Neurochem 2000; 75(1):274-281.
117. Quetglas S, Leveque C, Miquelis R et al. Ca2+-dependent regulation of synaptic SNARE complex assembly via a calmodulin- and phospholipid-binding domain of synaptobrevin. Proc Natl Acad Sci USA 2000; 97(17):9695-9700.
118. Quetglas S, Iborra C, Sasakawa N et al. Calmodulin and lipid binding to synaptobrevin regulates calcium-dependent exocytosis. EMBO J 2002; 21(15):3970-3979.

119. Turner KM, Burgoyne RD, Morgan A. Protein phosphorylation and the regulation of synaptic membrane traffic. Trends Neurosci 1999; 22(10):459-464.

120. Greengard P, Valtorta F, Czernik AJ et al. Synaptic vesicle phosphoproteins and regulation of synaptic function. Science 1993; 259(5096):780-785.

121. Pombo I, Martin-Verdeaux S, Iannascoli B et al. IgE receptor type I-dependent regulation of a Rab3D-associated kinase. A possible link in the calcium-dependent assembly of SNARE complexes. J Biol Chem 2001; 12:12.

122. Gerst JE. SNARE regulators: Matchmakers and matchbreakers. Biochim Biophys Acta 2003; 1641(2-3):99-110.

123. Hilfiker S, Pieribone VA, Nordstedt C et al. Regulation of synaptotagmin I phosphorylation by multiple protein kinases. J Neurochem 1999; 73(3):921-932.

124. Ozawa K, Yamada K, Kazanietz MG et al. Different isozymes of protein kinase C mediate feedback inhibition of phospholipase C and stimulatory signals for exocytosis in rat RBL-2H3 cells. J Biol Chem 1993; 268(4):2280-2283.

125. Nechushtan H, Leitges M, Cohen C et al. Inhibition of degranulation and interleukin-6 production in mast cells derived from mice deficient in protein kinase C beta. Blood 2000; 95(5):1752-1757.

126. Leitges M, Gimborn K, Elis W et al. Protein kinase C-delta is a negative regulator of antigen-induced mast cell degranulation. Mol Cell Biol 2002; 22(12):3970-3980.

127. Ludowyke RI, Peleg I, Beaven MA et al. Antigen-induced secretion of histamine and the phosphorylation of myosin by protein kinase C in rat basophilic leukemia cells. J Biol Chem 1989; 264(21):12492-12501.

128. Ludowyke RI, Scurr LL, McNally CM. Calcium ionophoreinduced secretion from mast cells correlates with myosin light chain phosphorylation by protein kinase C. J Immunol 1996; 157(11):5130-5138.

129. Shimazaki Y, Nishiki T, Omori A et al. Phosphorylation of 25-kDa synaptosome-associated protein. Possible involvement in protein kinase C-mediated regulation of neurotransmitter release. J Biol Chem 1996; 271(24):14548-14553.

130. Chung SH, Polgar J, Reed GL. Protein kinase C phosphorylation of syntaxin 4 in thrombin-activated human platelets. J Biol Chem 2000; 275(33):25286-25291.

131. Hepp R, Puri N, Hohenstein AC et al. Phosphorylation of SNAP-23 regulates exocytosis from mast cells. J Biol Chem 2005; 280(8):6610-6620.

132. Graham ME, Fisher RJ, Burgoyne RD. Measurement of exocytosis by amperometry in adrenal chromaffin cells: Effects of clostridial neurotoxins and activation of protein kinase C on fusion pore kinetics. Biochimie 2000; 82(5):469-479.

133. Fujita Y, Sasaki T, Fukui K et al. Phosphorylation of Munc-18/n-Sec1/rbSec1 by protein Kinase C. J Biol Chem 1996; 271:7265-7268.

134. Devries E, Koene HR, Vossen JM et al. Identification of an unusual fc gamma receptor IIIa (CD16) on natural killer cells in a patient with recurrent infections. Blood 1996; 88(8):3022-3027.

135. Reed GL, Houng AK, Fitzgerald ML. Human platelets contain SNARE proteins and a Sec1p homologue that interacts with syntaxin 4 and is phosphorylated after thrombin activation: Implications for platelet secretion. Blood 1999; 93(8):2617-2626.

136. Cabaniols JP, Ravichandran V, Roche PA. Phosphorylation of SNAP-23 by the novel kinase SNAK regulates t-SNARE complex assembly. Mol Biol Cell 1999; 10(12):4033-4041.

137. Chahdi A, Choi W, Kim Y et al. Serine/threonine protein kinases synergistically regulate phospholipase D1 and 2 and secretion in RBL-2H3 mast cells. Mol Immunol 2002; 38(16-18):1269.

138. Marash M, Gerst JE. t-SNARE dephosphorylation promotes SNARE assembly and exocytosis in yeast. EMBO J 2001; 20(3):411-421.

139. Marash M, Gerst JE. Phosphorylation of the autoinhibitory domain of the Sso t-SNAREs promotes binding of the Vsm1 SNARE regulator in yeast. Mol Biol Cell 2003; 14(8):3114-3125.

140. Ludowyke RI, Holst J, Mudge LM et al. Transient translocation and activation of protein phosphatase 2A during mast cell secretion. J Biol Chem 2000; 275(9):6144-6152.

141. Holst J, Sim AT, Ludowyke RI. Protein phosphatases 1 and 2A transiently associate with myosin during the peak rate of secretion from mast cells. Mol Biol Cell 2002; 13(3):1083-1098.

142. Wang X, Huynh H, Gjorloff-Wingren A et al. Enlargement of secretory vesicles by protein tyrosine phosphatase PTP-MEG2 in rat basophilic leukemia mast cells and Jurkat T cells. J Immunol 2002; 168(9):4612-4619.

143. Koffer A, Tatham P, Gomperts B. Changes in the state of Actin during the exocytotic reaction of permeabilized rat mast cell. J Cell Biol 1990; 111:919-927.

144. Ludowyke R, Kawasugi K, French P. PMA and calcium ionophore induce myosin and F-actin rearrangement during histamine secretion from RBL-2H3 cells. Cell Motil Cytoskeleton 1994; 29:354-365.

145. Oliver C, Sahara N, Kitani S et al. Binding of monoclonal antibody AA4 to gangliosides on rat basophilic leukemia cells produces changes similar to those seen with Fc epsilon receptor activation. J Cell Biol 1992; 116(3):635-646.

146. Nielsen EH. A filamentous network surrounding secretory granules from mast cells. J Cell Sci 1990; 96(Pt 1):43-46.

147. Pendleton A, Koffer A. Effects of latrunculin reveal requirements for the actin cytoskeleton during secretion from mast cells. Cell Motil Cytoskeleton 2001; 48(1):37-51.

148. Price DJ, Kawakami Y, Kawakami T et al. Purification of a major tyrosine kinase from RBL-2H3 cells phosphorylating Fc(epsilon)RI gamma-cytoplasmic domain and identification as the btk tyrosine kinase. Biochim Biophys Acta - Molecular Cell Research 1995; 1265(2-3):133-142.

149. Norman J, Price L, Ridley A et al. The small GTP-binding proteins, Rac and Rho, regulate cytoskeletal organization and exocytosis in mast cells by parallel pathways. Mol Biol Cell 1996; 7:1429-1442.

150. Guillemot JC, Montcourrier P, Vivier E et al. Selective control of membrane ruffling and actin plaque assembly by the Rho GTPases Rac1 and CDC42 in FcepsilonRI-activated rat basophilic leukemia (RBL-2H3) cells. J Cell Sci 1997; 110(Pt 18):2215-2225.

151. Frigeri L, Apgar JR. The role of actin microfilaments in the downregulation of the degranulation response in RBL-2H3 mast cells. J Immunol 1999; 162(4):2243-2250.

152. Hong-Geller E, Holowka D, Siraganian RP et al. Activated Cdc42/Rac reconstitutes Fcepsilon RI-mediated Ca2+ mobilization and degranulation in mutant RBL mast cells. Proc Natl Acad Sci USA 2001; 98(3):1154-1159.

153. Pivniouk VI, Snapper SB, Kettner A et al. Impaired signaling via the high-affinity IgE receptor in Wiskott-Aldrich syndrome protein-deficient mast cells. Int Immunol 2003; 15(12):1431-1440.

154. Kettner A, Kumar L, Anton IM et al. WIP regulates signaling via the high affinity receptor for immunoglobulin E in mast cells. J Exp Med 2004; 199(3):357-368.

155. Nielsen EH, Johansen T. Effects of dimethylsulfoxide (DMSO), nocodazole, and taxol on mast cell histamine secretion. Acta Pharmacol Toxicol (Copenh) 1986; 59(3):214-219.

156. Tasaka K, Mio M, Akagi M et al. Role of the cytoskeleton in Ca2+ release from the intracellular Ca store of rat peritoneal mast cells. Agents Actions 1991; 33(1-2):44-47.

157. Nishida K, Yamasaki S, Ito Y et al. Fc{epsilon}RI-mediated mast cell degranulation requires calcium-independent microtubule-dependent translocation of granules to the plasma membrane. J Cell Biol 2005; 170(1):115-126.

158. Smith AJ, Pfeiffer JR, Zhang J et al. Microtubule-dependent transport of secretory vesicles in RBL-2H3 cells. Traffic 2003; 4(5):302-312.

159. Clark RH, Stinchcombe JC, Day A et al. Adaptor protein 3-dependent microtubule-mediated movement of lytic granules to the immunological synapse. Nat Immunol 2003; 4(11):1111-1120.

160. Williams RM, Shear JB, Zipfel WR et al. Mucosal mast cell secretion processes imaged using three-photon microscopy of 5-hydroxytryptamine autofluorescence. Biophys J 1999; 76(4):1835-1846.

161. Ikonen E. Roles of lipid rafts in membrane transport. Curr Opin Cell Biol 2001; 13(4):470-477.

162. Simons K, Toomre D. Lipid rafts and signal transduction. Nat Rev Mol Cell Biol 2000; 1(1):31-39.

163. Simons K, Ikonen E. Functional rafts in cell membranes. Nature 1997; 387(6633):569-572.

164. Mukherjee S, Maxfield FR. Role of membrane organization and membrane domains in endocytic lipid trafficking. Traffic 2000; 1(3):203-211.

165. Tooze SA, Martens GJ, Huttner WB. Secretory granule biogenesis: Rafting to the SNARE. Trends Cell Biol 2001; 11(3):116-122.

166. Lafont F, Verkade P, Galli T et al. Raft association of SNAP receptors acting in apical trafficking in Madin-Darby canine kidney cells. Proc Natl Acad Sci USA 1999; 96(7):3734-3738.

167. Wang Y, Thiele C, Huttner WB. Cholesterol is required for the formation of regulated and constitutive secretory vesicles from the trans-Golgi network. Traffic 2000; 1(12):952-962.

168. Schmidt K, Schrader M, Kern HF et al. Regulated apical secretion of zymogens in rat pancreas. Involvement of the glycosylphosphatidylinositol-anchored glycoprotein GP-2, the lectin ZG16p, and cholesterol-glycosphingolipid-enriched microdomains. J Biol Chem 2001; 276(17):14315-14323.

169. Lang T, Bruns D, Wenzel D et al. SNAREs are concentrated in cholesterol-dependent clusters that define docking and fusion sites for exocytosis. EMBO J 2001; 20(9):2202-2213.

170. Chamberlain LH, Burgoyne RD, Gould GW. SNARE proteins are highly enriched in lipid rafts in PC12 cells: Implications for the spatial control of exocytosis. Proc Natl Acad Sci USA 2001; 98(10):5619-5624.

171. Pombo I, Rivera J, Blank U. Munc18-2/syntaxin3 complexes are spatially separated from syntaxin3-containing SNARE complexes. FEBS Lett 2003; 550(1-3):144-148.

CHAPTER 9

Acrosomal Exocytosis

Claudia Nora Tomes*

Why Is This Chapter in the Book? *Or* the Sperm Acrosome Reaction as a Model for Regulated Exocytosis

Sexual reproduction to perpetuate a given species occurs through fertilization, during which a diploid zygote is formed to produce a genetically distinct individual. To this end, the haploid sperm and haploid egg must collide to allow entry of the sperm head delivering the male chromatin into the egg cytoplasm. Both the male and female gametes undergo regulated exocytosis—termed the acrosome reaction and the cortical reaction respectively—at different times during their encounter. The success of fertilization depends on these exocytoses.

Exocytosis of the sperm's single vesicle—the acrosome—is a synchronized, all-or-nothing process that happens only once in the life of the cell and shares the basic fusion molecules with neuronal, endocrine, and all the other cells covered in this book. Acrosomal exocytosis (AE) depends on both Rab3 activation and neurotoxin-sensitive SNAREs; it also requires the efflux of calcium from inside the acrosome. Convergence of Rab- and toxin-sensitive SNARE-dependent pathways is a hallmark of AE that makes it an attractive mammalian model to study the different phases of the membrane fusion cascade. Furthermore, because nature has endowed sperm with a cellular specialization that gives them a single, irreversible chance to fertilize an egg, AE is more straightforward to dissect compared to fusion in other cell types, where the same substances are secreted again and again, requiring the membranes and fusion machinery to recycle multiple times.

The presence of secretory granules in the egg and the sperm highlights the essential nature of regulated exocytosis for the continued existence of multicellular species. Identifying and understanding the role and timing of the players involved are important for manipulating fertilization, either to enhance it or block it. The benefits of this knowledge are, however, not restricted to the reproduction field, but could be extended to more complex membrane fusion scenarios.

The Spermatozoon

The spermatozoon is the end product of the process of gametogenesis in the male, which occurs within the testicular seminiferous epithelium. Spermatogenesis involves a series of mitotic divisions of spermatogonial stem cells, two meiotic divisions by spermatocytes, extensive morphological remodeling of spermatids during spermiogenesis, and the release of free cells into the lumen of the seminiferous tubules by spermiation.[1] The result of spermatogenesis is a highly polarized cell, consisting of a head—that contains the nucleus and the acrosome in the apical region—and a flagellum—containing a 9 + 2 array of microtubules, sheath proteins, and mitochondria—which are joined at the neck (Fig. 1, left). The flagellum and the head are wrapped by the plasma membrane and contain little cytoplasm.

*Claudia Nora Tomes—Laboratorio de Biología Celular y Molecular, Instituto de Histología y Embriología (IHEM-CONICET), Facultad de Ciencias Médicas, CC 56, Universidad Nacional de Cuyo, 5500 Mendoza, Argentina. Email: ctomes@fcm.uncu.edu.ar

Molecular Mechanisms of Exocytosis, edited by Romano Regazzi. ©2007 Landes Bioscience and Springer Science+Business Media.

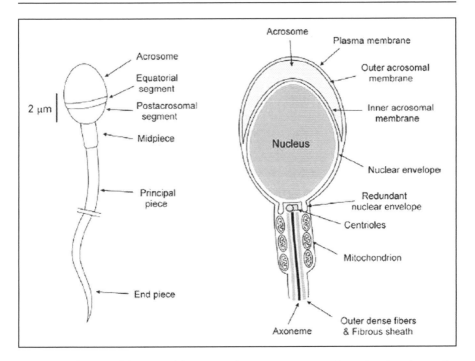

Figure 1. Left) general features of the mammalian spermatozoon. The major domains on the sperm head are the acrosomal and post acrosomal regions, and the equatorial segment. The flagellum can be divided in three regions: the middle piece, containing the mitochondria, the principal piece, containing the fibrous sheath, and the end piece. Right) schematic representation of a sagittal section, revealing the detailed structure of the head and midpiece. The plasma membrane overlies the outer acrosomal membrane, the inner acrosomal membrane, in turn, overlies the nuclear envelope. Mitochondria are helically wrapped around the outer dense fibers, which are adjacent to the axoneme. (Reprinted from: Darszon A et al. Int Rev Cytol 243:79-172, ©2005, with permission from Elsevier.[41])

Although in general spermatozoa share these structures, there are species-specific variations in the size and shape of the cells among mammals, lack of acrosome and flagellum in amoeboid cells, etc. As I will discuss later, the acrosome contains enzymes essential for fertilization, and the flagellum contains the energy sources and machinery necessary to produce motility. The roles of these components are to ensure delivery of the genetic material contained in the sperm nucleus to the egg, restoring the somatic chromosome number and initiating the development of a new individual exhibiting characteristics of the species.

After release from the seminiferous epithelium in the testis, sperm are transported through the epididymis, where numerous biochemical and functional modifications occur. These mature sperm are then stored within the cauda epididymis, in a functionally inactive state, until ejaculated or degraded in the lumen.[2] Mammalian sperm are unable to fertilize eggs until they achieve a functional reprogramming process, or capacitation, by migration through the female reproductive tract. Likewise, sperm of many nonmammalian species are activated by factors that are either associated with eggs or spawned with eggs. Capacitation consists of a variety of processes, including: the functional coupling of the signal transducing pathways that regulate the initiation of AE by physiological triggers; alterations in flagellar motility that may be required to penetrate the egg extracellular coat, the zona pellucida; and development of the capacity to fuse with eggs. These changes are accompanied by alterations in metabolism, membrane biophysical characteristics, changes in protein phosphorylation state, elevations of intracellular pH and calcium levels, and hyperpolarization of membrane potential.[3]

In mammals, fertilization itself begins in the oviduct with binding of capacitated, free-swimming sperm to the ovulated egg's zona pellucida, and ends a short time later with fusion of egg and sperm plasma membranes to form the zygote. Along the way, several recognizable events take place, including the sperm AE and penetration of the zona pellucida, and the egg cortical reaction and zona reaction. The latter is a built-in safety feature that results in alteration of the zona pellucida such that free-swimming sperm are unable to bind to fertilized eggs and hinder the normal development of the zygotes.[4]

The Acrosome

The acrosome is a relatively large, Golgi-derived, lysosome-like organelle that overlies the most anterior two-thirds of the nucleus in the apical region of the sperm head. It consists of the anteriorly located acrosomal cap and the posteriorly located equatorial segment. While the former is loaded with hydrolyzing enzymes, the enzyme content of the latter might be scant.[5] The acrosomal secretory products are concentrated and condensed, and the acrosomal granule is stored for a long period of time as the spermatid and sperm mature in the testis and epididymis, respectively. This highly condensed packaging format allows efficient storage of large amounts of secretory material in an osmotically inert form.[6] As is the case with other secretory granules, the acrosome has an electron-opaque content known as a dense core. In some species—e.g., guinea pig, mouse, rabbit, bovine, hamster, and hedgehog—the anterior acrosome has several different regions (subdomains) with varying degrees of electron density.[7]

The acrosome has all the characteristics attributed to secretory vesicles. Secretion from the acrosome is accelerated by extracellular stimuli, such as the cumulus extracellular matrix and its components, and the zona pellucida. Although the acrosome is sorrounded by a continuous membrane, the section overlying the nucleus is termed the inner acrosomal membrane, and the one underlying the plasma membrane, the outer acrosomal membrane (Fig. 1, right).[5] As for other granules, the membranes surrounding the acrosome can be removed by detergent treatment of spermatozoa without dissolving its condensed contents. This feature allows the definition of two biochemically and morphologically distinguishable compartments within the acrosome: the insoluble acrosomal matrix and the soluble proteins.[8] In this context, AE should not be described as a wholesale, instantaneous release of components from a fluid-filled sack. On the contrary, individual soluble proteins and insoluble acrosomal matrix components have their own characteristic patterns of release. Soluble factors are more readily released once AE has been initiated, whereas matrix components remain associated with the sperm for prolonged periods of time even after exocytosis. This suggests that compartmentalization of acrosomal contents may play a functional role in regulating the release of proteins during exocytosis and sperm-zona interaction.[7,8]

The Acrosome Reaction

Morphological Features

Exocytosis of the acrosome is a terminal morphological alteration that must occur prior to penetration of the zona pellucida. The acrosome reaction is a regulated exocytic process with many parallels to granule exocytosis in other cell types (see below). Yet, AE differs strikingly from other known exocytotic events in ways that can be summarized as follows: (i) sperm contain a single secretory vesicle; (ii) there are multiple fusion points between the outer acrosomal membrane and the overlying plasma membrane; (iii) exocytosis does not lead to a single fused membrane but to vesiculation and actual membrane loss; (iv) the AR is a singular event, with no known membrane recycling taking place following exocytosis (Table 1).

Morphologically, AE in eutherian mammals proceeds through a series of steps that begin with the punctate apposition of the sperm plasma membrane with the membrane directly

Table 1. Exocytosis of the sperm acrosome differs morphologically from that exhibited by more typical secretory cells

	Sperm	Other Cells
Number of secretory vesicles/cell	one	many
Number of fusion pores/vesicle	many	one
Rounds of exocytosis/cell	one	more than one
Membrane fusion	outer acrosomal and plasma membranes lost (vesiculation)	vesicle membrane incorporated to plasma membrane
Post-exocytosis membrane remodelling and recycling, and/or endocytosis	no	yes

adjacent to it, the outer acrosomal membrane, in the anterior region of the sperm head. The subsequent steps include swelling of the acrosomal contents, prior to or concomitant with fusion of the outer acrosomal membrane with the overlying plasma membrane at the points of apposition, followed by formation of pores at these points. In most cells, the pores widen, the granule contents are discharged, and granule membrane remodelling eventually occurs. In sperm, however, the pores widen, but since the outer acrosomal membrane is as large as the area of plasma membrane it is fusing with, the result of the pore widening is the fenestration of the fusion membranes and joining of pores to produce a reticulum of tubules and hybrid plasma-outer acrosomal membrane vesicles. In human sperm, for instance, the initial fusion process should be classified as fenestration, but continued fusion more closely resembles vesiculation. Acrosomal contents begin to be released at this point, with or without temporary retention of a shroud of fused and fenestrated membranes and some acrosomal contents. In some species the fenestrated cap or shroud has been shown to tether the reacted sperm to the zona pellucida. In species where the shroud has been observed, there is an insoluble component of the acrosome that stays with the cap. Finally, all of the shroud and contents are lost, so that the inner acrosomal membrane becomes the leading edge of the sperm (Fig. 2).[9,10]

Inducers, Timing and Location

Mammalian eggs are surrounded by the zona pellucida, an egg-specific extracellular matrix that is secreted by the growing oocyte and is a key regulator of fertilization. A somatic cell layer, the cumulus, also accompanies the egg into the oviduct at ovulation. The cumulus layer has both cellular and acellular components. The cellular component is a subset of follicular granulosa cells; its acellular component is a hyaluronic acid-rich extracellular matrix secreted by the cumulus cells.[11] During the course of fertilization, the acrosome undergoes exocytosis near the time that the sperm encounters the zona (Fig. 3). Although the actual site of AE is subject to debate, the prevailing dogma is that sperm must have completely intact acrosomes to adhere initially to the zona pellucida.

Initial sperm-zona pellucida binding in the mouse is mediated by ZP3, a constituent glycoprotein of the zona pellucida that binds to receptors in the anterior head of acrosome-intact sperm.[5,12] A sperm receptor for ZP3 has not yet been identified inequivocally despite the fact that several high affinity ZP3 binding proteins have been put forth as candidates.[13] ZP3 triggers AE in sperm bound to the zona pellucida. The ability of mouse ZP3 to serve as an acrosome

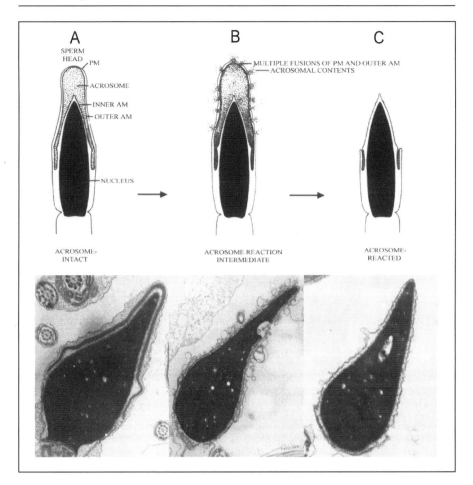

Figure 2. Schematic representation (top) and transmission electron micrographs (bottom) of the progression of the acrosome reaction viewed through sagittal sections of mouse (top) and human (bottom) sperm heads. A) Before the reaction; B) exocytosis in progress: multiple fusion points between the plasma and outher acrosomal membrane lead to vesiculation and release or exposure of acrosomal contents; C) the reaction is completed, inner acrosomal membrane becomes the leading cell membrane in the acrosomal region. PM: plasma membrane, AM: acrosomal membrane. (Top of the figure modified from: Wassarman PM et al. Mol Cell Endocrinol 234:95-103, ©2005, with permission from Elsevier.[125] Bottom of the figure modified from: Michaut M et al. Proc Natl Acad Sci USA 97:9996-10001, ©2000, with permission from the National Academy of Sciences, U.S.A.[99])

reaction inducer depends on the glycan moieties as well as on the polypeptide backbone of the molecule. Thus, small glycopeptides bound to capacitated spermatozoa but did not induce the acrosome reaction unless they were crosslinked on the sperm surface by anti-ZP3 antibodies.[14] These results suggest that ZP3 may be polyvalent with regard to adhesion oligosaccharide chains, and the resulting crosslinking and multimerization of sperm receptors is what initiates exocytosis.[3]

Although the zona can induce a reaction in most species, some acrosome reacted spermatozoa have been seen within the cumulus matrix before sperm arrive at the zona both in vivo (rabbit and hamster), and in vitro (human), perhaps induced by high concentrations of

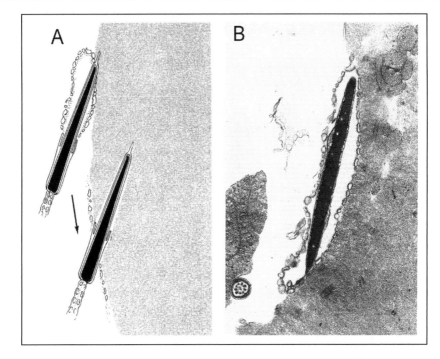

Figure 3. A) Schematic representation of eutherian spermatozoa as they interact with the zona pellucida. The acrosome reaction releases enzymes in part as soluble entities and in part complexed to insoluble acrosome matrix associated in some cases with the vesiculated shroud. The head intrudes into the zona, apparently devoid of acrosomal contents. (Reprinted from: Bedford JM. Biol Reprod 59:1275-1287, ©1998, with permission from the Society for the Study of Reproduction, Inc.[17]) B) Transmission electron micrograph of an oblique section of a reacted rabbit sperm head on the zona pellucida of an ovulated egg. Note the vesiculated shroud, which has lifted away from the inner acrosomal membrane. (Reprinted with permission from: Bedford JM. Ultrastructural changes in the sperm head during fertilization in the rabbit. Am J Anat 123:329-358, ©1968 Wiley-Liss, Inc., a subsidiary of John Wiley & Sons, Inc.)

progesterone.[9,15] Progesterone is present in the oviductal millieu, derived both from follicular fluid and from secretion by cumulus oophorus cells associated with eggs.[16] In some shrews, AE is always induced by the cumulus oophorus, with the zona seemingly unable to evoke it. It is possible that the behaviour of the acrosome within the cumulus differs among species or even between individual spermatozoa.[15]

Compared to other regulated exocytotic events, the acrosome reaction is a slow process. In human sperm, swelling and some fenestration are observed within 5-10 seconds after exposure to follicular fluid, with complete fenestration at 20 seconds and complete fusion of outer acrosomal-plasma membrane accomplished three minutes after inducing. The human sperm AR, as initiated by human follicular fluid, has an anterior initiation site and fusion moves posteriorly to the equatorial segment, generating vesicles of uniform size.[9] These hybrid vesicles composed of roughly equal proportions of plasma and outer acrosomal membranes form whether the acrosome reaction occurs spontaneously or is induced.

Purpose of the Acrosome Reaction

During fertilization in eutherian mammals, the spermatozoon must penetrate the zona pellucida to reach the oolema. Only sperm that have completed the acrosome reaction can succesfuly accomplish this task. First, they must adhere to the zona as hyperactivated spermatozoa in the acrosome-intact state for a period long enough to establish the primary binding, presumably mediated by ZP3. In contrast to metazoa and subtherian mammals, the eutherian zona surface is not visibly eroded by the soluble acrosomal enzymes released at the primary binding/reaction site.[15] After reacting, the sperm must remain attached (secondary binding) to the egg coat. Sperm proteins localized within the acrosomal granule and linked to the insoluble matrix/inner acrosomal membrane—such as acrosin—are thought to stabilize contact between the acrosome-reacted sperm and the zona pellucida. In other words, the acrosome is responsible for maintenance of binding by vigorous spermatozoa until a point of zona entry is established. From then on, penetration is accomplished by mechanical action derived from the strong flagellar beat of the sperm. It has been suggested that the acrosome reaction contributes to this late phase by changing the apical profile of the sperm head to favor penetration dependent on physical thrust.[17]

Membrane Fusion

Exocytosis is a highly regulated, multistage process consisting of multiple functionally definable stages, including recruitment, targeting, tethering, priming, and docking of secretory vesicles with the plasma membrane, followed by calcium-triggered membrane fusion.[18] During the last few years, a number of molecules involved in this complex cascade of events have been identified.[19-21] Despite the fact that exocytosis of the acrosomal vesicle is somewhat unique, sperm use the same conserved machinery and regulatory components as characterized in other secretory events. Thus, all of the molecules that are involved in signalling in sperm during AE were initially found in somatic cells and were already implicated in various membrane-fusion events. Therefore, those molecules are discussed at length elsewhere in this textbook and I will only briefly summarize their general properties and function in this chapter.

Methods to Study the Molecular Mechanisms Driving the Acrosome Reaction

The molecular mechanisms operating in the exocytosis of the acrosome are only now beginning to emerge. Traditionally, immunodetection of proteins in the acrosomal region was interpreted as an indicator of their role in exocytosis. This view was clearly inadequate, since presence and localization alone do not necessarily prove physiological relevance. Sperm's biggest drawback is their inability to synthesize proteins—thus ruling out the use of standard powerful techniques such as transfection and overexpression. In addition, the possibility of studying the role of macromolecules i.e., Rabs, SNAREs, etc is severely limited because such molecules are unable to cross the cell membrane. To overcome these limitations, we have applied in sperm a widespread methodology in the study of exocytosis in other cells, that of controlled plasma membrane permeabilization with streptolysin O (SLO). The permeable sperm preparations we have established are readily manipulated and reflective of the in vivo organization of the cell, and allow us to measure a functional, calcium-regulated acrosome reaction that is morphologically indistinguishable from that in nonpermeabilized cells.[22,23] Permeabilization allows penetration into the cell of exogenous factors—such as ions, neurotoxins, and neutralizing antibodies—that usually are not able to penetrate the plasma membrane and can be used to affect the function of target proteins. Furthermore, the SLO-permeabilized human sperm model is suitable for addition of recombinant proteins to replace, modulate and/or compete with the endogenous isoforms. SLO-permeabilization has recently been adopted by other laboratories for the study of the acrosome reaction,[24-26] and sperm motility.[27]

A few other laboratories have inferred the role of fusion proteins in AE simply by adding peptides to intact (i.e., nonpermeabilized) ram[28] and mouse[29] sperm, or antibodies to bovine[30] sperm in the absence or presence of calcium ionophores and observed an inhibition of the acrosome reaction. The explanation offered for those results was that initial fusion of the outer acrosomal membrane and the adjacent plasma membrane makes permeable pores through which macromolecules can enter into sperm and halt exocytosis before extensive content release takes place. This method is by no means as established a technology as SLO permeabilization to study secretion and presents the limitation of not being suitable for the study of regulatory steps that govern exocytosis before actual membrane fusion has begun.

Several signalling pathways serve the purpose of regulating more than one function in sperm. This is particularly true when it comes to capacitation and the acrosome reaction, since they are biologically intertwined. An additional advantage of the SLO-permeabilization approach resides in the possibility of isolating these two processes.

Calcium Signalling

Calcium is the most widely used intracellular messenger in cell signalling.[31-34] Unsurprisingly, one common theme in the regulation of exocytosis is the general role of intracellular calcium as a mediator of stimulus-secretion coupling. Thus, the elevation of calcium concentrations from resting (≤ 0.1 μM) to stimulated ($\sim$ 1-100 μM) levels is necessary and sufficient to initiate regulated exocytosis.[35,36]

In sperm, calcium is involved in virtually all physiological functions, including capacitation, hyperactivation, and the acrosome reaction. During the latter, two phases of solubilised zona pellucida- and progesterone-evoked calcium responses can be resolved with the use of ion-selective fluorescent probes. First a transient increase occurs, followed by a second phase in which increased calcium levels are sustained for the duration of the stimulation. The sustained calcium response develops slowly, requiring minutes to achieve maximal levels. AE only occurrs after a plateau is established.[37,38] The transient response exhibits the characteristics of a low voltage-activated, T-type calcium channel-mediated phenomenon. Transient calcium influx initiates a downstream cascade through the activation of phospholipase C isoforms, resulting in the generation of inositol-1, 4, 5-trisphosphate (IP₃). This second messenger mobilizes cal-

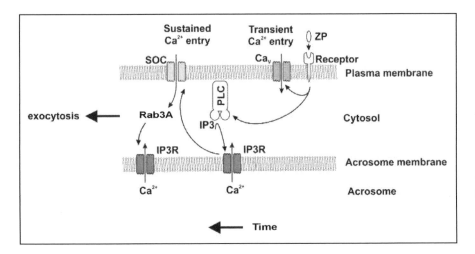

Figure 4. Diagrammatic representation of calcium peaks during AE. In the sperm head, the activation of ZP3 receptors on the plasma membrane results in a transient calcium influx from the extracellular millieu to the cytosol through voltage-gated channels (Ca$_v$), and the activation of phospholipase Cs (PLC). This results in the production of IP₃ and the subsequent opening of IP₃-sensitive calcium permeable channels (IP₃R) on the outer acrosomal membrane, and promotes the release of calcium from the intra-acrosomal pool. Depletion of this store produces the gating signal that opens SOCs in the plasma membrane and produces a sustained calcium entry into the cytosol. This drives activation of Rab3A and subsequently of all the fusion machinery. It also elicits a second efflux of calcium from the acrosome that drives exocytosis. (Modified with permission from: Felix R. Reproduction 129:251-262,[40] ©2005 Society for Reproduction and Fertility.)

cium from the intra-acrosomal store by binding to specific receptors. The emptying of this store generates a gating signal that couples it to the opening of store-operated channels (SOCs) in the plasma membrane, through a mechanism typically known as calcium-induced calcium release. SOC channels are responsible for the second—sustained—calcium phase (Fig. 4).[37,39-41] The mammalian homologue of the *Drosophila melanogaster* transient receptor potential channel Trp2 has emerged as a candidate for the zona pellucida-regulated SOC activation during AE in mouse sperm.[42] It has been suggested that AE requires sustained calcium influx to prevent precocious activation and loss of sperm function in cells with a single secretory granule.[43] Nevertheless, this pathway is not unique to sperm, since a similar mechanism of sequential cytoplasmic calcium increases via voltage-gated channels-phospholipase C activation-SOC channels has recently been described in insulin-secreting β-cells.[44]

The Acrosome Is a Calcium Store

Calcium efflux from membrane-bound vesicles is a prerequisite for transport in several systems such as the yeast vacuolar fusion model,[45,46] and the secretion of exocytotic granules.[47,48] The acrosomal granule had been postulated to be an internal reservoir of releasable calcium for several years. However, direct evidence for the existence of this pool was first acquired after developing a SLO plasma membrane permeabilization method. By using calcium sensitive probes, we could visualize directly the presence of a calcium pool inside the acrosome (Fig. 5B) and monitor the changes in the acrosomal pool under different experimental conditions that affect exocytosis.[49] It is noteworthy that, in nonpermeabilized sperm, a diffuse cytosolic staining is observed that precludes direct observation of the acrosomal pool (Fig. 5C-E). A calcium ionophore, and thapsigargin and cyclopiazonic acid (inhibitors of calcium-ATPases) strongly inhibit fusion, indicating that the luminal calcium pool must be actively maintained to keep sperm exocytosis competent.[49] Another laboratory has subsequently published similar evidence that the mouse acrosome is a calcium store, by loading nonpermeabilized sperm with the membrane permeant forms of calcium-sensitive fluorescent indicator dyes, and quenching all fluorescence in the cytosol with high manganese concentrations.[50]

According to the current model for calcium signalling during AE, the role of the intra-acrosomal calcium pool is limited to the opening of SOC channels in the plasma membrane. Incubating permeabilized sperm with calcium resembles the physiological situation of calcium influx through open SOC channels, because SLO permits free diffusion of ions. In consequence, permeabilization allows us to examine relatively late steps of the exocytic cascade, occuring after the sustained calcium influx, while bypassing earlier pathways whose end point is the opening of SOC channels. With this in mind, we were surprised to discover a direct correlation between depletion of an acrosomal calcium pool and inhibition of the acrosome reaction in permeabilized human sperm. In other words, exocytosis requires a release of calcium from the acrosome at a step downstream of SOC channels opening. Like the first efflux, the second utilizes IP_3-sensitive channels.[49] We have expanded the standard model of biphasic calcium signalling during sperm exocytosis to accommodate these new findings (Fig. 4). No-one had ever reported on the existence of this second intra-acrosomal calcium efflux phase in sperm, even less that exocytosis itself depends upon it! We have experimentally classified fusion factors as: (i) acting after SOC channel opening but before the second intra-acrosomal calcium efflux (Rab3A, α-SNAP, and NSF); and (ii) acting during or after the second intra-acrosomal release (SNARE proteins and synaptotagmin VI).[51]

Interestingly, AE depends on this calcium efflux from the acrosome even in the presence of relatively high external calcium concentrations, which are identical to cytosolic levels in permeabilized cells.[49,52] Why would a calcium efflux from the acrosome be necessary for AE under these conditions? Given the high luminal calcium concentrations, efflux could accomplish a drastic transient increase in calcium close to the acrosomal surface. This is consistent with a close spatial arrangement of the calcium release sites and their place of action.[53] During the fusion process, membranes are maintained in close proximity, creating a partially isolated

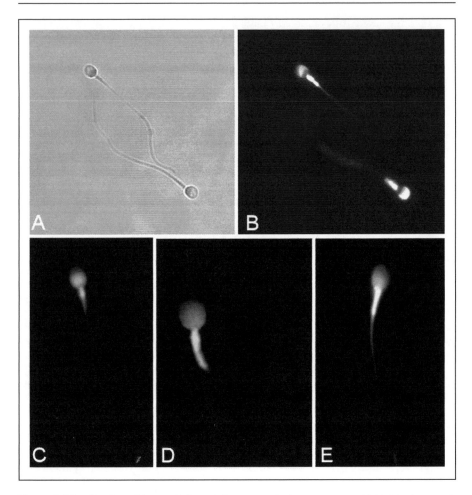

Figure 5. Visualization of intracellular calcium stores in human sperm. A) Light micrograph corresponding to the field depicted in panel (B); B-E) Fluo3-AM fluorescence images. A calcium accumulation is clearly observed in the acrosome and midpiece of permeabilized sperm (B), whereas a diffuse, whole head plus midpiece signal is detected in intact cells (C-E). Notice that the acrosomal store can only be visualized when sperm are permeabilized (Photographs provided by Gerardo De Blas).

microenvironment that could easily be characterized by highly localized, elevated calcium concentrations, suitable for specific—if yet undefined—fusion reactions. We favour the hypothesis that synaptotagmin is the main calcium target in this late exocytotic phase.

Calcium Sensing Proteins in Exocytosis

In the early stages of the secretory pathway, in constitutive exocytosis and in the endocytic pathway, vesicle fusion follows immediately after docking. In regulated exocytosis a trigger is required, so that vesicle fusion begins only when an appropriate signal arrives.[54] As I have discussed above, the signal is most frequently an elevation of the intracellular calcium concentration. Calcium must act at a very late stage on vesicles closely engaged with the plasma membrane and poised for immediate fusion, indicating that a late-acting calcium sensor is likely to be associated with the fusion machinery. A number of calcium-binding proteins thought to participate in exocytosis, based on functional analysis using a variety of techniques, have been described and reviewed elsewhere.[54] I will cover two of them, but not offer further details on the nature and behavior of the other proteins until evidence that they are present in sperm and involved in AE is gathered.

Synaptotagmin

Synaptotagmins are sensor proteins that detect the concentration of calcium reversibly by binding it and thereafter signalling to other targets. The transduction process relies on specific conformational changes in the sensor. Synaptotagmins bind phospholipids in a calcium-dependent manner. This property is endowed by their two C2 domains, C2A and C2B, which are homologous to the C2 domain in protein kinase C (PKC). Preceding the tandem C2 domains is a variable linker region connecting to the transmembrane anchor and a short intravesicular or extracellular amino terminus.[55] Distinct synaptotagmin isoforms have different calcium affinities, which could account for the different calcium requirements to trigger the exocytosis of various types of secretory vesicles.[56] In addition to calcium ions and phospholipids, synaptotagmins bind other synaptotagmins, calcium channels, isolated Q-SNAREs and those in the SNARE complex, β-SNAP, RIM, calmodulin, etc.[6,20,57,58]

Several synaptotagmin isoforms have been described in mammalian sperm. For instance synaptotagmin VI localizes to the acrosomal region in human,[59] synaptotagmin VIII in mouse,[60] synaptotagmin I in stallion,[61] and undefined isoforms in hamster, bull, rhesus monkey, mouse and human sperm.[30] Preincubation of SLO-permeabilized cells with antibodies to synaptotagmin VI,[59] and VIII[24] block calcium-initiated AE in human and mouse sperm respectively. Likewise, various peptides encompassing portions of cytoplasmic domains of synaptotagmins I, VI and VIII prevent the calcium-triggered acrosome reaction in both these species.[24,59,62] These results demonstrate that synaptotagmin is required for sperm exocytosis. Interestingly, AE elicited by GTP-γ-S and Rab3A in permeabilized human sperm proceeds at very low calcium concentrations. Yet, exocytosis continues to depend on synaptotagmin VI under these conditions, indicating that this protein acts after the Rab3A-elicited step.[59] Where does the calcium synaptotagmin binds to come from then? We believe it comes from the intra-acrosomal store. Consistent with this view, synaptotagmin VI displays a late role in the acrosome reaction, simultaneous with or downstream of the calcium efflux from the acrosome.[51] How does synaptotagmin participate in sperm outer acrosomal to plasma membrane fusion? We favor the hypothesis that synaptotagmin binds to the SNARE complex, perhaps in its loose trans configuration, contributing to its full zippering into the tight trans conformation. Others have suggested that the C2 domains oligomerize and form a bridge between the apposed membranes, in what would constitute a potentially direct role for synaptotagmin in fusion.[63,64]

Regulation of Synaptotagmin VI Activity by Phosphorylation

As I discussed earlier, the first calcium increase in the initial phase of the acrosome reaction activates phospholipase Cs, releasing IP_3 and diacylglycerol, a potent activator of several PKC isoforms. It has been proposed that PKC is a key component of the AE transduction pathway, although its role in the process is not yet completely understood.[65]

We have discovered that synaptotagmin VI is one of the substrates for PKC in sperm and that phosphorylation regulates its action on the acrosome reaction.[59,62] Both C2 domains interact with the endogenous machinery involved in exocytosis and these interactions appear to be disturbed by phosphorylation. A polybasic region on the fourth beta strand of the C2B domain is known as the effector domain since it has been implicated in the binding to the molecules enumerated above. This region contains two threonines that fit in a PKC phosphorylation consensus sequence. Phosphorylation in this polybasic region would strongly affect its overall charge, preventing interaction with effectors and rendering the motif inactive. In fact, mutations of these residues abolishing or mimicking phosphorylation influence the effect of recombinant C2B on the acrosome reaction.[62] Similar results were obtained by mutating a threonine in the polybasic region of the C2A domain. Using a specific antibody, we demonstrated that sperm synaptotagmin is phosphorylated in vivo on threonine 419 within the polybasic region of C2B. Phospho-synaptotagmin localizes to the acrosomal region of resting human sperm (Fig. 6A). The percentage of cells exhibiting this labelling diminishes upon triggering AE (in the presence of the permeant calcium chelator BAPTA acetoxymethyl ester [BAPTA-AM] to prevent membrane loss due to exocytosis). Because endogenous synaptotagmin must be dephosphorylated to be functional in exocytosis, we conclude that challenging sperm with acrosome reaction inducers must activate protein phosphatases that dephosphorylate synaptotagmin. Synaptotagmin dephosphorylation takes place after initiation of AE and prior to the efflux of calcium from the acrosome.[62] These findings are relevant for two reasons: first, they identify a substrate for PKC in sperm, the target residue phosphorylated, and the function of this post-translational modification. Second, very little was previously known about regulation of synaptotagmins by phosphorylation, and nothing about this particular site. Figure 6D depicts a model for PKC-mediated regulation of synaptotagmin activity in fusion.

Calmodulin

Several calcium binding proteins that lack C2 domains have also been implicated in granule exocytosis. Calmodulin binds calcium via its four EF-hand domains. Calcium-calmodulin has been invoked as a universal requirement for the membrane fusion process, based on observations from several constitutive membrane fusion events.[6,66] Calmodulin can function as a fast-responding calcium sensor when its concentrations are elevated to high micromolar levels.[66] It also has a well-established role in exocytosis in synapses through the activation of calmodulin-dependent protein kinase II.[67] Calmodulin has been suggested to exhibit a direct role in membrane fusion, through its binding to VAMP2 in vitro and in PC12 cells.[68,69] A late role for calmodulin has been reported in exocytosis of these and chromaffin cells.[66] Calmodulin is the sensor protein responding to calcium released from the vacuolar lumen in the first post-docking phase of yeast vacuolar fusion.[45] The V0 sector of the vacuolar-ATPase is a binding partner for calmodulin. Trans V0 complexes form subsequent to docking of vacuoles, and have been proposed to mature under the influence of calcium/calmodulin into the fusion pore itself.[70]

The presence of calmodulin in sperm has been known for many years and roles have been suggested in several functions (see the discussion in ref. 71). For instance, calmodulin antagonists inhibit the acrosome reaction in sea urchin,[72,73] mouse[74,75] and human[76] sperm. It has been suggested that they might do so by inhibiting T-type calcium currents. Accordingly, calmodulin antagonists prevent both the transient and sustained phases of cytosolic calcium increase elicited by progesterone.[76] If the only effect of calmodulin on AE were the modulation of calcium channels on the plasma membrane, we would not expect calcium ionophore-triggered AE to be affected by calmodulin antagonists. In fact, such antagonists no longer block A23187-stimulated AE, but rather stimulate it. Interestingly, in SLO-treated sperm (where all channel activity is nullified by permeabilization), these antagonists are able to elicit exocytosis by themselves. This exocytosis relies on a functional fusion machinery involving Rab and SNAREs. Consistent with these observations, calmodulin itself inhibits the acrosome reaction in permeabilized sperm.[76] In summary, it appears that calmodulin modulates more than one step during AE. In an early phase, it positively regulates the opening of calcium channels elicited by physiological inducers. Later on, calmodulin exhibits an inhibitory role. What its targets might be at this late phase is not yet known. We do know, however, that it is not Rab3A.[76,77]

Fusion Molecules/Stages: The Molecules and Mechanisms of Exocytosis

Calcium-triggered membrane fusion and exocytosis is widespread in neural, endocrine, exocrine, haemopoietic and perhaps all cells, where several distinct types of vesicles are employed. In all cases, membrane fusion is governed by conserved sets of proteins that are classified into a few protein families. These proteins include GTPases of the Rab family,[78,79] the sec1p/Munc-18 protein family,[80] and SNARE proteins.[20,21,57,81] The SNARE protein superfamily contains the syntaxin, SNAP-25 and synaptobrevin/VAMP protein families. Syntaxins and SNAP-25-like proteins were originally designated as t-SNAREs due to their localization on target membranes,

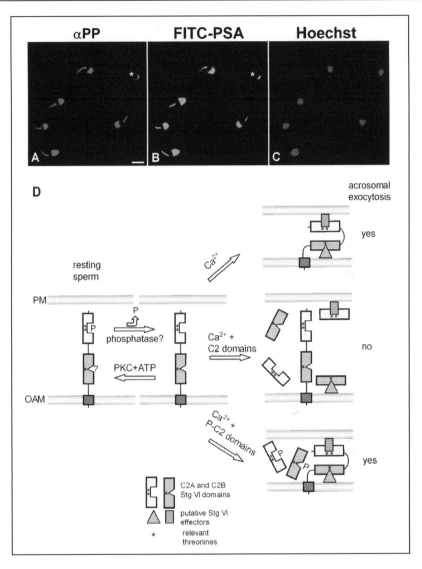

Figure 6. Human sperm synaptotagmin is phosphorylated in the polybasic C2B region. Sperm were triple stained with an anti-phosphosynaptotagmin antibody (A); FITC-PSA to differentiate between reacted and nonreacted sperm (B); and Hoechst 3258 to visualize all cells in the field (C). Scale bar = 5 μm. Phosphosynaptotagmin (PP) localizes to the acrosomal region, consistent with its role in AE. Accordingly, a spontaneously reacted sperm (asterisk in B) was negative for acrosomal PP staining (asterisk in A). D) model for PKC-mediated regulation of synaptotagmin VI C2 domains' interaction with putative membrane-bound effectors during AE. In resting sperm, synaptotagmin is phosphorylated and maintained in an inactive conformation. Sperm stimulation with calcium leads to dephosphorylation by a still unidentified phosphatase and substantial conformational changes that de-repress synaptotagmin's ability to interact with its effectors and accomplish exocytosis. Recombinant C2A and C2B domains compete with the endogenous synaptotagmin for the effectors, preventing AE. This effect is abolished by PKC catalyzed domain-phosphorylation. PM: plasma membrane; OAM: outer acrosomal membrane. (Reprinted from: Roggero CM et al. Dev Biol 285:422-435, ©2005, with permission from Elsevier.[62])

whereas the vesicular VAMPs were termed v-SNAREs. More recently, these proteins have been reclassified as R-SNAREs (arginine-containing SNAREs) or Q-SNAREs (glutamine-containing SNAREs), based on the identity of a highly conserved residue.[82] Syntaxin and SNAP-25 are Q-SNAREs whereas VAMP is an R-SNARE. The Q- and R-SNAREs form tight heterotrimeric complexes during the fusion process. These ternary complexes are highly resistant to denaturing reagents but undergo disassembly with the assistance of chaperone-like molecules known as NSF (an ATPase) and SNAPs (α, β and γ).[83] It is thought that the energy released during complex formation is transfered to the membranes via the proteins' membrane anchors and used to initiate membrane fusion by pulling the membranes together. In vitro, isolated sets of cognate SNARE proteins are able to drive liposome fusion.[84] "Flipped" SNAREs, whose topology is inverted so that the cytoplasmic domains are extracellularly oriented, accomplish cell-cell fusion between cells where these proteins are expressed.[85] These reports identify SNAREs as part of the minimal machinery for membrane fusion. Others, conversely, claim that SNAREs are dispensable for fusion.[86,87] Syntaxin and VAMP are anchored to the lipid bilayer by transmembrane regions found at their carboxy termini, whereas SNAP-25—which lacks a transmembrane region—is thought to be anchored by palmitoylation of a cysteine-rich sequence in the center of the molecule. Q-SNAREs and R-SNAREs contribute three and one helixes respectively to ternary complexes. When Q- and R-SNAREs reside on the same membrane, complexes are in a cis, fusion-incompetent, configuration. In contrast, when Q- and R-SNAREs reside on opposite membranes, complexes are in a trans, fusion-competent, configuration. It has been shown that SNARE proteins assembly begins at the amino terminus portion of the molecules progressing from partly (loose) to fully (tight) zippered complexes (see Fig. 1 in ref. 20).

In spite of what is thought to be a common set of molecular mechanisms involving Rab, Munc-18, synaptotagmin and SNARE proteins, modes of regulated vesicle exocytosis exhibit a great deal of variation in calcium sensitivity, rates of calcium-triggered fusion and latencies to fusion following a calcium trigger.[35] Furthermore, while there is consensus regarding exocytosis as a highly regulated, multistage process including targeting, priming, docking, and fusion of secretory vesicles with the plasma membrane, there has been an unfortunate divergence in terminology with respect to terms such as docking, priming and fusion. Confusion in the literature has arisen from the application of dissimilar standards to define membrane fusion steps.[67] Thus, seemingly opposite conclusions in terms of whether priming takes place before or after docking have arisen from the fact that morphological, functional or molecular criteria have been used to claim granules are or not docked (see ref. 6 and the discussion in ref. 88). Furthermore, continued neurotransmission requires cycles of exo- and endocytosis, so it can be difficult to define in such recycling systems exactly when postfusion becomes predocking. In general, the morphological distribution of vesicles and granules can be misleading, as this cannot easily be interpreted in relation to protein interactions known mainly from in vitro studies.

I summarize in Table 2 four of the current, prevalent exocytosis models and their molecular correlates. My intention is to compare—and if possible unify—models, rather than to present a comprehensive review of all the plausible mechanisms put forth by researchers in this field. Whether the models illustrate an actual divergence of mechanisms inherent in the transport steps, or simply of interpretations based on different measurement assays, or both, remains to be seen. In model I, docking is SNARE-independent (the docking machinery is either undefined or Rab3-related) and priming is a post-docking process involving assembly of trans SNARE complexes. Priming includes all the molecular rearrangements and ATP-dependent protein and lipid modifications that take place after initial docking of a secretory vesicle but before exocytosis, such that the influx of calcium is all that is needed to trigger nearly instantaneous release. In this context, priming converts docked vesicles into the slowly releasable pool (SRP) and later into the readily releasable pool (RRP). Both pools of vesicles exocytose, albeit with different kinetics and calcium sensitivities. Finally, α-SNAP/NSF display a post-fusion role, recycling the fusion machinery. A similar sequence of priming-docking-fusion-post-fusion (α-SNAP/NSF)—but without the prediction of two exocytotic pools—has been proposed for yeast exocytosis.[89] In model II SNAREs remain

Table 2. Fusion steps in secretory vesicle exocytosis and their proposed SNARE configuration correlates

Fusion Step	Loose, Reversible Attachment Between V and PM	Loose, Reversible Attachment Between V and PM	Tight Association Between V and PM	Tighter Association Between V and PM	Post Fusion	Ref.
SNAREs	*cis*	monomeric	loose *trans* complexes	tight *trans* complexes	*cis*	
Model I	–	docking	priming (1), SRP	priming (II), RRP	recycling NSF/α–SNAP	A
Model II	morphological docking	NSF/α-SNAP-sensitive	SRP	RRP (primed)		B
Model III (AE)	Rab	priming NSF/α-SNAP	docking	docking		C
Model IV		priming NSF/α-SNAP		tethering→docking Rab/SNAREs		D

A:35,143,155-158; B:114,159,160; C:18,51; D:48,86,90,161,162. Although Model IV as depicted is mainly derived from the yeast vacuolar fusion model, I maintained the mammalian fusion protein nomenclature for simplicity. SNARE proteins: VAMP in blue, syntaxin in red and SNAP-25 in green. A color version of this table is available online at http://www.Eurekah.com.

engaged in inactive cis complexes even after the vesicle is morphologically docked to the plasma membrane. Once again, priming follows docking, but here it involves the dissociation of cis SNARE complexes by α-SNAP/NSF, whereby these proteins exhibit a prefusion role. As we can see, the term "priming" means dissociation in Model II and exactly the opposite (association) in Model I. In Model IV, priming by α-SNAP/NSF might be concomitant with— or immediately precede—the complex process of tethering (loose membranes attachment) initiated by Rab, and takes place before trans SNARE mediated docking. As we will see in the rest of this chapter, the acrosome reaction fits nicely within model III. This model is molecularly similar to model II, but with a few differences, mainly in the nomenclature used to define each step. We favor the use of the term "tethering" to define the initial recognition and binding of the outer acrosomal and plasma membranes. By definition, this binding extends over a relatively long distance. May I point out, however, that we haven't yet developed a method to distinguish visually between tethered and nontethered acrosomes. Therefore, we employ a well established, molecular definition for tethering as being SNARE-independent, triggered by Rabs and accomplished through the recruiment of tethering factors.[90,91] Productive tethering is followed by membrane "docking"; a stronger—and spatially tighter—interaction of the two bilayers engaged in fusion. Again, we use a molecular, rather than a morphological, definition, whereby docking is achieved by trans SNARE pairing. We use the term "priming" to describe the activation of the fusion machinery. Priming includes—but is not limited to—the unpairing of cis SNARE complexes on both membranes by α-SNAP/NSF to make them available for productive, trans pairing and subsequent fusion. It follows from these definitions that both tethering and priming precede docking. Because in the acrosome reaction Rab3 functions prior to α-SNAP/NSF, we believe tethering—defined as a Rab3-elicited step and not necessarily as the complete physical association between membranes—precedes priming. Tethering, priming and docking take place before the second efflux of calcium from the acrosomal store. Models I and II derive from studies conducted in cells with different pools of secretory vesicles, such as neurons, PC12 and chromaffin cells. Models III and IV, better fit data obtained in systems where vesicle populations are thought to be homogeneous, such as sperm and yeast vacuoles. Model III is, nevertheless, favored by some authors to explain the role of α-SNAP/NSF in secretory granule exocytosis in chromaffin and PC12 cells.[18]

Rab3A

Rabs are small GTP-binding proteins that behave as molecular switches by cycling between inactive (GDP-bound, in the cytosol) and active (GTP-bound, membrane-associated) states.[79] Before fusion, the transport intermediate or secretory vesicle needs to recognize and become attached to the target membrane. As we have seen before, in addition to coordinating several sequential events during vesicular trafficking, Rabs are responsible for the recognition and physical attachment of the membranes that are going to fuse.[90,92] This association, or tethering, represents one of the earliest known events in membrane fusion. The concept emerging from numerous studies is that a large fibrous protein or protein complex is recruited by Rabs to different intracellular sites where they establish contact between cognate membranes. Tethering factors contain large coiled-coiled domains. It has been suggested that they may direct assembly of SNARE complexes at fusion sites.[93-95] It is through these tethering factor complexes that Rab3 has been implicated in post-tethering steps.[96]

Rab3 participates in regulated exocytosis of neurotransmitters and hormones.[79,97] One of the four isoforms, Rab3A, is present in the acrosomal region of human,[22] rat,[29] and mouse[98] sperm. Rab3A has been reported to stimulate human[22] and ram,[28] and to inhibit rat[29] sperm acrosome reaction. The Rab3A specific tool used in the latter two papers was a synthetic peptide encompasing the effector domain. We have used an identical peptide and also the full length, in vitro prenylated, and persistently activated with GTP-γ-S recombinant Rab3A (referred to as "Rab3A" from now on) to analyze the effects of this small GTPase on AE. Rab3A triggers sperm exocytosis to a magnitude similar to calcium. Predictably, extraction of the endogenous protein by GDI,[22,99] or blocking with a specific antibody[51,100] prevent AE. Rab3A does not require cytosolic calcium to elicit the acrosome reaction.[49]

Capacitation leads to an enhanced association of Rab3A with sperm membranes, presumably correlating with its activation in preparation for the acrosome reaction. Interestingly, acute removal of cholesterol from sperm membranes enhances both calcium-triggered exocytosis and Rab3A targeting to membranes.[100] When AE is elicited by calcium, Rab3A is activated, supposedly by boosting the exchange of GTP for GDP and the dissociation from GDI.[99]

Where in the fusion cascade involved in the acrosome reaction does sperm Rab3A play its part? When elicited by exogenous Rab3A, exocytosis depends on α-SNAP/NSF,[101] SNAREs,[102] synaptotagmin VI,[59] and an efflux of calcium from inside the acrosome,[49] coinciding with the "post-SOC-pre-second acrosomal calcium release" role described above. We believe Rab3A initiates the signalling cascade that will culminate in the tethering between the outer acrosomal and plasma membranes, and the priming of the fusion machinery by α-SNAP/NSF. Whether the actual initial, reversible, physical contact between these membranes happens before, or concomitantly with, cis-SNARE disassembly is something we do not yet know.

α-SNAP/NSF

SNAREs assemble into tightly packed helical bundles, the core complexes.[103] Disassembly requires the concerted action of the adaptor α-SNAP (Sec 17 in yeast) and the hexameric ATPase NSF (Sec 18 in yeast), whereby NSF uses energy from ATP hydrolysis to dissociate itself and SNAP from the SNAREs and to disassemble the core complexes. Interestingly, α-SNAP and NSF are not compartment-specific but participate in both constitutive secretion and in regulated exocytosis.[104] In the original SNARE hypothesis, disassembly of SNARE complexes would lead to membrane fusion.[105] That model has been extensively revised, and we know now that α-SNAP/NSF dissociate the cis—but not the trans[51,106]—SNARE complex. Sec 18, on the contrary, appears to disassemble both.[86] As I have discussed above, defining when SNARE complexes dissociate has proven difficult and controversial. Dissociation of the cis-SNARE complex by α-SNAP/NSF has been reported to occur at a late, post-fusion step to recycle the fusion machinery accumulating in the plasma membrane as a result of vesicle fusion. Alternatively, an early, prefusion, priming role for α-SNAP/NSF is strongly supported by evidence from studies in a variety of models (reviewed in refs. 20,57).

NSF has been detected in human, mouse, rat, bovine, murine and rhesus macaque sperm. Indirect immunofluorescence experiments revealed that it localizes to the acrosomal region.[61,99,107] Not surprisingly, α-SNAP exhibits a similar localization in human sperm[101] and mouse round spermatids.[108] Our knowledge that the acrosome reaction relies on NSF derives from its sensitivity to dominant mutants that cannot bind or hydrolyze ATP[99] and on the use of specific, inhibitory antibodies.[101] Likewise, an excess recombinant protein as well as antibodies to α-SNAP prevent the onset of AE.[101] These proteins display a relatively early role, prior to the second efflux of calcium from the acrosome.[101] We believe the step catalysed by α-SNAP/NSF constitutes the priming of the fusion machinery, which is coupled to the tethering between the outer acrosomal and plasma membranes initiated by Rab3A. As I have stated above, priming includes the dissociation of cis SNARE complexes on sperm membranes. Interestingly, the activity of sperm α-SNAP/NSF is not constitutive and appears to be completely dormant in resting cells. Upon sperm stimulation by calcium or Rab3A, the activity is induced and cis SNARE complexes disassembled.

SNAREs

SNAREs are required in multiple fusion events mandatory for cell survival even under resting conditions. The notion that VAMP, syntaxin and SNAP-25 have a direct function in exocytosis has received strong support from the identification of these proteins as the targets of clostridial neurotoxins. Tetanus toxin (TeTx) and seven structurally related botulinum neurotoxins (BoNT/A, B, C1, D, E, F and G) are potent inhibitors of secretory vesicle release due to their highly specific, zinc-dependent, proteolytic cleavage of SNARE proteins. BoNT/A and E cleave SNAP-25, BoNT/C cleaves syntaxin and, with much lower efficiency, SNAP-25. The remaining BoNTs, as well as TeTx, are specific for VAMP.[109] All SNAREs are sensitive to cleavage by neurotoxins only when not packed in tight heterotrimeric complexes.[110] The ratio of monomeric to assembled SNAREs depends on the type and physiological condition of the cell. Thus, while some studies indicate that most SNAREs are free in the plasma membrane,[111] others suggest that they are engaged in complexes.[112,113] Whatever the steady state configuration of SNAREs in neuroendocrine cells might be, exocytosis is blocked by neurotoxins, suggesting that SNAREs go through toxin-sensitive stages.[114,115]

Published data on the presence of members of the SNARE complex in sperm comprise all three protein homologues in sea urchin,[116-118] and mammals.[24,30,51,61,102,119] As I will detail below, we have demonstrated a requirement for all three members of the SNARE complex in the AR following several different approaches. We have also demonstrated that the onset of sperm's exocytosis relies on the productive assembly of ternary SNARE complexes. The first line of evidence comes from the use of specific antibodies, which prevent the acrosome reaction.

The second line of evidence comes from the use of BoNTs. Treatment with BoNT/A, E, F, B and C and TeTx resulted in a zinc-dependent inhibition of acrosomal release, indicating a need for toxin-sensitive members of all three SNARE families in the acrosome reaction, regardless of the stimulus applied.[51,102] Ours constitutes the first piece of evidence that a Rab-promoted fusion event can be effectively blocked by specific neurotoxins attacking SNAREs.

Dynamics of SNARE Assembly and Disassembly

SNARE Configuration in Resting Sperm

The dynamics of SNARE assembly and disassembly during membrane recognition and fusion is a central issue in intracellular trafficking, and much effort has been devoted to unraveling the mechanisms underlying regulated exocytosis. While a variety of critical observations have been made regarding microscopic, functional, electrophysiological, and biophysical processes, only a few have been reduced to a molecular explanation. This is mainly due to a paucity of integrated physiological models that has so far restricted our ability to assign molecular correlates to each stage of the fusion cascade. Regulated exocytosis in human sperm is a privileged model in terms of permitting the direct inspection—in great detail and with high resolution—of the dynamics of individual molecular species during exocytosis in a near in vivo configuration.

As I have already discussed, α-SNAP/NSF are required for the acrosome reaction. Since they only bind and dissociate cis SNARE complexes, their requirement suggested to us that sperm SNAREs might be engaged in such a configuration (remember that cis complexes are resistant to neurotoxin cleavage).[110] To test the veracity of this hypothesis, permeabilized sperm were incubated with the light chains of BoNT/E, BoNT/B, or TeTx. After incubation, toxins were inactivated by chelating zinc with N,N,N',N'-tetrakis (2-pyridymethyl) ethylenediamine (TPEN). AE was subsequently stimulated with calcium and secretion assessed. Although all three toxins are able to block exocytosis, they do not do so when zinc is chelated before sperm stimulation (Fig. 7). Because neurotoxins are proteases, their effect on sperm SNAREs is reflected in immunostaining. Immunolabelling with an anti-syntaxin1A antibody results in a clear immunolabelling in the acrosomal region of most cells (Fig. 8A). A similar pattern is revealed with an anti-VAMP2 antibody (Fig. 9A). Addition of BoNT/C does not decrease the immunofluorescence labelling, indicating that syntaxin is protected from toxin cleavage in resting spermatozoa (Fig. 8D). Likewise, VAMP2 is not susceptible to cleavage by BoNT/B (Fig. 9D) or TeTx (Fig. 9M). To verify that sperm activation triggers α-SNAP/NSF-catalyzed SNARE disassembly in a step prior to the calcium efflux from the intra-acrosomal pool, this store was depleted with BAPTA-AM, and sperm stimulated with extracellularly-applied calcium. The percentage of sperm with acrosomal syntaxin labelling decreased significantly, indicating that the protein became BoNT/C-sensitive after sperm activation (Fig. 8G, asterisks indicate sperm with intact acrosomes not stained for syntaxin1A). Likewise, VAMP2 is sensitized to BoNT/B (Fig. 9G, asterisks indicate sperm with intact acrosomes not stained for VAMP2) and TeTx (Fig. 9P, asterisks) when added prior to challenging with calcium. These data suggest that both R- and Q-SNAREs are protected from toxin cleavage in resting cells, instead of cycling through toxin-resistant and toxin-sensitive configurations as proposed in all other fusion systems. A corollary of this observation is that α-SNAP/NSF activity is not constitutive but rather induced by sperm stimulation. Two post-translational modifications of NSF affecting its activity on SNARE complex disassembly have recently been described.[120,121] Whether these or other mechanisms are involved in the regulation of NSF funcion on AE remains to be elucidated.

SNARE Configuration following Sperm Activation

As we have just discussed, cis SNARE complexes disassemble and pass through a toxin-sensitive configuration upon sperm stimulation. We also know that SNAREs are not protected before the second efflux of intra-acrosomal calcium;[51] hence only two configurations

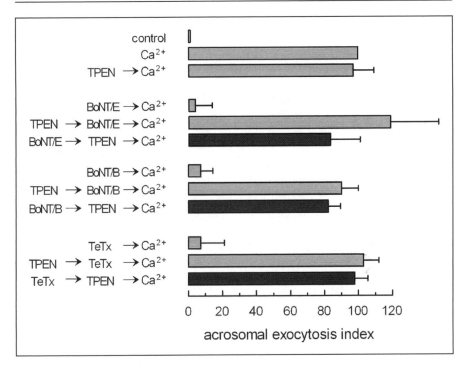

Figure 7. SNAREs are assembled in neurotoxin-resistant complexes in resting sperm. Exocytosis in permeabilized spermatozoa was abolished by the recombinant light chains of BoNT/E, BoNT/B and TeTx (neurotoxin→Ca^{2+}), reflecting the dependence of exocytosis on neurotoxin-sensitive Q- and R-SNARE proteins. Toxins were inactive when added in the presence of TPEN (TPEN→neurotoxin→Ca^{2+}). Surprisingly, they were also unable to cleave SNAREs prior to sperm stimulation, even when preincubated long enough to cut (neurotoxin→TPEN→Ca^{2+}, black bars). This means that VAMP and SNAP-25 are in a toxin-resistant configuration in resting sperm, and that challenging with calcium makes them transit through toxin-sensitive conformations. AE was assessed with FITC-PSA, data normalized with respect to the positive control (Ca^{2+}), to which we assign the 100% value, and expressed as acrosomal exocytosis index. (Figure reproduced from: De Blas GA et al. PLoS Biol 2005; 3:e323;[51] this is an open access article. ©2005 De Blas et al.)

are possible. Either they remain as monomers or reassemble partially in loose trans complexes. Due to the high propensity of SNAREs to form unproductive cis complexes in vitro, it is not possible to directly study the nature of the partial or total trans complexes involved in exocytosis by common biochemical methods such as immunoprecipitation or SDS-PAGE analysis.[122] In this regard, differential sensitivity to BoNT/B and TeTx constitutes an independent approach to distinguish between monomeric VAMP and that engaged in loose trans complexes. These toxins cleave the same peptide bond, exposed in both configurations.[123] Interestingly, TeTx binds to the N-terminal whereas BoNT/B binds the C-terminal portions of VAMP coil domain. Since SNARE complex assembly begins at the N-terminus, the TeTx-recognition site is hidden in loose SNARE complexes while the BoNT/B recognition site is exposed. In other words, TeTx can only cleave monomeric VAMP while BoNT/B also cuts VAMP loosely assembled in SNARE complexes.[124] BoNT/B—but not TeTx—is capable of inhibiting AE when the system is stimulated with calcium or Rab3A and allowed to reach the intra-acrosomal calcium-sensitive step.[51] Therefore we can establish that: (i) SNAREs are in a loose trans configuration before the second intra-acrosomal calcium release; and (ii) assembly of trans SNARE complexes is calcium independent, unlike what has been described for exocytosis Models I and

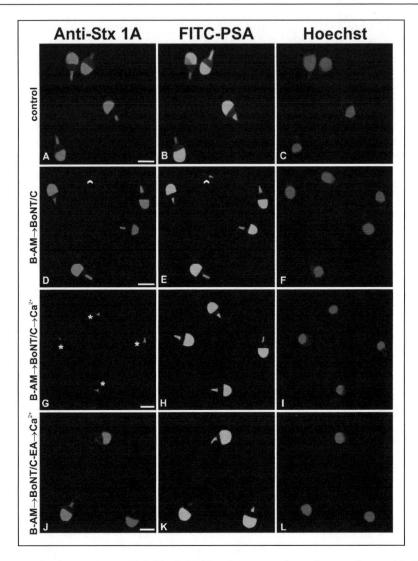

Figure 8. Syntaxin1A is assembled in BoNT/C-resistant complexes that are disassembled by sperm activation. Sperm were incubated with the recombinant light chain of BoNT/C. The cells were then fixed and triple stained with an anti-syntaxin1A antibody (that recognizes an epitope that is trimmed by the toxin, red, left panels), FITC-PSA (to differentiate between reacted and intact sperm, green, central panels), and Hoechst 33258 (to visualize all cells in the field, blue, right panels). Notice the acrosomal staining revealed by the anti-syntaxin antibody. This pattern is expected for a protein that participates in sperm exocytosis and is not observed in spontaneously reacted sperm (arrowheads in D and E) because they have lost the membranes where syntaxin is anchored. BoNT/C has no effect on resting sperm (compare panels A-C with D-F). However, labelling in sperm stimulated with 10 μM calcium (in the presence of BAPTA-AM to prevent exocytosis; observe that PSA staining is not affected) was significantly reduced by the toxin (asterisks, panel G). In contrast, the same experimental condition in the presence of a protease-inactive toxin (BoNT/C-EA) had no effect (panels J-L). Bars = 5 μm. (Figure reproduced from: De Blas GA et al. PLoS Biol 2005; 3:e323;[51] this is an open access article. ©2005 De Blas et al.) A color version of this figure is available online at http://www.Eurekah.com.

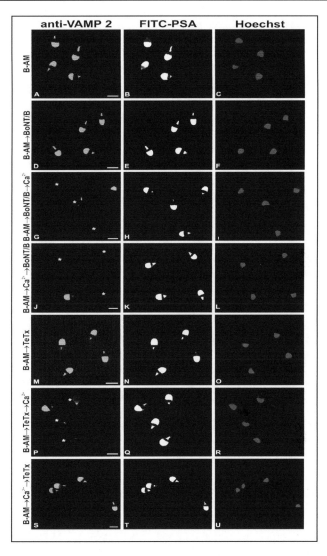

Figure 9. VAMP2 is engaged in loose trans complexes after sperm stimulation and before the second efflux of intra-acrosomal calcium. Sperm were incubated with the recombinant light chain of BoNT/B or TeTx. BAPTA-AM was used in all cases to chelate intra-acrosomal calcium and prevent the VAMP2-containing membrane loss inherent to the acrosome reaction (monitored in central panels). The cells were then triple stained with an anti-VAMP2 antibody (notice the acrosomal pattern, red, left panels), FITC-PSA (to differentiate between reacted and intact sperm, green, central panels), and Hoechst 33258 (to visualize all cells in the field, blue, right panels). BoNT/B and TeTx had no effect on resting sperm (compare panels D-F and M-O with A-C). However, labelling in sperm stimulated with calcium was significantly reduced by the toxins (asterisks, panels G and P). In contrast, when cells were first allowed to arrive at the intra-acrosomal calcium sensitive step and then treated with toxins, only BoNT/B caused a significant decrease in the VAMP2 labeling (compare panels J-L with panels S-U). Bars = 5 μm. (Figure reproduced from: De Blas GA et al. PLoS Biol 2005; 3:e323;[51] this is an open access article. ©2005 De Blas et al.) A color version of this figure is available online at http://www.Eurekah.com.

II (Table 2). As shown in Figure 8, immunofluorescence is a powerful technique to monitor SNARE assembly status as reflected by their sensitivity to cleavage by neurotoxins. Differential sensitivity of VAMP2 to TeTx and BoNT/B when engaged in loose complexes shows—by VAMP2 immunostaining—that these complexes form after sperm stimulation. As I have discussed earlier, a significant decrease in the percentage of cells exhibiting VAMP2 acrosomal staining is observed when toxin-loaded sperm are challenged with calcium (asterisks in Fig. 9G and P). In contrast, when toxins are added after calcium (i.e., after AE has progressed to the intra-acrosomal calcium-sensitive step), VAMP2 labelling is attenuated by BoNT/B (Fig. 9J, asterisks) but not by TeTx (Fig. 9S). This pattern of VAMP2 sensitivity to BoNT/B coupled to resistance to TeTx implies its engagement in loose trans SNARE complexes. To summarize, functional[51] and immunofluorescence (Fig. 9) data demonstrate that, following sperm stimulation, SNAREs are engaged in loose trans complexes awaiting the release of intra-acrosomal calcium that will trigger the final steps of AE. In the yeast vacuolar fusion model, docking by trans SNARE pairing produces an unknown signal leading to the transient opening of calcium channels on the vacuolar membrane. The resulting calcium efflux activates targets that catalyze bilayer mixing.[45,46] In sperm, on the contrary, productive trans SNARE pairing does not appear to be a prerequisite for intra-acrosomal calcium efflux, given that Rab3A is able to mobilize intravesicular calcium in cells treated with BoNT/E.[49] We are currently devoting our efforts to elucidate what—other than mere docking—is the signal that opens the IP_3-sensitive calcium channels on the outer acrosomal membrane to give origin to the second efflux.

Model

Model III applied to AE is depicted in Figure 10 where, for simplicity, only SNAREs are shown. Initially, SNAREs are locked in inactive cis complexes on plasma and outer acrosomal membranes. Rab3A is activated upon calcium entrance into the cytoplasm, initiating the tethering of the acrosome to the plasma membrane and the activation of α-SNAP/NSF, which disassemble cis SNARE complexes on both membranes. Monomeric SNAREs are free to assemble in loose trans complexes, causing the irreversible docking of the acrosome to the plasma membrane. At this point, calcium is released from inside the acrosome through IP_3-sensitive calcium channels to trigger the final steps of membrane fusion, which require SNAREs (presumably in tight trans complexes) and synaptotagmin.

Signalling Cascades

I have so far described the factors that induce the acrosome reaction and those that accomplish it in the end. In between, there are a number of intracellular signalling molecules that carry the message from the inducer bound to the sperm surface to the membranes that ultimately fuse. These include several signal-transducing components, second messengers, G proteins, phospholipases, lipids, protein kinases and phosphatases, etc. The subject has been recently covered in the literature.[3,5,40,111,125-132] Since signalling pathways studies are, for the most part, not concerned with membrane fusion, I will not consider them further in this chapter, with the following two exceptions.

Protein Kinases and Phosphatases

Protein phosphorylation is a post-translational modification that allows the cell to control various cellular processes. The phosphorylation state of phosphoproteins is controlled by the activity of protein kinases and phosphatases. Mature spermatozoa have their own idiosyncrasies as highly specialized cells: they are highly compartmentalized, transcriptionally inactive, and unable to synthesize new proteins. Therefore, their reliance on protein phosphorylation as a means of modulating their function is greater than in other cells.[133] For instance, we have shown regulation of synaptotagmin function by PKC-mediated phosphorylation on threonine residues within the C2A and C2B domains. Western blotting of sperm populations during capacitation and/or after exposure to AE inducers has been widely used in the reproductive biology field, giving rise to a catalogue of phosphorylated proteins that is unfortunately not

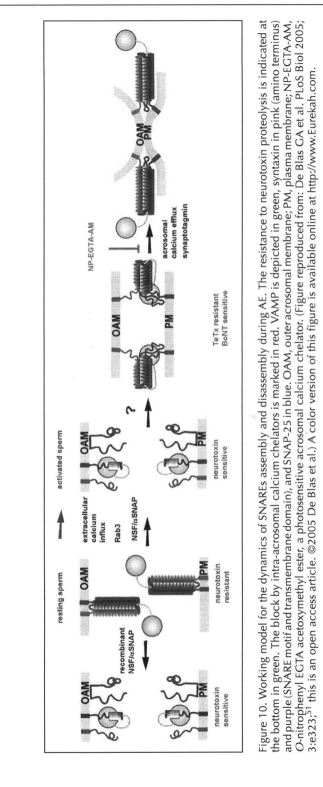

Figure 10. Working model for the dynamics of SNAREs assembly and disassembly during AE. The resistance to neurotoxin proteolysis is indicated at the bottom in green. The block by intra-acrosomal calcium chelators is marked in red. VAMP is depicted in green, syntaxin in pink (amino terminus) and purple (SNARE motif and transmembrane domain), and SNAP-25 in blue. OAM, outer acrosomal membrane; PM, plasma membrane; NP-EGTA-AM, O-nitrophenyl EGTA acetoxymethyl ester, a photosensitive acrosomal calcium chelator. (Figure reproduced from: De Blas GA et al. PLoS Biol 2005; 3:e323;[51] this is an open access article. ©2005 De Blas et al.) A color version of this figure is available online at http://www.Eurekah.com.

very reproducible and, with a few exceptions, conveys little information about the identity or function of these proteins.

Protein tyrosine phosphorylation had been implicated in mammalian sperm AE since treatment with inhibitors can block zona pellucida-[134,135] and progesterone-[136-139] induced exocytosis. As I have already discussed, this approach is not ideally suited to distinguish between changes ocurring specifically during AE, and not during capacitation. We reasoned that if this post-translational modification plays a role during the acrosome reaction, then changes in tyrosine phosphorylation when we stimulate sperm to undergo exocytosis shoud be easily detected. Indeed, challenging sperm with calcium doubles the percentage of cells depicting phosphotyrosine staining on the head (Fig. 11), indicating that the initiation of signalling cascades leading to AE induces tyrosine kinase activity in sperm.[52] Head tyrosine phosphorylation takes place in two waves, the first shortly after cytoplasmic calcium increase and the second in a later step, following the second intra-acrosomal calcium efflux. Our functional assay shows that calcium- and Rab3A-induced protein tyrosine phosphorylation is essential for human sperm AE, and again, that there are late acting kinases on whose activity exocytosis depends. Sperm must contain kinases that presumably require a specific microenvironment for optimal activity, with high and/or localized calcium concentrations achieved upon efflux of intra-acrosomal calcium. Alternatively, a component activated by intra-acrosomal calcium release, instead of calcium itself, could be responsible for the activation of these late acting kinases.[52] Is this a signature of sperm's exocytotic cascade or does it represent a widespread, and as of yet unexplored, phenomenon in membrane fusion?

Although there has been progress in the last decade in dissecting tyrosine phosphorylation-related signaling pathways that are crucial for sperm function, surprisingly little attention has been paid to the dynamics of the process. Protein tyrosine phosphatase activity is of particular relevance since it can potentially control the strength of phosphotyrosine-related pathways. So far, the only protein tyrosine phosphatase described in sperm is the nontransmembrane PTP1B.[52] Inhibition of tyrosine phosphatase activity blocks AE and enhances tyrosine phosphorylation during the restricted time frame when sperm are primed to undergo the AR.[52] In summary, it would appear that sperm possess one or more active tyrosine kinases and that phosphoprotein accumulation is normally prevented by vigorous concomitant phosphatase activity. The fact that both tyrosine kinase and phosphatase activities are necessary for acrosomal exocytosis is intriguing. One possible explanation is that one or more tyrosine residues on the proteins required for exocytosis need to cycle between a phosphorylated and a dephosphorylated states to function properly. Alternatively, certain tyrosine residues on protein members of the fusion machinery might need to be phosphorylated while others might need to be dephosphorylated to elicit exocytosis. Interestingly, serine/threonine - but not tyrosine - dephosphorylation have been shown to participate in the late phases of membrane fusion in yeast.[140-142]

cAMP/Epac

In many cell types, an increase in intracellular cyclic adenosine 3', 5'-monophosphate (cAMP) concentration regulates calcium-triggered exocytosis.[143,144] In most cells, however, an elevation of cAMP alone in the absence of a calcium rise is not sufficient to trigger exocytosis.[6] Yet, a limited range of cells use cAMP as a major trigger for exocytosis. The cAMP dependent pathways coexist with calcium-dependent ones for exocytosis in these cells, and it is likely that they use a common final SNARE-dependent mechanism. cAMP-triggered exocytosis in parotid gland—one of the best studied—requires PKA and VAMP2. The involvement of other SNAREs or other components of the general fusion machinery in cAMP-triggered exocytosis has not, until now, been investigated,[6] with the exception of the pathway cAMP-Epac2-Rim-Rab3-calcium sensor.[145-148]

In sperm, various activation pathways required to achieve egg-fertilizing ability depend on the intracellular rise of cAMP. Of particular interest here are those concerned with the acrosome

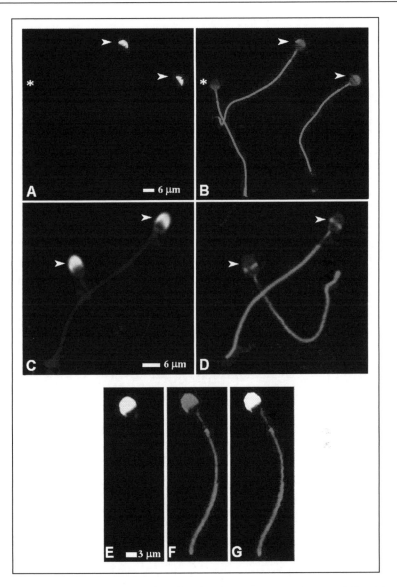

Figure 11. Localization of tyrosine phosphorylated proteins in SLO-permeabilized human sperm. After permeabilization with SLO, human sperm samples were fixed, permeabilized and dual labelled with an anti-phosphotyrosine antibody (red staining, B, D, F) and FITC-PSA (green staining, A, C, E). Two labeling patterns are detected on the sperm head, acrosomal (F and arrowheads in B) and equatorial (arrowheads in D). All sperm present strong labeling in the principal piece of the flagellum. Only acrosome-intact cells display tyrosine phosphory-lation in the acrosomal region or in the equatorial segment, whereas acrosome reacted sperm (asterisk in A) do not (asterisk in B). The yellow acrosomal signal in panel G indicates overlap between the green and red signals seen in panels E and F. Shown are fluorescence micrographs recorded with a confocal microscope. (Reprinted from: Tomes CN et al. Dev Biol 265:399-415, ©2003, with permission from Elsevier.[52]) A color version of this figure is available online at http://www.Eurekah.com.

reaction.[126,149] It has been known for some time that reagents that increase intracellular cAMP levels induce exocytosis in mammalian sperm.[150] One of the mechanisms proposed to explain these observations involved PKA activation by cAMP, leading to the opening of calcium channels.[126] We have recently unveiled an alternative mechanism—not related to PKA-induced opening of calcium channels—by which cAMP triggers exocytosis in human sperm. This cAMP-dependent step is located downstream of the influx of calcium into the cytosol through SOC channels and is independent of PKA, suggesting that another cAMP target is involved. We have identified Epac, a small guanine nucleotide exchange factor for the small GTPase Rap,[151,152] as the cAMP target responsible for all its post-SOC opening effects in sperm exocytosis. Like calcium, cAMP—via Epac activation—has the ability to drive the whole cascade of events necessary to bring exocytosis to completion, including tethering and docking of the acrosome to the plasma membrane, priming of the fusion machinery, mobilization of intravesicular calcium, and ultimately, bilayer mixing and fusion. In fact, AE elicited by cAMP proceeds in the virtual absence of cytosolic calcium, but all the effects of calcium on the acrosome reaction appear to be mediated by cAMP/Epac.[153,154]

Concluding Remarks

Sperm are attractive cells. Their study has drawn the interest of researchers from all over the globe for decades. Initially came the description and cataloguing of the overall shape and properties of these cells in many different species, together with their genesis and development in the testis. Later on, the study of the acrosome reaction took off owing to ultrastructural analysis that achieved exquisite levels of detail. Those early studies described and characterized the acrosome reaction as we know it today from a morphological point of view. Biochemical analysis ensued, identifying ligands and signalling pathways whose end point was exocytosis. Somehow, the unveiling of the molecular mechanisms involved in membrane fusion itself lagged behind all this progress. This picture has changed dramatically in the last few years, due to an explosion in our knowledge of the many proteins required for exocytosis and its regulation and the discovery that very similar versions of these proteins play the same roles in virtually all membrane fusion models. Sperm are not the exception to this rule.

The universality of mechanisms underlying exocytosis has greatly simplified our thinking and means that these need not be studied in detail in all cell types, but future work can concentrate on the analysis of favorable secretory cell models. For instance, researchers are very actively searching for ways to manipulate exocytosis—to regulate the insulin-producing beta cells of the pancreas to prevent diabetes or to get neurotransmitters released in the brain at the right time or concentration. AE is a relatively simple model to address these issues in addition to its inherent role in regulation of fertility.

Aknowledgements

Thanks to Luis Mayorga for helpful discussions, critical reading of this manuscript and assistance preparing the figures. Thanks also to Sean Patterson for critical reading of this chapter and valuable suggestions. I have cited reviews when possible; I apologize to those whose work I was not able to cite owing to space restrictions.

References

1. Eddy EM, O'Brien DA. The Spermatozoon. In: Knobil E, Neill JD, eds. The Physiology of Reproduction. 2nd ed. New York: Raven Press Ltd., 1994:29-77.
2. Jones R. Sperm survival versus degradation in the mammalian epididymis: A hypothesis. Biol Reprod 2004; 71:1405-1411.
3. Evans JP, Florman HM. The state of the union: The cell biology of fertilization. Nat Cell Biol 2002; 4(Suppl):s57-s63.
4. Wassarman PM. Mammalian fertilization: Molecular aspects of gamete adhesion, exocytosis, and fusion. Cell 1999; 96:175-183.
5. Yanagimachi R. Mammalian fertilization. In: Knobil E, Neill JD, eds. The Physiology of Reproduction. 2nd ed. New York: Raven Press, 1994:189-317.

6. Burgoyne RD, Morgan A. Secretory granule exocytosis. Physiol Rev 2003; 83:581-632.
7. Yoshinaga K, Toshimori K. Organization and modifications of sperm acrosomal molecules during spermatogenesis and epididymal maturation. Microsc Res Tech 2003; 61:39-45.
8. Kim KS, Foster JA, Gerton GL. Differential release of guinea pig sperm acrosomal components during exocytosis. Biol Reprod 2001; 64:148-156.
9. Yudin AI, Gottlieb W, Meizel S. Ultrastructural studies of the early events of the human sperm acrosome reaction as initiated by human follicular fluid. Gamete Res 1988; 20:11-24.
10. Storey BT. Interactions between gametes leading to fertilization: The sperm's eye view. Reprod Fertil Dev 1995; 7:927-942.
11. McLeskey SB, Dowds C, Carballada R et al. Molecules involved in mammalian sperm-egg interaction. Int Rev Cytol 1998; 177:57-113.
12. Kerr CL, Hanna WF, Shaper JH et al. Characterization of zona pellucida glycoprotein 3 (ZP3) and ZP2 binding sites on acrosome-intact mouse sperm. Biol Reprod 2002; 66:1585-1595.
13. Snell WJ, White JM. The molecules of mammalian fertilization. Cell 1996; 85:629-637.
14. Leyton L, Saling P. Evidence that aggregation of mouse sperm receptors by ZP3 triggers the acrosome reaction. J Cell Biol 1989; 108:2163-2168.
15. Bedford JM. Enigmas of mammalian gamete form and function. Biol Rev Camb Philos Soc 2004; 79:429-460.
16. Contreras HR, Llanos MN. Detection of progesterone receptors in human spermatozoa and their correlation with morphological and functional properties. International Journal of Andrology 2001; 24:246-252.
17. Bedford JM. Mammalian fertilization misread? Sperm penetration of the eutherian zona pellucida is unlikely to be a lytic event. Biol Reprod 1998; 59:1275-1287.
18. Burgoyne RD, Morgan A. Analysis of regulated exocytosis in adrenal chromaffin cells: Insights into NSF/SNAP/SNARE function. Bioessays 1998; 20:328-335.
19. Jahn R, Sudhof TC. Membrane fusion and exocytosis. Annu Rev Biochem 1999; 68:863-911.
20. Sollner TH. Regulated exocytosis and SNARE function. Mol Membr Biol 2003; 20:209-220.
21. Jahn R, Lang T, Sudhof TC. Membrane fusion. Cell 2003; 112:519-533.
22. Yunes R, Michaut M, Tomes C et al. Rab3A triggers the acrosome reaction in permeabilized human spermatozoa. Biol Reprod 2000; 62:1084-1089.
23. Diaz A, Dominguez I, Fornes MW et al. Acrosome content release in streptolysin O permeabilized mouse spermatozoa. Andrologia 1996; 28:21-26.
24. Hutt DM, Baltz JM, Ngsee JK. Synaptotagmin VI and VIII and syntaxin 2 are essential for the mouse sperm acrosome reaction. J Biol Chem 2005; 280:20197-20203.
25. Hernandez-Gonzalez EO, Lecona-Valera AN, Escobar-Herrera J et al. Involvement of an F-actin skeleton on the acrosome reaction in guinea pig spermatozoa. Cell Motil Cytoskeleton 2000; 46:43-58.
26. Kitamura K, Tanaka H, Nishimune Y. Haprin, a novel haploid germ cell-specific RING finger protein involved in the acrosome reaction. J Biol Chem 2003; 278:44417-44423.
27. Johnson LR, Moss SB, Gerton GL. Maintenance of motility in mouse sperm permeabilized with streptolysin O. Biol Reprod 1999; 60:683-690.
28. Garde J, Roldan ERS. Rab3-peptide stimulates exocytosis of the ram sperm acrosome via interaction with cyclic AMP and phospholipase A_2 metabolites. FEBS Lett 1996; 391:263-268.
29. Iida H, Yoshinaga Y, Tanaka S et al. Identification of Rab3A GTPase as an acrosome-associated small GTP-binding protein in rat sperm. Dev Biol 1999; 211:144-155.
30. Ramalho-Santos J, Moreno RD, Sutovsky P et al. SNAREs in mammalian sperm: Possible implications for fertilization. Dev Biol 2000; 223:54-69.
31. Clapham DE. Calcium signaling. Cell 1995; 80:259-268.
32. Berridge MJ, Lipp P, Bootman MD. The versatility and universality of calcium signalling. Nat Rev Mol Cell Biol 2000; 1:11-21.
33. Berridge MJ, Lipp P, Bootman MD. Signal transduction. The calcium entry pas de deux. Science 2000; 287:1604-1605.
34. Nowycky MC, Thomas AP. Intracellular calcium signaling. J Cell Sci 2002; 115:3715-3716.
35. Martin TF. Tuning exocytosis for speed: Fast and slow modes. Biochim Biophys Acta 2003; 1641:157-165.
36. Neher E. Vesicle pools and Ca^{2+} microdomains: New tools for understanding their roles in neurotransmitter release. Neuron 1998; 20:389-399.
37. Florman HM, Arnoult C, Kazam IG et al. A perspective on the control of mammalian fertilization by egg- activated ion channels in sperm: A tale of two channels. Biol Reprod 1998; 59:12-16.
38. Kirkman-Brown JC, Punt EL, Barratt CL et al. Zona pellucida and progesterone-induced Ca^{2+} signaling and acrosome reaction in human spermatozoa. J Androl 2002; 23:306-315.
39. Trevino CL, Serrano CJ, Beltran C et al. Identification of mouse trp homologs and lipid rafts from spermatogenic cells and sperm. FEBS Lett 2001; 509:119-125.
40. Felix R. Molecular physiology and pathology of Ca^{2+}-conducting channels in the plasma membrane of mammalian sperm. Reproduction 2005; 129:251-262.

41. Darszon A, Nishigaki T, Wood C et al. Calcium channels and Ca^{2+} fluctuations in sperm physiology. Int Rev Cytol 2005; 243:79-172.
42. Jungnickel MK, Marrero H, Birnbaumer L et al. Trp2 regulates entry of Ca^{2+} into mouse sperm triggered by egg ZP3. Nat Cell Biol 2001; 3:499-502.
43. O'Toole CM, Arnoult C, Darszon A et al. Ca^{2+} entry through storeoperated channels in mouse sperm is initiated by egg ZP3 and drives the acrosome reaction. Mol Biol Cell 2000; 11:1571-1584.
44. Thore S, Dyachok O, Gylfe E et al. Feedback activation of phospholipase C via intracellular mobilization and store-operated influx of Ca^{2+} in insulin-secreting β-cells. J Cell Sci 2005; 118:4463-4471.
45. Peters C, Mayer A. Ca^{2+}/calmodulin signals the completion of docking and triggers a late step of vacuole fusion. Nature 1998; 396:575-580.
46. Merz AJ, Wickner WT. Trans-SNARE interactions elicit Ca^{2+} efflux from the yeast vacuole lumen. J Cell Biol 2004; 164:195-206.
47. Scheenen WJ, Wollheim CB, Pozzan T et al. Ca^{2+} depletion from granules inhibits exocytosis. A study with insulin-secreting cells. J Biol Chem 1998; 273:19002-19008.
48. Thirion S, Troadec JD, Pivovarova NB et al. Stimulus-secretion coupling in neurohypophysial nerve endings: A role for intravesicular sodium? Proc Natl Acad Sci USA 1999; 96:3206-3210.
49. De Blas G, Michaut M, Trevino CL et al. The intraacrosomal calcium pool plays a direct role in acrosomal exocytosis. J Biol Chem 2002; 277:49326-49331.
50. Herrick SB, Schweissinger DL, Kim SW et al. The acrosomal vesicle of mouse sperm is a calcium store. J Cell Physiol 2005; 202:663-671.
51. De Blas GA, Roggero CM, Tomes CN et al. Dynamics of SNARE assembly and disassembly during sperm acrosomal exocytosis. PLoS Biol 2005; 3:e323.
52. Tomes CN, Roggero CM, De Blas G et al. Requirement of protein tyrosine kinase and phosphatase activities for human sperm exocytosis. Dev Biol 2004; 265:399-415.
53. Bootman MD, Lipp P, Berridge MJ. The organisation and functions of local Ca^{2+} signals. J Cell Sci 2001; 114:2213-2222.
54. Burgoyne RD, Morgan A. Calcium sensors in regulated exocytosis. Cell Calcium 1998; 24:367-376.
55. Mikoshiba K, Fukuda M, Ibata K et al. Role of synaptotagmin, a Ca^{2+} and inositol polyphosphate binding protein, in neurotransmitter release and neurite outgrowth. Chem Phys Lipids 1999; 98:59-67.
56. Sugita S, Shin OH, Han W et al. Synaptotagmins form a hierarchy of exocytotic Ca^{2+} sensors with distinct Ca^{2+} affinities. EMBO J 2002; 21:270-280.
57. Li L, Chin LS. The molecular machinery of synaptic vesicle exocytosis. Cell Mol Life Sci 2003; 60:942-960.
58. Sudhof TC, Rizo J. Synaptotagmins: C2-domain proteins that regulate membrane traffic. Neuron 1996; 17:379-388.
59. Michaut M, De Blas G, Tomes CN et al. Synaptotagmin VI participates in the acrosome reaction of human spermatozoa. Dev Biol 2001; 235:521-529.
60. Hutt DM, Cardullo RA, Baltz JM et al. Synaptotagmin VIII is localized to the mouse sperm head and may function in acrosomal exocytosis. Biol Reprod 2002; 66:50-56.
61. Gamboa S, Ramalho-Santos J. SNARE proteins and caveolin-1 in stallion spermatozoa: Possible implications for fertility. Theriogenology 2005; 64:275-291.
62. Roggero CM, Tomes CN, De Blas GA et al. Protein kinase C-mediated phosphorylation of the two polybasic regions of synaptotagmin VI regulates their function in acrosomal exocytosis. Developmental Biology 2005; 285:422-435.
63. Davis AF, Bai J, Fasshauer D et al. Kinetics of synaptotagmin responses to Ca^{2+} and assembly with the core SNARE complex onto membranes. Neuron 1999; 24:363-376.
64. Bai J, Earles CA, Lewis JL et al. Membrane-embedded synaptotagmin penetrates cis or trans target membranes and clusters via a novel mechanism. J Biol Chem 2000; 275:25427-25435.
65. Breitbart H, Naor Z. Protein kinases in mammalian sperm capacitation and the acrosome reaction. Rev Reprod 1999; 4:151-159.
66. Burgoyne RD, Clague MJ. Calcium and calmodulin in membrane fusion. Biochim Biophys Acta 2003; 1641:137-143.
67. Burgoyne RD, Morgan A. Ca^{2+} and secretory-vesicle dynamics. Trends Neurosci 1995; 18:191-196.
68. Quetglas S, Iborra C, Sasakawa N et al. Calmodulin and lipid binding to synaptobrevin regulates calcium-dependent exocytosis. EMBO J 2002; 21:3970-3979.
69. de Haro L, Ferracci G, Opi S et al. Ca^{2+}/calmodulin transfers the membrane-proximal lipid-binding domain of the v-SNARE synaptobrevin from cis to trans bilayers. PNAS 2004; 101:1578-1583.
70. Peters C, Bayer MJ, Buhler S et al. Trans-complex formation by proteolipid channels in the terminal phase of membrane fusion. Nature 2001; 409:581-588.
71. Marin-Briggiler CI, Jha KN, Chertihin O et al. Evidence of the presence of calcium/calmodulin-dependent protein kinase IV in human sperm and its involvement in motility regulation. J Cell Sci 2005; 118:2013-2022.

72. Guerrero A, Darszon A. Egg jelly triggers a calcium influx which inactivates and is inhibited by calmodulin antagonists in the sea urchin sperm. Biochim Biophys Acta 1989; 980:109-116.
73. Sano K. Inhibition of the acrosome reaction of sea urchin spermatozoa by a calmodulin antagonist, N-(6-aminohexyl)-5-chloro-1-naphthalenesulfonamide (W-7). J Exp Zool 1983; 226:471-473.
74. Bendahmane M, Lynch C, Tulsiani DR. Calmodulin signals capacitation and triggers the agonist-induced acrosome reaction in mouse spermatozoa. Arch Biochem Biophys 2001; 390:1-8.
75. Lopez-Gonzalez I, Vega-Beltran JL, Santi CM et al. Calmodulin antagonists inhibit T-type Ca^{2+} currents in mouse spermatogenic cells and the zona pellucida-induced sperm acrosome reaction. Dev Biol 2001; 236:210-219.
76. Yunes R, Tomes C, Michaut M et al. Rab3A and calmodulin regulate acrosomal exocytosis by mechanisms that do not require a direct interaction. FEBS Letters 2002; 525:126-130.
77. Coppola T, Perret-Menoud V, Luthi S et al. Disruption of Rab3-calmodulin interaction, but not other effector interactions, prevents Rab3 inhibition of exocytosis. EMBO J 1999; 18:5885-5891.
78. Novick P, Zerial M. The diversity of Rab proteins in vesicle transport. Curr Opin Cell Biol 1997; 9:496-504.
79. Zerial M, McBride H. Rab proteins as membrane organizers. Nat Rev Mol Cell Biol 2001; 2:107-117.
80. Pevsner J. The role of Sec1p-related proteins in vesicle trafficking in the nerve terminal. J Neurosci Res 1996; 45:89-95.
81. Chen YA, Scheller RH. SNARE-mediated membrane fusion. Nat Rev Mol Cell Biol 2001; 2:98-106.
82. Fasshauer D, Sutton RB, Brunger AT et al. Conserved structural features of the synaptic fusion complex: SNARE proteins reclassified as Q- and R-SNAREs. Proc Natl Acad Sci USA 1998; 95:15781-15786.
83. May AP, Whiteheart SW, Weis WI. Unraveling the mechanism of the vesicle transport ATPase NSF, the N-Ethylmaleimide-sensitive factor. J Biol Chem 2001; 276:21991-21994.
84. Weber T, Zemelman BV, McNew JA et al. SNAREpins: Minimal machinery for membrane fusion. Cell 1998; 92:759-772.
85. Hu C, Ahmed M, Melia TJ et al. Fusion of cells by flipped SNAREs. Science 2003; 300:1745-1749.
86. Ungermann C, Sato K, Wickner W. Defining the functions of trans-SNARE pairs. Nature 1998; 396:543-548.
87. Tahara M, Coorssen JR, Timmers K et al. Calcium can disrupt the SNARE protein complex on sea urchin egg secretory vesicles without irreversibly blocking fusion. J Biol Chem 1998; 273:33667-33673.
88. Xu T, Ashery U, Burgoyne RD et al. Early requirement for α-SNAP and NSF in the secretory cascade in chromaffin cells. EMBO J 1999; 18:3293-3304.
89. Grote E, Carr CM, Novick PJ. Ordering the final events in yeast exocytosis. J Cell Biol 2000; 151:439-452.
90. Pfeffer SR. Transport-vesicle targeting: Tethers before SNAREs. Nat Cell Biol 1999; 1:E17-E22.
91. Waters MG, Hughson FM. Membrane tethering and fusion in the secretory and endocytic pathways. Traffic 2000; 1:588-597.
92. Gonzalez Jr L, Scheller RH. Regulation of membrane trafficking: Structural insights from a Rab/effector complex. Cell 1999; 96:755-758.
93. Ungermann C, Langosch D. Functions of SNAREs in intracellular membrane fusion and lipid bilayer mixing. J Cell Sci 2005; 118:3819-3828.
94. Coppola T, Magnin-Luthi S, Perret-Menoud V et al. Direct Interaction of the Rab3 Effector RIM with Ca^{2+} Channels, SNAP-25, and Synaptotagmin. J Biol Chem 2001; 276:32756-32762.
95. Johannes L, Doussau F, Clabecq A et al. Evidence for a functional link between Rab3 and the SNARE complex. J Cell Sci 1996; 109:2875-2884.
96. Weimer RM, Jorgensen EM. Controversies in synaptic vesicle exocytosis. J Cell Sci 2003; 116:3661-3666.
97. Coppola T, Frantz C, Perret-Menoud V et al. Pancreatic β-cell protein granuphilin binds Rab3 and Munc-18 and controls exocytosis. Mol Biol Cell 2002; 13:1906-1915.
98. Ward CR, Faundes D, Foster JA. The monomeric GTP binding protein, Rab3a, is associated with the acrosome on mouse sperm. Mol Reprod Dev 1999; 53:413-421.
99. Michaut M, Tomes CN, De Blas G et al. Calcium-triggered acrosomal exocytosis in human spermatozoa requires the coordinated activation of Rab3A and N-ethylmaleimide-sensitive factor. Proc Natl Acad Sci USA 2000; 97:9996-10001.
100. Belmonte SA, Lopez CI, Roggero CM et al. Cholesterol content regulates acrosomal exocytosis by enhancing Rab3A plasma membrane association. Developmental Biology 2005; 285:393-408.
101. Tomes CN, De Blas GA, Michaut MA et al. α-SNAP and NSF are required in a priming step during the human sperm acrosome reaction. Mol Hum Reprod 2005; 11:43-51.
102. Tomes CN, Michaut M, De Blas G et al. SNARE complex assembly is required for human sperm acrosome reaction. Dev Biol 2002; 243:326-338.

103. Sutton RB, Fasshauer D, Jahn R et al. Crystal structure of a SNARE complex involved in synaptic exocytosis at 2.4 A resolution. Nature 1998; 395:347-353.
104. Hay JC, Scheller RH. SNAREs and NSF in targeted membrane fusion. Curr Opin Cell Biol 1997; 9:505-512.
105. Sollner T, Bennett MK, Whiteheart SW et al. A protein assembly-disassembly pathway in vitro that may correspond to sequential steps of synaptic vesicle docking, activation, and fusion. Cell 1993; 75:409-418.
106. Weber T, Parlati F, McNew JA et al. SNAREpins are functionally resistant to disruption by NSF and α-SNAP. J Cell Biol 2000; 149:1063-1072.
107. Ramalho-Santos J, Schatten G. Presence of N-ethyl maleimide sensitive factor (NSF) on the acrosome of mammalian sperm. Arch Androl 2004; 50:163-168.
108. Ramalho-Santos J, Moreno RD, Wessel GM et al. Membrane trafficking machinery components associated with the mammalian acrosome during spermiogenesis. Exp Cell Res 2001; 267:45-60.
109. Pellizzari R, Rossetto O, Schiavo G et al. Tetanus and botulinum neurotoxins: Mechanism of action and therapeutic uses. Philos Trans R Soc Lond B Biol Sci 1999; 354:259-268.
110. Hayashi T, McMahon H, Yamasaki S et al. Synaptic vesicle membrane fusion complex: Action of clostridial neurotoxins on assembly. EMBO J 1994; 13:5051-5061.
111. Darszon A, Beltran C, Felix R et al. Ion transport in sperm signaling. Dev Biol 2001; 240:1-14.
112. Lang T, Margittai M, Holzler H et al. SNAREs in native plasma membranes are active and readily form core complexes with endogenous and exogenous SNAREs. J Cell Biol 2002; 158:751-760.
113. Rickman C, Meunier FA, Binz T et al. High affinity interaction of syntaxin and SNAP-25 on the plasma membrane is abolished by botulinum toxin E. J Biol Chem 2004; 279:644-651.
114. Xu T, Binz T, Niemann H et al. Multiple kinetic components of exocytosis distinguished by neurotoxin sensitivity. Nat Neurosci 1998; 1:192-200.
115. Chen YA, Scales SJ, Patel SM et al. SNARE complex formation is triggered by Ca^{2+} and drives membrane fusion. Cell 1999; 97:165-174.
116. Schulz JR, Wessel GM, Vacquier VD. The exocytosis regulatory proteins syntaxin and VAMP are shed from sea urchin sperm during the acrosome reaction. Dev Biol 1997; 191:80-87.
117. Schulz JR, Sasaki JD, Vacquier VD. Increased association of synaptosome-associated protein of 25 kDa with syntaxin and vesicle-associated membrane protein following acrosomal exocytosis of sea urchin sperm. J Biol Chem 1998; 273:24355-24359.
118. Schulz JR, De La Vega-Beltran J, Beltran C et al. Ion channel activity of membrane vesicles released from sea urchin sperm during the acrosome reaction. Biochemical and Biophysical Research Communications 2004; 321:88-93.
119. Katafuchi K, Mori T, Toshimori K et al. Localization of a syntaxin isoform, syntaxin 2, to the acrosomal region of rodent spermatozoa. Mol Reprod Dev 2000; 57:375-383.
120. Huynh H, Bottini N, Williams S et al. Control of vesicle fusion by a tyrosine phosphatase. Nat Cell Biol 2004; 6:831-839.
121. Matsushita K, Morrell CN, Cambien B et al. Nitric oxide regulates exocytosis by S-nitrosylation of N-ethylmaleimide-sensitive factor. Cell 2003; 115:139-150.
122. Chen YA, Scales SJ, Scheller RH. Sequential SNARE assembly underlies priming and triggering of exocytosis. Neuron 2001; 30:161-170.
123. Schiavo G, Matteoli M, Montecucco C. Neurotoxins affecting neuroexocytosis. Physiol Rev 2000; 80:717-766.
124. Hua SY, Charlton MP. Activity-dependent changes in partial VAMP complexes during neurotransmitter release. Nat Neurosci 1999; 2:1078-1083.
125. Wassarman PM, Jovine L, Qi H et al. Recent aspects of mammalian fertilization research. Mol Cell Endocrinol 2005; 234:95-103.
126. Breitbart H. Signaling pathways in sperm capacitation and acrosome reaction. Cell Mol Biol (Noisy -le-grand) 2003; 49:321-327.
127. Baldi E, Luconi M, Bonaccorsi L et al. Signal transduction pathways in human spermatozoa. J Reprod Immunol 2002; 53:121-131.
128. Gadella BM, Rathi R, Brouwers JF et al. Capacitation and the acrosome reaction in equine sperm. Anim Reprod Sci 2001; 68:249-265.
129. Herrero MB, de Lamirande E, Gagnon C. Nitric oxide is a signaling molecule in spermatozoa. Curr Pharm Des 2003; 9:419-425.
130. Neill AT, Vacquier VD. Ligands and receptors mediating signal transduction in sea urchin spermatozoa. Reproduction 2004; 127:141-149.
131. Fraser LR, Adeoya-Osiguwa S, Baxendale RW et al. First messenger regulation of mammalian sperm function via adenylyl cyclase/cAMP. J Reprod Dev 2005; 51:37-46.
132. Brewis IA, Moore HD. Molecular mechanisms of gamete recognition and fusion at fertilization. Hum Reprod 1997; 12:156-165.
133. Urner F, Sakkas D. Protein phosphorylation in mammalian spermatozoa. Reproduction 2003; 125:17-26.

134. Leyton L, LeGuen P, Bunch D et al. Regulation of mouse gamete interaction by a sperm tyrosine kinase. Proc Natl Acad Sci USA 1992; 89:11692-11695.
135. Pukazhenthi BS, Wildt DE, Ottinger MA et al. Inhibition of domestic cat spermatozoa acrosome reaction and zona pellucida penetration by tyrosine kinase inhibitors. Mol Reprod Dev 1998; 49:48-57.
136. Tesarik J, Moos J, Mendoza C. Stimulation of protein tyrosine phosphorylation by a progesterone receptor on the cell surface of human sperm. Endocrinology 1993; 133:328-335.
137. Luconi M, Bonaccorsi L, Krausz C et al. Stimulation of protein tyrosine phosphorylation by platelet- activating factor and progesterone in human spermatozoa. Mol Cell Endocrinol 1995; 108:35-42.
138. Kirkman-Brown JC, Lefievre L, Bray C et al. Inhibitors of receptor tyrosine kinases do not suppress progesterone-induced $Ca^{2+}{}_i$ signalling in human spermatozoa. Mol Hum Reprod 2002; 8:326-332.
139. Meizel S, Turner KO. Chloride efflux during the progesterone-initiated human sperm acrosome reaction is inhibited by lavendustin A, a tyrosine kinase inhibitor. J Androl 1996; 17:327-330.
140. Marash M, Gerst JE. t-SNARE dephosphorylation promotes SNARE assembly and exocytosis in yeast. EMBO J 2001; 20:411-421.
141. Peters C, Andrews PD, Stark MJ et al. Control of the terminal step of intracellular membrane fusion by protein phosphatase 1. Science 1999; 285:1084-1087.
142. Bryant NJ, James DE. The Sec1p/Munc18 (SM) protein, Vps45p, cycles on and off membranes during vesicle transport. J Cell Biol 2003; 161:691-696.
143. Seino S, Shibasaki T. PKA-dependent and PKA-independent pathways for cAMP-regulated exocytosis. Physiol Rev 2005; 85:1303-1342.
144. Holz GG. Epac: A new cAMP-binding protein in support of glucagon-like peptide-1 receptor-mediated signal transduction in the pancreatic beta-cell. Diabetes 2004; 53:5-13.
145. Fujimoto K, Shibasaki T, Yokoi N et al. Piccolo, a Ca^{2+} sensor in pancreatic β-cells. Involvement of cAMP-GEFII.Rim2.Piccolo complex in cAMP-dependent exocytosis. J Biol Chem 2002; 277:50497-50502.
146. Kashima Y, Miki T, Shibasaki T et al. Critical role of cAMP-GEFII—Rim2 complex in incretin-potentiated insulin secretion. J Biol Chem 2001; 276:46046-46053.
147. Ozaki N, Shibasaki T, Kashima Y et al. cAMP-GEFII is a direct target of cAMP in regulated exocytosis. Nat Cell Biol 2000; 2:805-811.
148. Shibasaki T, Sunaga Y, Fujimoto K et al. Interaction of ATP sensor, cAMP sensor, Ca^{2+} sensor, and voltage-dependent Ca^{2+} channel in insulin granule exocytosis. J Biol Chem 2004; 279:7956-7961.
149. Lefievre L, Jha KN, de Lamirande E et al. Activation of protein kinase A during human sperm capacitation and acrosome reaction. J Androl 2002; 23:709-716.
150. De Jonge CJ, Han HL, Lawrie H et al. Modulation of the human sperm acrosome reaction by effectors of the adenylate cyclase/cyclic AMP second-messenger pathway. J Exp Zool 1991; 258:113-125.
151. de Rooij J, Zwartkruis FJ, Verheijen MH et al. Epac is a Rap1 guanine-nucleotide-exchange factor directly activated by cyclic AMP. Nature 1998; 396:474-477.
152. Kawasaki H, Springett GM, Mochizuki N et al. A family of cAMP-binding proteins that directly activate Rap1. Science 1998; 282:2275-2279.
153. Branham MT, Zarelli V, Tomes C. AMPc induce exocitosis acrosomal por la vía Epac/Rap. Buenos Aires Medicina 2004; 64(Suppl II):220.
154. Branham MT, Mayorga LS, Tomes CT. Calcium-induced acrosomal exocytosis requires cAMP acting through a protein Kinase A-independent, Epac-mediated pathway. J Biol Chem 2006; 281:8656-8666.
155. Rizo J, Sudhof TC. SNAREs and Munc18 in synaptic vesicle fusion. Nat Rev Neurosci 2002; 3:641-653.
156. Lonart G, Sudhof TC. Assembly of SNARE core complexes prior to neurotransmitter release sets the readily releasable pool of synaptic vesicles. J Biol Chem 2000; 275:27703-27707.
157. Voets T. Dissection of three Ca^{2+}-dependent steps leading to secretion in chromaffin cells from mouse adrenal slices. Neuron 2000; 28:537-545.
158. Littleton JT, Barnard RJ, Titus SA et al. SNARE-complex disassembly by NSF follows synaptic-vesicle fusion. Proc Natl Acad Sci USA 2001; 98:12233-12238.
159. He P, Southard RC, Chen D et al. Role of α-SNAP in promoting efficient neurotransmission at the crayfish neuromuscular junction. J Neurophysiol 1999; 82:3406-3416.
160. Banerjee A, Kowalchyk JA, DasGupta BR et al. SNAP-25 is required for a late postdocking step in Ca^{2+}-dependent exocytosis. J Biol Chem 1996; 271:20227-20230.
161. Wickner W. Yeast vacuoles and membrane fusion pathways. EMBO J 2002; 21:1241-1247.
162. Wang L, Merz AJ, Collins KM et al. Hierarchy of protein assembly at the vertex ring domain for yeast vacuole docking and fusion. J Cell Biol 2003; 160:365-374.

CHAPTER 10

Nonsecretory, Regulated Exocytosis:
A Multifarious Mechanism Employed by Cells to Carry Out a Variety of Functions

Emanuele Cocucci and Jacopo Meldolesi*

Abstract

R egulated exocytosis is most often identified as the late step of protein and neurotransmitter secretion, consisting in the fusion between the membrane of secretory granule/vesicle and the plasma membrane. However, a large number of nonsecretory processes, often playing key roles in cellular function, are also based on regulated exocytosis. We propose to call them nonsecretory regulated exocytoses. Here we illustrate a number of these exocytoses taking place in cells where they play particularly important roles. Specifically, we deal with the circulation of glutamatergic receptors, AMPA and NMDA receptors, at the postsynaptic compartment (usually dendritic spines); with the growth of neuronal axons, made possible by the exocytosis of "plasmalemma precursor vesicles"; with the exocytosis-mediated wound healing, by which cells repair mechanical lesions which otherwise would be lethal; with the exocytosis of vesicles, including the enlargeosomes, at the varicosities of neurosecretory cells, necessary for neurite growth and for the uptake of myo-inositol; with the traffic of vesicles rich in the Glut4 transporter in adipocytes, which is regulated by insulin and defective in diabetes mellitus; with the H^+/K^+ pump in the parietal cells of the stomach; with the aquaporin2 channel in the duct cells of the kidney; with the compensatory plasmalemma enlargement in phagocytizing macrophages and macropinocytizing dendritic cells. At the end we discuss the competence for regulated exocytosis of specialized subfamilies of endosomes: multivesicular bodies; recycling endosomes targeted to the cleavage furrow, that may be necessary for cytokinesis at the end of mitosis; phagosomes and micropinosomes. Nonsecretory exocytoses are not peculiar properties of specialized cells only. Rather, they are widely expressed and several of them take place concomitantly in single cells. The coordination of these processes, among themselves and with secretory exocytoses, is a form of cellular complexity that remains to be investigated.

Introduction

The concept of regulated exocytosis, consisting in the stimulation-induced fusion with the plasmalemma of the membrane of specialized secretory organelles, was introduced in the mid fifties by George Palade as the mechanism responsible for the process of protein secretion (Palade, 1956).[1] This mechanism explained numerous observations made by cytologists in the previous decades, the first being the disappearance after animal feeding of the large and dense zymogen granules typical of resting pancreatic acinar cells.[2] Shortly thereafter regulated exocytosis, recognized at synapses[3] and in a variety of other secretory systems,[4-6]

*Corresponding Author: Jacopo Meldolesi—Vita-Salute San Raffaele University, DIBIT, via Olgettina 58, 20132 Milan, Italy. Email: meldolesi.jacopo@hsr.it

Molecular Mechanisms of Exocytosis, edited by Romano Regazzi. ©2007 Landes Bioscience and Springer Science+Business Media.

was proposed as the general mechanism by which proteins and other hydrophilic products, often of key physiological importance, are released in bulk by the cells in response to adequate cell stimulation. At the moment this concept is so widely and strongly established that in many papers the nomenclature of discharge or release of secretion products, and even that of secretion, have been used instead of regulated exocytosis to indicate the specific membrane fusion process (for examples see refs. 7-9).

However, regulated exocytosis occurs not only during secretion but also in a variety of other processes. Among these is lysosomal exocytosis, recognized in leucocytes and macrophages already in the sixties and now known to occur in many, possibly all types of cells when exposed to appropriate stimulation.[10] Similar to secretion, however, lysosomal exocytosis results in the discharge of specific materials segregated within the organelle lumen, in this case lysosomal enzymes. From this point of view, therefore it can be envisaged as a form of secretory exocytosis and therefore will not be considered here. In contrast, the processes we intend to discuss, defined as nonsecretory regulated exocytoses,[11] are aimed not to the discharge of segregated material, but only to the transfer of membrane patches, from specific organelles to the plasmalemma. Changes of the plasma membrane, including the temporary intermixing of its own components with the components of vesicle membranes, are known to take place also during secretory exocytosis. These events, however, do not occur as independent processes, but rather as the inevitable consequences of secretory exocytoses, destined to be rapidly corrected by specific endocytosis. Nevertheless, a few important proteins not involved in secretion (channels, receptors), integrated in the membrane of synaptic vesicles or granules, have been shown to reach their final destination at the cell surface by secretory exocytosis (for examples see refs. 12-14). This type of translocation, however, is not an independent process characterized by peculiar timing and regulation, but a sort of hitch hiking of individual proteins in the membrane of secretion carriers. In contrast, in regulated nonsecretory exocytoses the surface translocation of membrane components is the key event, and its regulation can occur independently of that of secretory events taking place in the same cell.

The idea of regulated traffic of membrane taking place to and from the plasma membrane independently of secretion is not new. Important examples were demonstrated, beginning in the eighties/early nineties. In all previous studies, however, the interest was focused exclusively on specialized processes taking place in single or a few types of cells, such as proton secretion in the principal cells of the stomach,[15] water resorption in the duct cells of the kidney[16] and glucose uptake in adipocytes.[17] For many years, therefore, these membrane traffic events were envisaged as unique properties of individual cell types, with little impact in general cell biology. Recently such an impact has increased. Several regulated exocytic processes, already known in specialized cells, have been shown to occur also in other cell types, and to induce translocation either of the same protein or of analogues of the same family. Moreover, the number of recognized nonsecretory exocytoses has greatly increased, and is still increasing at fast rate. Some of them have a role in new, others in classical processes, such as phagocytosis and cytokinesis,[11] that until recently had not been thought to include regulated exocytic events. Finally, the molecular properties of these nonsecretory exocytoses, in particular the differences among themselves and with respect to the secretory processes, have become to emerge. We conclude that this field, already very interesting at the moment, is destined to gain momentum in relation not only to cell biology and pathology (for example, in diabetes mellitus, diabetes insipidus, gastric ulcers, some forms of muscle distrophy and so on), but also of biotechnological applications.[11]

In our previous contribution, published last year, the most important examples of nonsecretory exocytosis were classified in two groups: those aimed at the translocation to the cell surface of specific proteins and those aimed at the expansion of the cell surface area, as needed when intense endocytosis or specific processes (cellularization, differentiation, wound healing etc) take place.[11] The known cellular and molecular properties of those processes were mentioned, however only briefly. Here we will focus specifically on those aspects, considered in the cells where nonsecretory exocytoses are particularly important and/or multifarious. In addition, we will discuss the state of a family of organelles, the endosomes, believed until recently to be incompetent for regulated exocytosis, some members of which have been found to undergo the process, at least in specific cells after adequate stimulation.

Synaptic Transmission

Changes of receptor surface density is one of the basic mechanism by which cells modulate their responses to external stimuli. In neurons this mechanism is more sophisticated than in other cells because most membrane receptors, after reaching the surface, need to cluster at discrete domains, for example at the postsynaptic membrane, in order to participate in key processes such as synaptic transmission. The demonstration, both in the hippocampus and the cerebral cortex, that the receptor-channel necessary for glutamatergic transmission, the AMPA receptor (AMPAR), lacking in a fraction of the specific synapses, is however reintroduced after a few min of stimulation due to a Ca^{2+}-dependent, calmodulin-mediated process of vesicle exocytosis, explained an electrophysiological observation that had remained mysterious, the existence of the so called silent synapses. The latter are normal in the presynaptic compartment where a normal number of synaptic vesicles are exocytized upon stimulation, with ensuing glutamate release as in functional synapses. In silent synapses, however, no rapid response is activated because the only glutamate receptor-channel exposed, the NMDA receptor (NMDAR), requires for opening the concomitant depolarization of the postsynaptic membrane, a physiological consequence of AMPAR activation. The traffic of the latter receptor, in and out the postsynaptic membrane, does therefore regulate transmission, playing a key role in synaptic plasticity. The latter process, by which synapses do retain messages, is regarded as the cellular basis of learning and memory (for reviews see refs. 18, 19).

Several aspects of the AMPAR trafficking in the hippocampal and cerebral cortex neurons deserve attention. The receptor-rich exocytic organelles are small vesicles distributed in the proximity of the postsynaptic compartment, usually the dendritic spines. So far they have not been isolated, therefore details about their composition are still unknown. However their exocytosis is known to be blocked by tetanus toxin, which works by cleaving the vesicular SNARE, VAMP2.[7] The plasma membrane SNAP25 is also involved. Because of these two proteins, the exocytosis of the AMPAR-rich vesicle seems to resemble that of synaptic vesicles. However, the appearance of newly exocytized AMPAR at the post-synaptic membrane does not occur in msec, as transmitter release, but requires a few min. Such a delay may be only apparent. The exocytosis of receptor-rich vesicles, in fact, cannot occur directly at the postsynaptic membrane, where the dense cytoskeleton (postsynaptic density) affects circulation of vesicles, but at some distance (Fig. 1). Once exocytized, the receptor needs therefore to diffuse in the plane of the plasmalemma in order to reach its functional site, where further diffusion is precluded by its direct binding to a complex composed by two postsynaptic density proteins, PSD95 and SAP97.[20] The regulative step of the vesicle exocytosis has been shown to depend on the direct phosphorylation of the GluR1 subunit (at the serine 845) by the Ca^{2+}-calmodulin-dependent protein kinase II (CamKII),[21] which is also concentrated at the postsynaptic compartment (Fig. 1).

AMPAR is not the only receptor trafficking at synapses. In the hippocampus the cytokine tumor necrosis factor α, which modulates positively the exocytosis of AMPAR-rich vesicles, induces a decreased exocytosis of the vesicles rich in the inhibitory receptor, the GABA-A receptor. The two effects cooperate to increase the excitability of neurons.[22] Also NMDAR is not a static component, rather, it does circulate from cytoplasmic organelles to the postsynaptic membrane and viceversa.[23] Interestingly NMDAR, which at nascent synapses is mostly located in relatively large structures that immunocytochemistry suggests to be, at least in part, endosomes, circulates much faster than AMPAR. Also the exocytosis of NMDAR appears different from that of AMPAR since its main controller appears to be not Ca^{2+} but protein kinase C (PKC) (Fig. 1).[24] Moreover, the SNAP SNARE involved in the fusion of NMDA-rich vesicles is SNAP23,

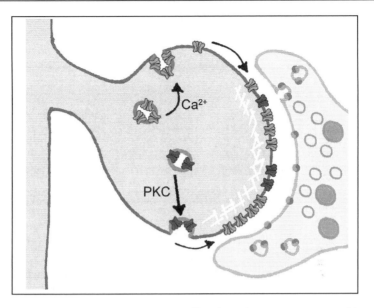

Figure 1. The multifarious exocytic membrane traffic taking place at synapses. In addition to the classical transmitter release exocytoses of clear and dense-core vesicles, taking place at the presynaptic terminals (right), at least three distinct types of nonsecretory exocytoses have been demonstrated at glutamatergic synapses, inducing the surface appearance of the PI transporter (presynaptic, HMIT, ●) and of the two main ionotropic receptors, AMPA and NMDA receptors (post-synaptic, and , respectively). The latter, which occur at the lateral face of the spines (and not at their final destination, the true post-synaptic membrane, where membrane traffic is precluded by the specific cytoskeletal density of PSD, yellow), are controlled by different signals: Ca^{2+} (via CamKII) for the AMPA receptor; protein kinase C for the NMDA receptor. A color version of this figure is available online at http://www.Eurekah.com.

not SNAP25.[25] Other brain areas, other receptors and other regulatory pathways can also exhibit properties analogous to those described for hippocampal AMPAR. For example, a calmodulin-dependent, clostridial toxin-sensitive exocytosis of AMPAR-rich vesicles induced by application of pain occurs in lumbar neurons of the spinal cord and is responsible for the hyperalgesia;[26] the vanilloid receptor (TRPV1) of dorsal root ganglion neurons is localized in VAMP2 positive vesicles and undergoes botulinum toxin-sensitive exocytosis in response to application of a variety of stimuli working via activation of PKC[27] and/or PI3kinase and Srk kinase.[28] Finally, at presynaptic and neurite terminals of differentiated PC12 cells, rapid (<1 min) exocytosis of small vesicles distinct from the secretory clear and dense vesicles, induced by a variety of stimuli (depolarization, activation of PKC, increased $[Ca^{2+}]_i$), results in the translocation of the H^+/myo-inositol symporter, a molecule essential for the neuronal uptake of the sugar necessary for both phosphoinositide metabolism and the secretory exocytic process (Fig. 1).

Axonal Growth and Wound Healing

Axonal growth, one of the key processes in embryonic brain development and nerve regeneration, requires considerable expansion of the neuronal plasma membrane depending to a large extent to exocytosis of intracellular organelles. For quite sometime this membrane enlargement was believed to be a by product of the exocytosis of secretory organelles, clear and/or dense-core vesicles, however this is not the case. Growing axons are believed to contain nonsecretory vesicles competent for exocytosis, defined as plasmalemma precursor vesicles, which may also serve for the growth of dendrites.[29] The precise nature of these vesicles is still

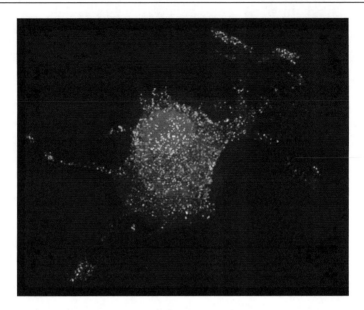

Figure 2. PC12 cells differentiated by NGF express enlargeosomes close to the plasma membrane at both the soma and varicosities. Merged 3D renderings of deconvoluted wide-field microscopy images stack of a differentiated PC12 cell immunolabeled for markers of secretory organelles (synaptophysin, green, localized in both clear and dense vesicles), enlargeosomes (desmoyokin/Ahnak, red) and nucleus (DAPI, blue). Notice that the two types of organelles do not coincide and therefore are distinct from each other; and that they are distributed both in the cell body and in the varicosities at the tip of the fibers. In terms of intracellular distribution the enlargeosomes appear close to the plasma membrane more often than the secretory vesicles. Enlargement: 1250X. A color version of this figure is available online at http://www.Eurekah.com.

largely unknown, however some interesting properties have emerged. Their SNARE appears to be not VAMP2, as in secretory and other exocytic vesicles, but Ti-VAMP (also known as VAMP7), which is insensitive to tetanus toxin.[30] Moreover, at least some of these vesicles appear rich in receptors for axonal guidance factors. Thus, their exocytosis increases the outgrowth potential of the axon.[31] Exocytoses governing axonal growth are stimulated by many factors, such as insulin and IGF1 (but not BDNF), working through various intracellular signals: cAMP and protein kinase A[32,31] Ca^{2+}-calmodulin and CamKII[33,34] PI3K and Akt.[35]

Another process dependent on regulated exocytosis, which occurs not only in axons but also in many, if not all cells, is the healing of cell membrane wounds. This process is important because cells are wounded quite often, especially when exposed to mechanical stress. The repair occurs spontaneously, however only when the margins of the wounds come in direct contact with each other. This cannot occur unless the surface area of the cell is increased. A form of exocytosis that appears to participate in wound healing is lysosomal exocytosis.[36] However, when this form is blocked pharmacologically, repair occurs anyway.[37] Additional organelles, therefore, must be involved. Among these are the enlargeosomes small juxta-plasmalemma vesicles expressed by a variety of cells which, after appropriate $[Ca^{2+}]_i$ rises, are exocytized very quickly (<1 sec) by a tetanus toxin-insensitive process[38,39] (Fig. 2; see also "Neurosecretory Cells" section). The capacitance increases observed previously by patch clamping in various nonsecretory cells (CHO and 3T3 fibroblasts) after strong stimulation[40,41] might also be due to exocytosis of enlargeosomes. In skeletal muscle a defect in wound healing occurs in case of mutations of dysferlin, a peripheral protein of a subplasmalemma vesicle competent for regulated exocytosis.[42] Other organelles, defined all together as the "emergency response team", appear to be also involved in plasmalemma repair.[43]

Neurosecretory Cells

When exposed to NGF, the best known neurosecretory cell line, PC12, differentiates with acquisition of a neuronal-like phenotype. For this reason PC12 cells are extensively investigated also as a neuronal model. Until recently the exocytoses investigated in these cells were those of the two secretory vesicles: the numerous dense, catecholamine-containing granules, which resemble the dense-core vesicles of neurons and the chromaffin granules of the adrenal medulla, and the clear, acetylcholine-containing vesicles, that have never been identified with certainty at the cellular level. Our studies on enlargeosomes, carried out primarily in a PC12 clone defective of neurosecretion, have shown that those nonsecretory exocytic organelles, distinct from the secretory vesicles, are scarce in nondifferentiated PC12 but increase markedly during differentiation, accumulating in the varicosities of neurites (Fig. 2).[38] Enlargeosomes appear as unique organelles. Their membrane, resistant to nonionic detergents, appears to be composed primarily by membrane rafts; their exocytosis resembles the clear vesicle exocytosis in its fast rate (<1sec from the rise of $[Ca^{2+}]_i$) and yet it is insensitive to tetanus toxin and does not involve the SNAREs of neurosecretion, VAMP2, SNAP25 and syntaxin1; their endocytosis, which does involve neither clathrin nor caveolin, appears distinct also from the other forms of endocytosis described so far.[38,39]

Enlargeosomes, however, are not the only nonsecretory exocytic organelle present in PC12. A specific marker, IRAP, documents the expression of the exocytic vesicles usually positive for Glut4 (see "Adipocytes, Kidney Duct Cells and Gastric Parietal Cells" section). Even more interesting, differentiated PC12 express, primarily at their varicosities, the vesicles specific for the myo-inositol transporter already presented in neurons (see "Synaptic Transmission" section) (Fig. 1).[44] The exocytosis of these vesicles, triggered by Ca^{2+} and PKC activation, but not by cAMP, is fast as that of enlargeosomes, however their endocytosis is dynamin-dependent, while that of enlargeosomes is not. The two organelles, therefore, appear distinct from each other. We conclude that, in addition to the two secretory organelles, which are discharged at different rates in response to different rises of $[Ca^{2+}]_i$,[45] at least two types of nonsecretory organelles are trafficking in differentiated PC12 cells, one enabling the cell to pick up myo-inositol, which is necessary for membrane fusion, another to control the surface area. All these vesicles respond to similar signals, although possibly under not identical conditions. The coordination of these events represents an exciting subject from future studies.

Adipocytes, Kidney Duct Cells and Gastric Parietal Cells

Each of these cell types is characterized by a major nonsecretory exocytic system which has attracted interest since long time, initially for physiological and pathological reasons, more recently also from the cellular and molecular points of view. By their continuous and regulated membrane trafficking, from the cytoplasm to the plasma membrane and viceversa, these systems control the surface expression of critical molecules: the glucose transporter Glut4, that mediates the insulin action on glucose metabolism; the aquaporin2 channel, that regulates reuptake of water from the kidney ducts; and the H^+/K^+ pump, which is responsible for the acidic environment of the gastric lumen, respectively. The organelles and their exocytoses have been discussed in detail in long lists of papers, including numerous recent reviews (see, among others, refs. 46-49, for aquaporin2; see ref. 50, for the gastric H^+/K^+ pump), as well as in our previous opinion paper (see ref. 11). Here we will concentrate only on a few molecular and functional aspects considered in parallel in the three systems.

Similarities among the Glut4, aquaporin and H^+/K^+ pump exocytic systems are numerous. First, their exocytoses do not follow quickly the application of stimuli but require a few min to get started even if the competent vesicles (in the case of gastric parietal cells the larger, so called tubulovesicles) are distributed in the proximity of the plasma membrane; second, these exocytoses use VAMP2 as the vesicular SNARE, thus they are all sensitive to tetanus toxin. The SNAREs of the plasma membrane are either SNAP23 and syntaxin 4 (aquaporin2 and Glut4) or SNAP25 and syntaxin3 (H^+/K^+ pump); third, the post-exocytic endocytosis is in all cases clathrin-dependent;

fourth, the endocytized vesicles reacquire competence for exocytosis within relatively short times, without travelling back to the Golgi complex or TGN. As far as the intracellular messengers responsible for the stimulation of exocytosis, aquaporin and H^+/K^+ pump vesicles respond to cAMP while Glut4 is regulated by the cascade of events triggered by stimulation of the insulin receptor: PI3 kinase, akt, PKCζ. Interestingly, in addition to the classical PKA pathway, with phosphorylation of relevant proteins, the exocytic effect of the cyclic nucleotide appears to be mediated by alternative mechanism(s), possibly based on the GTP/GDP exchange at specific G proteins.[51] The existing similarities suggested the vesicles of the three nonsecretory systems to be of a single type, with differences dependent on the different gene expression of various cells. This however does not seem to be the case. In fully differentiated 3T3-L1 adipocytes, Glut4 and aquaporin2 are distributed to distinct cytoplasmic vesicles which undergo exocytosis in response to different stimulations, i.e., by insulin in the first case; by increase of cAMP in the second.[52]

Macrophages and Dendritic Cells

In the course of their activity these cells, and other cell types as well, undergo major endocytic processes, i.e., phagocytosis and macropinocytosis, by which large fractions of the plasma membrane are transferred to the cytoplasm. Since the cells cannot reduce their surface area below a certain level, these processes necessarily require compensatory exocytoses to take place, concomitantly or even before phagosome or macropinosome sealing. Clear evidence demonstrates that this is indeed the case. An indirect assay of plasma membrane area, based on the changes of the surface tension in phagocytizing neutrophils, has revealed the plasma membrane area to increase up to 80% after stimulation of phagocytosis most likely due to exocytoses of cytoplasmic vesicles initiated even before sealing (Fig. 3).[53] In phagocytizing macrophages the exocytized vesicles seem to be of at least two types, one characterized by the SNARE VAMP3,[54] the other by VAMP7;[55] in dendritic cells the organelles exocytized to compensate for macropinocytosis are most likely enlargeosomes.[56]

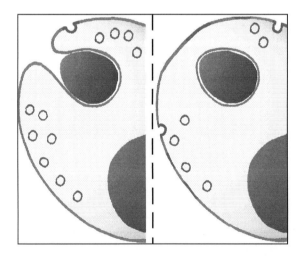

Figure 3. Nonsecretory exocytoses compensate for the large plasma membrane internalization taking place during phagocytosis. When macrophages and leukocytes phagocytize solid particles, for example bacteria, they internalize in bulk large plasma membrane sheets, thus giving rise to phagosomes. Extensive evidence demonstrates that this internalization does not result in a restriction of the cell surface because it is compensated by the Ca^{2+}-dependent exocytosis of vesicles, initiated even before the sealing of the phagosome (left) and continued thereafter (right). The nonsecretory vesicles involved have not been isolated yet. They may be of two types, using for membrane fusion either VAMP3 (cellubrevin) or VAMP7 (TiVAMP).

Exocytosis of Endosomes

During the last few years the number of endosome classes has increased from one to several, and at the moment is still growing. The classical acidic organelles are still intensely investigated, and numerous new details have been clarified concerning their origin from coated vesicles; their constitutive traffic from the early (or sorting) to the perinuclear recycling compartment; their constitutive recycling back to the plasma membrane; their alternative pathway, important especially for the endocytized materials, from early endosomes to lysosomes (see ref. 57). In addition new types of endosomes, distinct from the classical type in almost all properties: markers, lumenal pH (neutral), intracellular pathways, mechanisms of fusion etc, are attracting great attention (see refs. 58-60). The question about endosomes we intend to discuss concerns their exocytosis: Is it constitutive in all cases or regulated in some of them? Although many, most likely the vast majority of endosomes are indeed incompetent for regulated exocytosis, several exceptions have been reported.

The first example, known since several years, is the regulated exocytosis of multivesicular bodies, organelles generated by the internalization of many small vesicles from the membrane of large "mature" endosomes, taking place late in the pathway from endosomes to lysosomes. The ensuing release of the segregated vesicles (the so called exosomes), is important for various functions among which antigen presentation, transferrin receptor removal from reticulocytes, release of prion protein and aggregation of platelets (see refs. 61,62). However, multivesicular bodies are proximal to lysosomes. Their regulated exocytosis, therefore, could be considered as a form of prelysosomal (see "Introduction" section) and not bona fide endosomal exocytosis.

Other examples concern specialized forms of endosomes. The most impressive is cellularization of the drosophila eggs, a process that converts, within approximately 2hr, a single mega-cell exhibiting over 6,000 nuclei aligned below the plasma membrane into an equal number of discrete cells. In this case the enlargement of the plasma membrane is obtained by addition of intracellular membranes with formation of invaginations between adjacent nuclei, beginning with the establishment of intercellular adherens junctions. Extensive studies, carried out also by the use of dominant-negative constructs, have revealed the involvement of various proteins, including Rab11, which is considered a marker of recycling endosomes, and dynamin, known to participate both in the clathrin-dependent and in other forms of endocytosis. Whether these proteins are essential only for the membrane traffic necessary to drive vesicles to their site of insertion, or whether they are also markers of the vesicles undergoing exocytosis, is still discussed. Whatever the answer, the traffic of vesicles positive for recycling endosome markers is for sure the predominant membrane transfer necessary for cellularization (see ref. 63).

A similar example is mitotic cytokinesis (Fig. 4). Traditionally, the various phases of mitosis have received much more attention than cytokinesis, the process by which daughter cells separate from each other. For long time, in fact, the latter was believed to depend only on the contraction of the cytoskeleton. Now it is clear that, to make cell abscission possible, a considerable number of cytoplasmic vesicles needs to undergo exocytosis near the cleavage furrow. That this indeed occurs is documented also by the local appearance of a few surface markers distinct from the rest of the plasma membrane (see ref. 64). These exocytoses need to be preceded and accompanied by an intense program of coordinate membrane traffic. Endocytosis, no longer occurring from the beginning of mitosis, is resumed during later stages,[65] and large numbers of recycling endosomes are targeted to the cleavage furrow (Fig. 4). Also this membrane traffic appears regulated by Rab11, assembled in a complex with another protein, FIP3,[66] and working together with the small G protein Arf6.[64,67] All these markers suggested the recycling endosomal nature of the exocytized vesicles. Recently, however, a peculiar family of exocytic vesicles, characterized by the SNAREs known to be involved in cytokinesis, VAMP8 and syntaxin2,[68,69] have been shown to cluster at the midbody of only one of the two partners undergoing cytokinesis (Fig. 4), and to interact in a complex with a local protein, centriolin, and a SNAREbinding protein, snapin.[70] These vesicles exhibit no endosome markers. Rather,

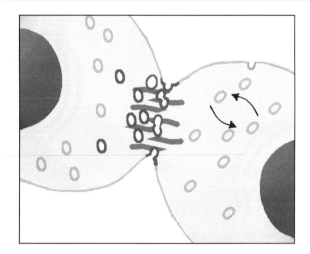

Figure 4. Cytokinesis, the last step of mitosis, is due to the exocytosis of nonsecretory vesicles. For long time the separation of the two cells at the end of telophase was believed to be due only to the contraction of the cytoskeleton, with ensuing squeezing of the cytoplasmic isthmus (midbody) connecting the two cells. It is now clear that the separation could not occur without nonsecretory exocytoses precisely focused at the midbody plasma membrane. The vesicles widely believed to sustain the process are recycling endosomes (green), recognized by specific markers (Rab12, FIP3, Arf6), which are known to undergo an intense traffic from a perinuclear area to the midbody and viceversa (arrows). Recently, however, a population of specific vesicles (blue) have been proposed to cluster near one of the two cells involved, and to undergo regulated exocytosis (Ca^{2+} dependent?) in a coordinate fashion, rapidly resulting in a sealing membrane patch in both cells. A color version of this figure is available online at http://www.Eurekah.com.

they possess an at least potential secretory competence because they accumulated, and released by exocytosis, a GFP-tagged construct devoid of retention and retrieval sequences and thus unable to proceed to endosomes and lysosomes. Moreover, disassembly of the vesicle-associated complex by mutation of either centriolin or snapin induced disruption of the midbody with block of cytokinetic abscission, confirming the key role of the vesicles in the latter process. The relationship between the new vesicles and the intense circulation of endosomes (Fig. 4) remains to be explained.[70] Whatever the nature of the vesicles, their fusion at the end of cytokinesis takes place rapidly, probably in response to a Ca^{2+} signal,[71] confirming the regulated nature of their exocytic system.

A final mention concerns phagosomes and macropinosomes. The first have been shown to respond to $[Ca^{2+}]_i$ increases, with release of free radicals and, at later stages, of digestion residues.[72,73] However, soon after their generation phagosomes fuse with lysosomes, thus their exocytoses cannot be distinguished from that of the latter organelles. Regulated exocytosis of macropinosomes is different, inasmuch as fusion of these organelles with lysosomes occurs only long time after their generation. On the other hand, exocytosis of even early macropinosomes can be very rapid and extensive upon strong $[Ca^{2+}]_i$ rise. Therefore, macropinosomes can be considered as true nonsecretory exocytic organelles.[56] Whether this is an example of endosome exocytosis is however questionable. Although extensively fused with conventional endosomes, macropinosomes are not identical to them. Competence for exocytosis appears therefore typical of the hybrid organelle, which is certainly related to endosomes, but does not coincide with them.

Conclusion

Nonsecretory regulated exocytoses are present, at least transiently, in most, if not all cells. Functionally, they play key roles in a long list of processes, some of which (such as glucose homeostasis, gastric acidification, urine volume) specifically investigated since long; and others (such as cytokinesis, embryonic development, synaptic transmission, neuronal growth, wound healing) also intensely investigated; however, until recently with no much attention to nonsecretory regulated exocytoses. Finally, it is not hard to predict that additional processes involving nonsecretory exocytoses will be discovered in the near future.

The organelles competent for these exocytic processes are most often small vesicles. Because of this phenotype, common to many other mini-organelles of different functional significance, they remain indistinguishable in conventional electron micrographs and are recognized only when adequate markers become available. Most of these vesicles remain to be isolated in pure subcellular fractions, and therefore their biochemical investigation is still at a preliminary stage. Nevertheless we now know that a fraction of the small vesicles, especially of those scattered in the superficial layers of the cytoplasm, are not involved in other, better known processes such as the traffic to and from the TGN, constitutive exocytosis and various forms of endocytosis, but are ready to be discharged by nonsecretory exocytosis regulated by intracellular messengers.

Although largely overlooked for long time, nonsecretory exocytoses are neither rare nor minor processes. In terms of number they may be even more numerous than their more established counterparts, the secretory regulated exocytoses. The latter, in fact, occur mostly in specialized cells equipped with only one or two secretory organelles, while specific nonsecretory organelles competent for regulated exocytosis can be present not only in specialized, but also in apparently nonspecialized cells. Similar to their secretory counterparts, the nonsecretory organelles are often distributed strategically within the cell. For example, in tiny microdistricts, such as the presynaptic compartments, the growth cones and the varicosities of neurosecretory cells, regulated secretory and nonsecretory exocytoses coexist and operate, involved the first in the discharge of products, the others in the regulation of intracellular events and in the expansion of the surface membrane. So far, the coordination of the two types of regulated exocytic processes, nonsecretory and secretory, has been considered only in a few cases and investigated only in small part. Therefore it is not hard to predict that exciting developments, revealing new aspects of cellular complexity, will take place in the next few years.

Acknowledgments

We thank Evelina Chieregatti for helpful discussions, Gabriella Racchetti for her important role in the experimental work and Francesca Floriani for the editorial assistance. The original work reported in this review was supported by grants from the European Community (APOPIS-LSHM-CT-2003-503330), the Italian Ministry of Education and Research(FIRB) and Telethon Fondazione ONLUS (GGP30234).

References

1. Palade GE. Intracisternal granules in the exocrine cells of the pancreas. J Biophys Biochem Cytol 1956; 2:417-242.
2. Heidenhein RPH. Handbuch der Physiologie. Leipzig, Vogel: 1883:V:173.
3. De Robertis E. Submicroscopic morphology and function of the synapse. Exp Cell Res 1958; 14:347-369.
4. De Robertis ED, Sabatini DD. Submicroscopic analysis of the secretory process in the adrenal medulla. Fed Proc 1960; 19:70-78.
5. Farquhar MG. Origin and fate of secretory granules in cells of the anterior pituitary gland. Trans NY Acad Sci 1961; 23:346-351.
6. Ito S, Winchester RJ. The fine structure of the gastric mucosa in the bat. J Cell Biol 1963; 16:541-577.
7. Lu W, Man H, Ju W et al. Activation of synaptic NMDA receptors induces membrane insertion of new AMPA receptors and LTP in cultured hippocampal neurons. Neuron 2001; 29:243-254.

8. Chou CL, Yip KP, Michea L et al. Regulation of aquaporin-2 trafficking by vasopressin in the renal collecting duct. Roles of ryanodine-sensitive Ca^{2+} stores and calmodulin. J Biol Chem 2000; 275:36839-36846.
9. Bezzerides VJ, Ramsey IS, Kotecha S et al. Rapid vesicular translocation and insertion of TRP channels. Nat Cell Biol 2004; 6:709-720.
10. Blott EJ, Griffiths GM. Secretory lysosomes. Nat Rev Mol Cell Biol 2002; 3:122-131.
11. Chieregatti E, Meldolesi J. Regulated exocytosis: New organelles for nonsecretory purposes. Nat Rev Mol Cell Biol 2005; 6:181-187.
12. Passafaro M, Rosa P, Sala C et al. N-type Ca^{2+} channels are present in secretory granules and are transiently translocated to the plasma membrane during regulated exocytosis. J Biol Chem 1996; 271:30096-30104.
13. Oynebraten I, Barois N, Hagelsteen K et al. Characterization of a novel chemokine-containing storage granule in endothelial cells: Evidence for preferential exocytosis mediated by protein kinase A and diacylglycerol. J Immunol 2005; 175:5358-5369.
14. Guan JS, Xu ZZ, Gao H et al. Interaction with vesicle luminal protachykinin regulates surface expression of delta-opioid receptors and opioid analgesia. Cell 2005; 122:619-631.
15. Forte JG, Forte TM, Black JA et al. Correlation of parietal cell structure and function. J Clin Gastroenterol 1983; 5:17-27.
16. Nielsen S, Chou CL, Marples D et al. Vasopressin increases water permeability of kidney collecting duct by inducing translocation of aquaporin-CD water channels to plasma membrane. Proc Natl Acad Sci USA 1995; 92:1013-1017.
17. Bell GI, Murray JC, Nakamura Y et al. Polymorphic human insulin-responsive glucose-transporter gene on chromosome 17p13. Diabetes 1989; 38:1072-1075.
18. Malinow R, Malenka RC. AMPA receptor trafficking and synaptic plasticity. Annu Rev Neurosci 2002; 25:103-126.
19. Sheng M, Hyoung Lee S. AMPA receptor trafficking and synaptic plasticity: Major unanswered questions. Neurosci Res 2003; 46:127-134.
20. Cai C, Li H, Rivera C et al. Interaction between SAP97 and PSD-95, two Maguk proteins involved in synaptic trafficking of AMPA receptors. J Biol Chem 2005, (Epub ahead of print).
21. Oh MC, Derkach VA, Guire ES et al. Extrasynaptic membrane trafficking regulated by GluR1 serine 845 phosphorylation primes AMPA receptors for LTP. J Biol Chem 2005, (Epub ahead of print).
22. Stellwagen D, Beattie EC, Seo JY et al. Differential regulation of AMPA receptor and GABA receptor trafficking by tumor necrosis factor-alpha. J Neurosci 2005; 25:3219-3228.
23. Wenthold RJ, Prybylowski K, Standley S et al. Trafficking of NMDA receptors. Annu Rev Pharmacol Toxicol 2003; 43:335-358.
24. Lan JY, Skeberdis VA, Jover T et al. Protein kinase C modulates NMDA receptor trafficking and gating. Nat Neurosci 2001; 4:382-390.
25. Washbourne P, Liu XB, Jones EG et al. Cycling of NMDA receptors during trafficking in neurons before synapse formation. J Neurosci 2004; 24:8253-8264.
26. Galan A, Laird JM, Cervero F. In vivo recruitment by painful stimuli of AMPA receptor subunits to the plasma membrane of spinal cord neurons. Pain 2004; 112:315-323.
27. Morenilla-Palao C, Planells-Cases R, Garcia-Sanz N et al. Regulated exocytosis contributes to protein kinase C potentiation of vanilloid receptor activity. J Biol Chem 2004; 279:25665-25672.
28. Zhang X, Huang J, McNaughton PA. NGF rapidly increases membrane expression of TRPV1 heat-gated ion channels. EMBO J 2005; 24:4211-4223.
29. Martinez-Arca S, Coco S, Mainguy G et al. A common exocytotic mechanism mediates axonal and dendritic outgrowth. J Neurosci 2001; 21:3830-3838.
30. Martinez-Arca S, Alberts P, Zahraoui A et al. Role of tetanus neurotoxin insensitive vesicle-associated membrane protein (TI-VAMP) in vesicular transport mediating neurite outgrowth. J Cell Biol 2000; 149:889-900.
31. Bouchard JF, Moore SW, Tritsch NX et al. Protein kinase A activation promotes plasma membrane insertion of DCC from an intracellular pool: A novel mechanism regulating commissural axon extension. J Neurosci 2004; 24:3040-3050.
32. Song HJ, Poo MM. Signal transduction underlying growth cone guidance by diffusible factors. Curr Opin Neurobiol 1999; 9:355-363.
33. Pfenninger KH, Laurino L, Peretti D et al. Regulation of membrane expansion at the nerve growth cone. J Cell Sci 2003; 116:1209-1127.
34. Wayman GA, Kaech S, Grant WF et al. Regulation of axonal extension and growth cone motility by calmodulin-dependent protein kinase I. J Neurosci 2004; 24:3786-3794.

35. Laurino L, Wang XX, de la Houssaye BA et al. PI3K activation by IGF-1 is essential for the regulation of membrane expansion at the nerve growth cone. J Cell Sci 2005; 118:3653-3662.
36. Jaiswal JK, Andrews NW, Simon SM. Membrane proximal lysosomes are the major vesicles responsible for calcium-dependent exocytosis in nonsecretory cells. J Cell Biol 2002; 159:625-635.
37. Cerny J, Feng Y, Yu A et al. The small chemical vacuolin-1 inhibits Ca(2+)-dependent lysosomal exocytosis but not cell resealing. EMBO Rep 2004; 5:883-888.
38. Borgonovo B, Cocucci E, Racchetti G et al. Regulated exocytosis: A novel, widely expressed system. Nat Cell Biol 2002; 4:955-962.
39. Cocucci E, Racchetti G, Podini P et al. Enlargeosome, an exocytic vesicle resistant to nonionic detergents, undergoes endocytosis via a nonacidic route. Mol Biol Cell 2004; 15:5356-5368.
40. Coorssen JR, Schmitt H, Almers W. Ca^{2+} triggers massive exocytosis in Chinese hamster ovary cells. EMBO J 1996; 15:3787-3791.
41. Ninomiya Y, Kishimoto T, Miyashita Y et al. Ca^{2+}-dependent exocytotic pathways in Chinese hamster ovary fibroblasts revealed by a caged-Ca^{2+} compound. J Biol Chem 1996; 271:17751-17754.
42. Bansal D, Miyake K, Vogel SS et al. Defective membrane repair in dysferlin-deficient muscular dystrophy. Nature 2003; 423:168-172.
43. McNeil PL, Kirchhausen T. An emergency response team for membrane repair. Nat Rev Mol Cell Biol 2005; 6:499-505.
44. Uldry M, Steiner P, Zurich MG et al. Regulated exocytosis of an H+/myo-inositol symporter at synapses and growth cones. EMBO J 2004; 23:531-540.
45. Kasai H, Kishimoto T, Liu TT et al. Multiple and diverse forms of regulated exocytosis in wild-type and defective PC12 cells. Proc Natl Acad Sci USA 1999; 96:945-949.
46. Watson RT, Kanzaki M, Pessin JE. Regulated membrane trafficking of the insulin-responsive glucose transporter 4 in adipocytes. Endocr Rev 2004; 25:177-204.
47. Ishiki M, Klip A. Minireview: Recent developments in the regulation of glucose transporter-4 traffic: New signals, locations, and partners. Endocrinology 2005; 146:5071-5078.
48. Brown D. The ins and outs of aquaporin-2 trafficking. Am J Physiol Renal Physiol 2003; 284:F893-901.
49. Valenti G, Procino G, Tamma G et al. Minireview: Aquaporin 2 trafficking. Endocrinology 2005; 146:5063-5070.
50. Yao X, Forte JG. Cell biology of acid secretion by the parietal cell. Annu Rev Physiol 2003; 65:103-131.
51. Seino S, Shibasaki T. PKA-dependent and PKA-independent pathways for cAMP-regulated exocytosis. Physiol Rev 2005; 85:1303-1342.
52. Procino G, Caces DB, Valenti G et al. Adipocytes support cAMP-dependent translocation of Aquaporin 2 (AQP2) from intracellular sites distinct from the insulin-responsive GLUT4 storage compartment. Am J Physiol Renal Physiol 2005, (Epub ahead of print).
53. Herant M, Heinrich V, Dembo M. Mechanics of neutrophil phagocytosis: Behavior of the cortical tension. J Cell Sci 2005; 118:1789-1797.
54. Bajno L, Peng XR, Schreiber AD et al. Focal exocytosis of VAMP3-containing vesicles at sites of phagosome formation. J Cell Biol 2000; 149:697-706.
55. Braun V, Fraisier V, Raposo G et al. TI-VAMP/VAMP7 is required for optimal phagocytosis of opsonised particles in macrophages. EMBO J 2004; 23:4166-4176.
56. Falcone S, Cocucci E, Podini P et al. Macropinocytosis triggers regulated exocytosis in human dendritic cells. EMBO-FEBS-ESF Workshop on Membrane Dynamics in Endocytosis, Sant Feliu de Guixols (Spain) 2005.
57. Maxfield FR, McGraw TE. Endocytic recycling. Nat Rev Mol Cell Biol 2004; 5:121-132.
58. Parton RG, Richards AA. Lipid rafts and caveolae as portals for endocytosis: New insights and common mechanisms. Traffic 2003; 4:724-738.
59. Nabi IR, Le PU. Caveolae/raft-dependent endocytosis. J Cell Biol 2003; 161:673-677.
60. Rajendran L, Simons K. Lipid rafts and membrane dynamics. J Cell Sci 2005; 118:1099-1102.
61. Stahl PD, Barbieri MA. Multivesicular bodies and multivesicular endosomes: The "ins and outs" of endosomal traffic. Sci STKE 2002; PE32.
62. de Gassart A, Geminard C, Hoekstra D et al. Exosome secretion: The art of reutilizing nonrecycled proteins? Traffic 2004; 5:896-903.
63. Lecuit T. Junctions and vesicular trafficking during Drosophila cellularization. J Cell Sci 2004; 117:3427-3433.
64. Matheson J, Yu X, Fielding AB et al. Membrane traffic in cytokinesis. Biochem Soc Trans 2005; 33:1290-1294.
65. Schweitzer JK, Burke EE, Goodson HV et al. Endocytosis resumes during late mitosis and is required for cytokinesis. J Biol Chem 2005; 280:41628-41635.

66. Wilson GM, Fielding AB, Simon GC et al. The FIP3-Rab11 protein complex regulates recycling endosome targeting to the cleavage furrow during late cytokinesis. Mol Biol Cell 2005; 16:849-860.
67. Fielding AB, Schonteich E, Matheson J et al. Rab11-FIP3 and FIP4 interact with Arf6 and the exocyst to control membrane traffic in cytokinesis. EMBO J 2005; 24:3389-3399.
68. Conner SD, Wessel GM. Syntaxin is required for cell division. Mol Biol Cell 1999; 10:2735-2743.
69. Low SH, Li X, Miura M et al. Syntaxin 2 and endobrevin are required for the terminal step of cytokinesis in mammalian cells. Dev Cell 2003; 4:753-759.
70. Gromley A, Yeaman C, Rosa J et al. Centriolin anchoring of exocyst and SNARE complexes at the midbody is required for secretory-vesicle-mediated abscission. Cell 2005; 123:75-87.
71. Shuster CB, Burgess DR. Targeted new membrane addition in the cleavage furrow is a late, separate event in cytokinesis. Proc Natl Acad Sci USA 2002; 99:3633-3638.
72. Shimada O, Ishikawa H, Tosaka-Shimada H et al. Exocytotic secretion of toxins from macrophages infected with Escherichia coli O157. Cell Struct Funct 1999; 24:247-253.
73. Di A, Krupa B, Bindokas VP et al. Quantal release of free radicals during exocytosis of phagosomes. Nat Cell Biol 2002; 4:279-285.

CHAPTER 11

Adaptation of the Secretory Machinery to Pathophysiological Conditions

Abderrahmani Amar*

Abstract

Regulated exocytosis is a fundamental and common feature of all secretory cells specialized in the release of essential bioactive substances. This process is tightly controlled to adapt the amount of released products in response to the variable physiological cues. However it occurs that inadequate, excessive and/or prolonged exposure to environmental demands become stressful and desensitizes or irreversibly damage the cell secretory competence. The insulin-releasing β-cells of the islets of Langerhans represent a well characterized model of adaptation of the secretory response. In this system, while physiological elevation of glucose or lipids concentration can acutely stimulate insulin release, prolonged exposure to these secretagogues can have deleterious effects on the cells contributing to severe insulin secretion defects in diabetic patients. New emerging data provide evidence that prolonged exposure to elevated glucose concentration is associated with a reduction in the expression of a set of genes essential for exocytosis. This chapter discusses this phenomenon, which occurs at transcriptional levels and leads to a decline in the secretory function. These glucose effects appear to be mediated by a sustained rise in cAMP levels, which in turn induces expression of the inducible cAMP early repressor (ICER), a transcription factor known to be activated in response to stress. Based on these findings, this chapter highlights the role of ICER in the control of the exocytotic apparatus challenged with a stressful environment.

Evidences for Adaptation of the Secretory Machinery to Physiopathologic Conditions: The Pancreatic β-Cell Model

Regulated exocytosis is a complex process that comprises the sequential participation of numerous subcellular mechanisms. Hormone secretion is dictated by extracellular stimuli that are transduced into activation/deactivation of signalling pathways and genes expression. This will ultimately determine the precise availability of hormone to be released. A well characterized secretory system is represented by the β cells of the islets of Langerhans from endocrine pancreas in which the molecular basis of insulin release and its regulation has been extensively studied. The endocrine pancreas is permanently subjected to dynamic changes in response to variations in blood glucose plasma levels, leading to an increase in insulin secretion from β cells. This physiological task needs to be finely tuned because insulin is not only required for glucose homeostasis and survival, but can also be lethal within minutes if inappropriately high concentrations are released. Thus, to match insulin secretion to the physiological demands, the

*Correspondence Author: Abderrahmani Amar—Service of Internal Medicine and Department of Cell Biology and Morphology, Rue du Bugnon 9, CH-1005, Lausanne, Switzerland. Email: Amar.Abderrahmani@unil.ch

Molecular Mechanisms of Exocytosis, edited by Romano Regazzi. ©2007 Landes Bioscience and Springer Science+Business Media.

differentiated β cell displays a sophisticated degree of plasticity that allows adaptation at various ranges of time, preventing shortage and excesses of insulin in the circulation.

The Machinery of Insulin Exocytosis

Proinsulin, the precursor of insulin, is synthesized in the endoplasmic reticulum and undergoes a series of maturation steps, starting in the Golgi. The product is then packaged into secretory granules or large dense-core vesicles that gradually acidify, allowing further processing to become insulin.[1] These granules are found throughout the cytosol and eventually translocate to the plasma membrane. The ultimate fusion of the granule with the plasma membrane is triggered by Ca^{2+} and controlled by a complex network of protein-protein and protein-lipid interactions that are similar to membrane fusion events occurring in other secretory cells. A vast body of molecular and physiological data accumulated over the past years has led to a unifying model for membrane fusion.[2] It is therefore not surprising that many of the proteins involved in the regulation of neurotransmitter release have also been identified in the pancreatic β-cell and demonstrated to participate in insulin secretion (for review see ref. 3). These proteins include the soluble *N*-ethylmaleimide-sensitive fusion protein [NSF] attachment protein [SNAP] receptors (SNAREs), synaptobrevin-2 (VAMP2), syntaxin-1A, and 25-kDa synaptosomal-associated protein (SNAP-25). It is now clear that a ternary complex consisting of these proteins plays a central role in insulin exocytosis, by tethering the granules to the plasma membrane.[4] During the last few years many of the proteins playing a key role in the regulation of the assembly of the SNARE complex have been identified. These proteins include Munc18 and complexin-1 and members of the Synaptotagmin family that couple Ca^{2+} changes to vesicle fusion.[3,5-7] In addition, Rab3a and Rab27a, two members of the Rab GTPase family that associate with insulin-containing secretory granules, and their effectors RIM2, MyRIP/Slac2c, Noc2 and Granuphilin/Slp4 have been found to be necessary for proper regulation of insulin release (Fig. 1).[8-13]

Coupling Glucose Stimulation to Insulin Exocytosis

Under condition of prolonged physiological stimulation, insulin secretion is biphasic. The first phase corresponds to the Ca^{2+}-dependent release of granules already docked to the plasma membrane, belonging to the "readily releasable pool".[3,14,15] This pool account for less than 1% of the granules and can be depleted in less than a second (for review see ref. 14). The second phase or the sustained phase reflects the relatively slow recruitment of granules from a reserve

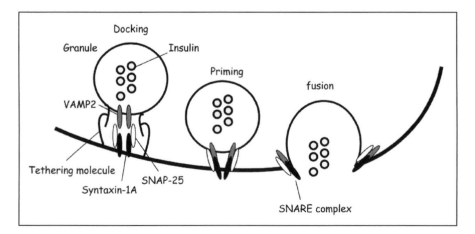

Figure 1. Schematic presentation of exocytosis. Exocytosis of insulin-containing granules implies vesicle recruitment, docking, priming and fusion.

pool, probably to some extent their translocation to the plasma membrane.[3,14] After docking at the release site the granules undergo an energy-dependent process (priming) that prepares them for fusion. Several sequential steps are necessary to render the granule release-competent, and numerous ideas have been put forward to explain the ATP-dependence of priming. These include the ATP-dependent synthesis of phosphoinositides and protein kinase-mediated protein phosphorylation.[16] Exocytosis of insulin-containing granules is regulated through a complex metabolic network in which the increased flux of glucose elevates the intracellular ATP/ADP ratio.[3,17-20] This results in the closure of the ATP-sensitive potassium channel (KATP channels) and subsequent cellular depolarization.[21,22] In turn, depolarization causes the influx of extracellular Ca^{2+} by the activation of the L-type voltage dependent Ca^{2+} channels triggering the initial fusion of insulin-containing granules with the plasma membrane.[23] Although Ca^{2+} influx is a an essential prerequisite for glucose-induced insulin exocytosis, other intracellular signals generated by sugar metabolism such as phospholipids, diacylglycerol and cAMP can potentiate the secretory response (Fig. 2).[24-26]

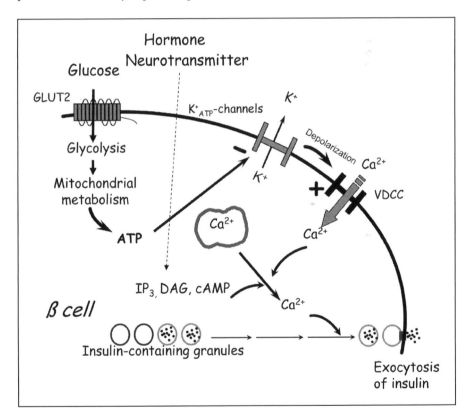

Figure 2. Model of insulin secretion. Glucose-induced insulin secretion and its potentiation constitute the principal mechanism of insulin release. Glucose is transported by the glucose transporter (GLUT2) into the pancreatic β-cell. Metabolism of glucose increases ATP production (and the ATP-to-ADP ratio), which in turn induce subsequently the closure of the K_{ATP} channels, resulting in membrane depolarization and opening of the voltage-dependent Ca^{2+} channels (VDCCs). Finally, the Ca^{2+} influx triggers insulin granule exocytosis. Insulin granule exocytosis is also regulated by hormones and neurotransmitters, which generate intracellular signals such as cAMP, diacylglycerol (DAG), and inositol trisphosphate (IP_3).

Loss of β-Cell Secretory Properties in Type 2 Diabetes

Failure in the β-cell secretory capacity is a characteristic feature of all forms of diabetes. In Type 2 Diabetes (T2D), the most prominent form of the disease, diminished sensitivity of insulin target tissues and reduced physical activity are major contributors of metabolic unbalance, but diabetes develops only when β-cells fail to compensate for the increased insulin demand. It is now believed that this β-cell inadequacy results from a combination of secretory dysfunction and insufficient β-cell mass caused by prolonged exposure to glucose and lipids in β-cells. The glucose and lipid toxicity concept postulates that β-cells are devised to function within a narrow range of plasma glucose and lipid levels and that even modest but chronic increases in the concentration of these nutrients lead to an unfavourable environment, which causes alteration in β-cell function. Several reports have now provided evidence for abnormalities of the secretory machinery, possibly accounting for β-cell dysfunction in T2D. A marked reduction in expression of secretory vesicle (*v*)-SNARE (VAMP-2 and Synaptotagmin) and target (*t*) membrane *t*-SNARE (syntaxin1A and SNAP-25) have been observed in islets from patients with T2D.[27] This finding has been confirmed in rat models of T2D. Among these T2D models are the Goto-Kakizaki (GK) rats that offer similarities to genetic predisposition to diabetes in humans. In these rats, not only the signalling coupling glucose metabolism to K_{ATP} channels is impaired, but the secretory machinery of β-cells is also defective.[24-26,28-31] Thus, the number of docked granules is reduced and the first phase of insulin release is reduced.[32] These defects were associated with a consistent reduction in several proteins required for β-cell exocytosis including the plasma membrane associated *t*-SNAREs syntaxin1A and SNAP-25.[33] Importantly, recovery of normal levels of these *t*-SNAREs was able to partially restore insulin secretion.[33] In islets of GK rats, diminution in the expression of several other proteins participating in the insulin secretory process including the *v*-SNARE VAMP2; synaptotagmin III, cysteine string protein (CSP), Munc-18, α-SNAP and NSF has also been reported.[34] However, because of the variation in many metabolic parameters and of genetic factors, it is difficult to attribute the loss in the expression of these genes to direct effects of environmental factors such as glucose and/or lipids.

Decline in the Expression of Exocytotic Genes in Cultured Insulin-Secreting Cells

Prolonged exposure of cultured insulin-secreting cells to elevated glucose concentrations lead to secretory defects including excessive insulin release under basal conditions and diminished secretory capacity in response to stimuli.[35] Thus, these experimental conditions offer an adequate in vitro model that mimic glucose-induced β-cell dysfunction without the confounding genetic and metabolic variables found in diabetic patients and GK rats. As expected, incubation of rat insulin-secreting INS-1E cells for two days in the presence of 20 mM glucose resulted in impairment in the secretory process in response to glucose.[36] K$^+$ is also used as an insulin secretagogue because it induces a direct depolarization of the β-cell membrane downstream of ATP production and K$^+$-ATP channel closure. Importantly, the insulin secretory response elicited by depolarizing K$^+$ concentrations is also defective in cells cultured for 48 hours in the presence of high glucose concentration.[36] Therefore, these data indicate possible alterations in signals coupling calcium entry to activation of the exocytotic machinery. In view of this finding, the expression levels of several key genes involved in insulin exocytosis was investigated. Northern blot, western blot and immunocytochemistry analyses indicated a drastic decline in the amount of the Rab GTPases Rab3a and Rab27a and of their effectors, Granuphilin and Noc2.[36] These effects of glucose occured in a time- and a dose-dependent manner.[36] In agreement with these data, the loss of expression in these four genes was confirmed in rat primary islets cultured for 72 and 96 hours at 33 mM of glucose. Unexpectedly, the expression of the SNAREs Syntaxin-1, SNAP-25 and VAMP-2 were unchanged in cells cultured at high glucose concentration. Interestingly, impaired expression of Rab3a, Rab27a, Granuphilin and Noc2 was reversible.[36] In fact, 36 hours of incubation in 2 mM glucose after two days of exposure to 20 mM glucose was sufficient to recover normal protein levels and to restore exocytosis.[36] Thus, these in vitro models provide functional evidence of an adaptation of the exocytotic machinery to chronic hyperglycemia. Although the mechanism that silences expression of exocytotic genes in GK rats is still unknown, the observed insulin secretory defects may also be the result of direct effects of chronic hyperglycemia. More in vivo experiments should be performed to confirm the decline in exocytotic genes in response to chronic elevation of plasma glucose concentration. The diabetic model of partially pancreatectomized rats would be ideally suited to address this problem. This approach would offer a better defined model to study the effects of diabetic environment without the confounding genetic variables of GK rats.

Although chronic hyperglycemia is undoubtedly a major contributor of β-cell dysfunction, it becomes apparent that abnormally elevated concentrations of circulating nonesterifed fatty acids (NEFA) often accompany the progression of T2D.[37-39] A number of in vitro and in vivo studies have shown that chronic exposure to NEFA adversely affects pancreatic β-cell function, leading to inhibition of glucose-stimulated insulin secretion.[37,39,40] An important observation is that incubation of islet β-cells or mouse insulin-secreting MIN6 cells for 48 h with NEFA (particularly 1 mmol/l of palmitate) causes an increase in SNAP-25 protein levels that is associated with a loss of insulin secretion in response to secretagogues.[41,42] Considering the role of SNAP-25 in the tight control of the exocytotic machinery, any alteration in its expression or function could have an impact on the secretory granules fate.[43] In view of the effects of prolonged exposure of glucose described above, the expression levels of other exocytotic genes such as Rab and Rab effectors genes will need to be carefully examined also in cells chronically cultured in the presence of high NEFA concentration.

Transcriptional Effects of the cAMP Signalling Pathways: Role of the Inducible cAMP Early Repressor (ICER)

Data obtained from the INS1-E and islets cells cultured at high glucose concentration have highlighted the importance of the cyclic adenosine monophosphate (cAMP) signalling cascade in mechanisms responsible for the loss of exocytotic proteins. The concomitant decline in mRNA levels of Rab3a, Rab27a, Granuphilin and Noc2 suggests that a common mechanism modulates their expression at transcriptional and/or post-transcriptional levels. The impaired activity of the Granuphilin promoter in INS-1E cells cultured with supraphysiological concentration of glucose argued in favour of a transcriptional modification.[36] To identify the molecular mechanism underlying this process a close inspection of promoter regions of genes impaired by glucose has been done.[36] This analysis in complement with a developed approach based on the genome wide scan revealed the presence of elements responsive to cAMP (CRE) TGACGTCA within *Rab3a, Rab27a, Granuphilin* and *Noc2* genes.[36,44] Under physiological conditions the cAMP signalling pathway is known to enhance the first and second phase of insulin secretion. When a meal is digested cAMP levels are normally raised by incretins such as gastric inhibitory polypeptide/glucose-dependent insulinotropic peptide (GIP), glucagon-like peptide-1 (GLP-1) and by glucose.[45-47] In this case, it is thought that the effects of cAMP in regulated exocytosis are mediated by activation of cAMP-dependent protein kinase A (PKA), a heterotetrameric molecule composed of two regulatory and two catalytic subunits. This leads to phosphorylation of exocytotic proteins.[48] A cAMP signalling pathway independent from PKA has also been shown to participate in the stimulation of insulin secretion by directly affecting the activity of the exocytotic machinery.[49-51] In contrast to these short term-effects, a rise in cAMP levels due to persistent exposure to supraphysiological concentration of glucose is associated with a loss in glucose-induced insulin secretion.[52] In these experimental conditions incubation of INS1-E cells to cAMP-raising agents such as forskolin and IBMX for 48 hours mimics the glucose effects, including the loss in granuphilin promoter activity and the decline in Rab3a, Rab27a, Granuphilin and Noc2 expression.[36] Furthermore, inhibition of PKA prevents the loss of the four exocytotic genes caused by glucose or cAMP-raising agents.[36]

The Inducible cAMP Early Repressor (ICER): A Powerful Repressor

The inducible cAMP early repressors (ICERs), are transcription factors that belong to the cAMP-responsive element modulator (CREM) family and are targets of cAMP/PKA pathways.[53] ICER isoforms are generated by the use of an alternative P2 promoter, lying near the 3' end of the *CREM* gene.[53,54] In contrast to other CRE-binding proteins (CREBs), ICER lacks the activation domain and thereby functions as a transcriptional repressor.[53,54] ICER expression is normally transiently induced over a time frame of several hours following intracellular elevation of cAMP. The induction of this repressor requires the activation of CREBs, which consist of a family of both activators and repressors of transcription.[55,56] After a rise in cAMP concentration these transcription factors are rapidly phosphorylated by PKA thereby converting CREB to a powerful activator.[57,58] Phosphorylated CREB binds to four CRE elements in the P2 promoter and activates ICER expression.[53] ICER can also bind to its P2 promoter, therefore repressing its own expression and allowing cAMP signalling to return to the basal state (Fig. 3).[53] This negative feedback loop seems to be altered after chronic exposure to glucose and sustained induction of ICER in INS1-E and islet cells cultured with high glucose parallels the loss of Rab3a, Rab27a, Granuphilin and Noc2.[36] Gel retardation assays demonstrated the ability of ICER to bind to all CRE motifs found in the promoter region of these genes. ICER contains DNA binding and leucine zipper domains, but not the N-terminal transactivation domain and thus, functions as an endogenous inhibitor of gene transcription. Forced β-cell-directed expression of ICER I in mice leads to severe Diabetes characterized by a reduction in glucose-induced secretion of insulin.[59] In line with this observation, overexpression of ICER in INS1-E cells represses expression of Rab3a, Rab27a, Granuphilin and Noc2 and results in a significant decrease in a pronounced reduction in hormone release in the presence of secretagogues.[36] Silencing of ICER with an antisense plasmid prevents the decrease in the expression of the four proteins caused by glucose and Forskolin/IBMX. Tagen together, these data provide evidence that sustained induction of ICER is responsible for pathological consequences elicited by chronic hyperglycemia in insulin-secreting cells. Indeed, abnormally elevated levels of this transcriptional repressor have been observed in islets chronically to exposed to palmitate and in islets from GK rats.[40,60] Therefore, it seems very likely that diminished expression of Rab3a, Rab27, Granuphilin and Noc2 can contribute to the loss of glucose-induced insulin secretion in T2D.

Induction of ICER in other Secretory Cells in Response to Stress: Potential Implication for the Secretory Machinery

The cAMP signalling pathway is also a major mediator for secretion of several other hormones from anterior pituitary cells including, β-endorphin, growth hormone (GH), prolactin (PRL) or adenocorticotropic hormone (ACTH).[61-63] This hormonal release plays a pivotal role in coordinating the physiologic adaptation of the hypothalamo-pituitary-adrenal axis to stress.[59,64-67] In this case, stress is defined as physiological changes that accompany increased physical and psychological demands.[68] Increasing evidence suggests that ICER plays a regulatory role in hormone release in pituitary cells. Consistently, targeted suppression of CREM gene in mice, leading to the loss of ICER expression, result in chronic increase in plasma β-endorphin.[69] A model of pituitary corticotroph AtT20 cell line overexpressing ICER has been developed to measure the effects of ICER on hormone secretion.[70] Ectopic expression of ICER blocks the synthesis of ACTH at post-translational levels and alters stimulus-induced secretion of this hormone.[70] It can be assumed that this loss of function is at least in part the result of perturbations in the function of the secretory apparatus. Consistent with this hypothesis, is the fact that hormone release from pituitary cells share with β-cells part of the components of the exocytotic machinery including Rab3a and Rab27a.[71-75] Because these genes are targets of ICER, it can be speculated that in pituitary cells overexpressing ICER the loss in the expression of these proteins could contribute to the defect in the secretory response. Hormone release from pituitary cells is also triggered by adrenergic receptors that are able to raise

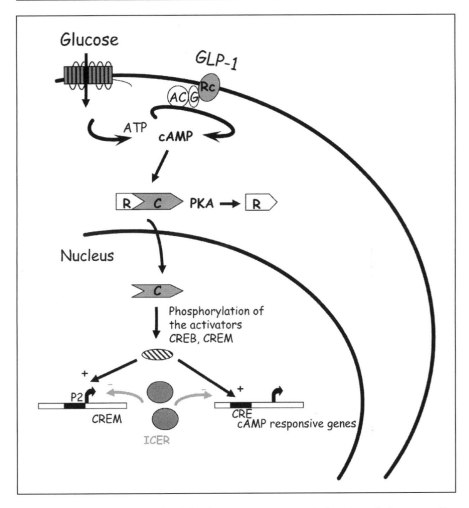

Figure 3. Scheme of ICER inducibility by cAMP. A sustained elevation of glucose and/or activation of membrane receptors (RC) (e.g., GLP-1 receptor, Glucagon) stimulate the activity of the membrane-associated adenylyl cyclase (AC) (ref. 45,52). This converts ATP to cAMP which causes the dissociation of the inactive tetrameric protein kinase A (PKA) complex into the active catalytic subunits (C) and the regulatory subunits (R). Catalytic subunits migrate into the nucleus where they phosphorylate transcriptional activators such as CREB and CREM, which in turn become activated. These factors then interact with the cAMP responsive elements (CRE) found in the promoters of cAMP-responsive genes to activate transcription and in the CREM intronic promoter (P2), which drives ICER expression. ICER proteins, in turn, can bind CREs and repress transcription from promoters containing these elements and from the CREM P2 promoter that contains four DNA elements capable of binding ICER.

intracellular cAMP levels. Prolonged agonist-mediated activation of these receptors leads to strong activation of ICER in mouse pituitary gland.[69]

Ovaries produce estradiol (E2), an estrogen hormone that controls the expression and secretion of PRL and GH.[76] Long-time exposure of cultured pituitary cells to E2 inhibits stimulated-release of GH.[77] It is established that long-term E2 administration produces PRL-cell adenomas and perturbed PRL release.[78] Interestingly, it has been observed that sustained culture in the presence of E2 causes a decrease in *t*-SNAREs and Rab3a expression levels in rat pituitary glands.[79] E2 treatment is known to induce the cAMP signalling cascade, suggesting a potent role of ICER in the E2-mediated reduction of Rab3a in pituitary cells.[80,81] In addition, in adrenal cortex of hypophysectomized rats, ICER can be rapidly and strongly induced after ACTH administration.[82] Increase in the plasma concentration of ACTH is one of the first events that take place after stress.[83]

Expression of ICER mRNA has been shown to be gradually increased by noradrenalin during the circadian rhythm in brain structures such as pineal and pituitary glands.[54,84,85] It is well documented that hormonal release from pituitary cells exhibits rhythmic patterns.[86,87] Interestingly, Rab3a is critical for short- and long-term synaptic plasticity and circadian motor activity in brain.[88] Thus, these data indicate that ICER may play a pivotal role in modulating hormone release and directing the synaptic plasticity.

Conclusion and Perspective

This chapter has highlighted the fact that the secretory apparatus can be directly modified in response to stressful environment. Based on β-cell models, one transcriptional mechanism has been deciphered that can account for the decline in the expression of key components of the β-cell secretory machinery. This involves the transcriptional repressor ICER, which selectively silences expression of a set of genes directly involved in exocytosis. Typically, elevation in ICER expression is observed in neuronal and neuroendocrine cells in response to different stressors (cognitive and physical) or to a relative excess and prolonged, environmental demands. In these cases, this process is physiological and is part of an adaptive mechanism that allows secretory cells to return to basal state. Therefore, it can be assumed that transient induction of ICER is primarily a regulatory process that enables β-cells to adjust insulin secretion after a meal. In contrast, persistent activation of ICER caused by chronic hyperglycemia may represent a pathogenic effect susceptible to worsen the insulin secretory defects and thereby contributing to the progression of diabetes. Measurement of ICER expression levels in islets of partially pancreatectomized diabetic rats may be useful to support this hypothesis. The β-cell cAMP-mediated repression of *Rab3a*, *Rab27a*, *granuphilin* and *Noc2* genes raises the speculation that cAMP pathways might also up-regulate expression of these genes on a rapid time scale to potentiate the acute stimulation of insulin secretion. Indeed, as evoked herein, rapid activation of CREB transcriptional activators by cAMP precedes induction of ICER.[53] CREBs bind to CRE where they can activate transcription of genes. A forthcoming investigation would be to examine the transcription rates of these four ICER target genes over a shorter time frame. Such findings would have relevant consequences for understanding the physiological regulation of the secretory machinery. However, it is still unclear how the loss of expression in *t*- or *v*-SNAREs occurs in β-cells of diabetic patients and GK rats as well as the up-regulation of SNAP-25 by NEFA. In the first case, considering the available in vitro data, it seems unlikely that expression levels of SNAREs components are affected by hyperglycemia or by NEFA. Thus, studies are required to address whether these modulations are provoked by diabetic environment or by genetics. An increase in SNAP-25 levels in the brain of patients with Down's Syndrome has been demonstrated to be mediated by an increase in protein stability.[89] Such mechanism could account for the increase of SNAP-25 by NEFA.[41,42] Consequently, it will be important to search in a systematic manner e.g., by proteome analysis for potential translational and post translational modifications of exocytotic proteins in β-cells challenged by a stressful environment.

Experiments performed in β-cells challenged for a long time with diabetic milieu suggest the implication of transcription factors in the changes in exocytotic gene expression. Chronic hyperglycemia and elevated free fatty acids have been recognized to elevate the expression of CAAT enhancer binding protein (C/EBPβ) and to mediate activation of the nuclear factor-κB (NF-κB).[90-95] C/EBPβ inhibits transcription of the insulin gene.[92] NF-κB is considered as an essential cell mediator acting at the crossroads of life and death in both neurons and insulin-secreting cells.[96-99] In fact NF-κB is activated by phosphorylation and subsequent degradation of inhibitor κB (IκB), which renders NF-κB free to enter the nucleus and modulate genes expression.[100] A recent work shows that the islet-cell specific silencing of NF-κB in mice leads to impaired insulin secretion in response to glucose and arginine.[101] This loss of function has been associated with perturbed expression in Rab3C and Rab3D, two members of the Rab3 GTPase family that are implicated in stimulus-induced insulin exocytosis.[9,101] Future studies will have to clarify the role of C/EBPβ and NF-κB activation on the expression of genes encoding the components of the exocytotic apparatus.

While transcriptional regulation seems extremely important, alterations in the function of secretory machinery in response to pathophysiological conditions could also occur at the post-transcriptional and post-translational levels. The recent identification of microRNAs as major regulators of gene expression, offers a promising field to investigate the mechanism that regulate expression of genes at post-transcriptional levels under stressful environment (for review see ref. 102). Although relatively few studies have been carried out on post-translational modification, a case can be illustrated by the neuronal-specific synapsin I. Synapsin I is an important molecule that is believed to regulate the association of synaptic vesicles to the microtubule cytoskeleton.[103,104] Activation of synapsin I is dependent on phosphorylation. It has been reported that phosphorylation sites of synapsin I are changed in transgenic mice model of Huntington's disease, leading to a reduced activity of synapsin I.[105]

Finally, a careful examination of transcriptional, translational and post-translational modifications of components of the exocytotic apparatus may become a fascinating area of future research in view of better understanding the molecular pathogenesis of various neurological and endocrine disorders.

Acknowledgements

The author is indebted to Prof. Gérard Waeber for his invaluable aid, to Guy Niederhauser for excellent technical assistance and is grateful to Dr. Margo Rumerio and Dr. Thierry Roger for the critical reading of the manuscript. The work was supported by the Swiss National Science Foundation grant (3100A0-105425), the Placide Nicod and Octav Botnar Foundations.

References

1. Hutton JC. Insulin secretory granule biogenesis and the proinsulin-processing endopeptidases. Diabetologia 1994; 37(Suppl 2):S48-S56.
2. Calakos N, Scheller RH. Synaptic vesicle biogenesis, docking, and fusion: A molecular description. Physiol Rev 1996; 76(1):1-29.
3. Lang J. Molecular mechanisms and regulation of insulin exocytosis as a paradigm of endocrine secretion. Eur J Biochem 1999; 259(1-2):3-17.
4. Sollner T, Whiteheart SW, Brunner M et al. SNAP receptors implicated in vesicle targeting and fusion. Nature 1993; 362(6418):318-324.
5. Easom RA. Beta-granule transport and exocytosis. Semin Cell Dev Biol 2000; 11(4):253-266.
6. Rorsman P, Renstrom E. Insulin granule dynamics in pancreatic beta cells. Diabetologia 2003; 46(8):1029-1045.
7. Abderrahmani A, Niederhauser G, Plaisance V et al. Complexin I regulates glucose-induced secretion in pancreatic {beta}-cells. J Cell Sci 2004; 117(Pt 11):2239-2247.
8. Coppola T, Frantz C, Perret-Menoud V et al. Pancreatic beta-cell protein granuphilin binds Rab3 and Munc-18 and controls exocytosis. Mol Biol Cell 2002; 13(6):1906-1915.
9. Iezzi M, Escher G, Meda P et al. Subcellular distribution and function of Rab3A, B, C, and D isoforms in insulin-secreting cells. Mol Endocrinol 1999; 13(2):202-212.
10. Waselle L, Coppola T, Fukuda M et al. Involvement of the Rab27 binding protein Slac2c/MyRIP in insulin exocytosis. Mol Biol Cell 2003; 14(10):4103-4113.
11. Yi Z, Yokota H, Torii S et al. The Rab27a/granuphilin complex regulates the exocytosis of insulin-containing dense-core granules. Mol Biol Cell 2002; 22(6):1858-1867.
12. Ozaki N, Shibasaki T, Kashima Y et al. cAMP-GEFII is a direct target of cAMP in regulated exocytosis. Nat Cell Biol 2000; 2(11):805-811.
13. Cheviet S, Waselle L, Regazzi R. Noc-king out exocrine and endocrine secretion. Trends Cell Biol 2004; 14(10):525-528.

14. Barg S. Mechanisms of exocytosis in insulin-secreting B-cells and glucagon-secreting A-cells. Pharmacol Toxicol 2003; 92(1):3-13.
15. Hay JC, Martin TF. Resolution of regulated secretion into sequential MgATP-dependent and calcium-dependent stages mediated by distinct cytosolic proteins. J Cell Biol 1992; 119(1):139-151.
16. Klenchin VA, Martin TF. Priming in exocytosis: Attaining fusion-competence after vesicle docking. Biochimie 2000; 82(5):399-407.
17. Ashcroft FM, Proks P, Smith PA et al. Stimulus-secretion coupling in pancreatic beta cells. J Cell Biochem 1994; 55(Suppl):54-65.
18. Detimary P, Dejonghe S, Ling Z et al. The changes in adenine nucleotides measured in glucose-stimulated rodent islets occur in beta cells but not in alpha cells and are also observed in human islets. J Biol Chem 1998; 273(51):33905-33908.
19. Gembal M, Gilon P, Henquin JC. Evidence that glucose can control insulin release independently from its action on ATP-sensitive K+ channels in mouse B cells. J Clin Invest 1992; 89(4):1288-1295.
20. Maechler P, Wang H, Wollheim CB. Continuous monitoring of ATP levels in living insulin secreting cells expressing cytosolic firefly luciferase. FEBS Lett 1998; 422(3):328-332.
21. Inagaki N, Gonoi T, Clement JP et al. Reconstitution of IKATP: An inward rectifier subunit plus the sulfonylurea receptor. Science 1995; 270(5239):1166-1170.
22. Sakura H, Ammala C, Smith PA et al. Cloning and functional expression of the cDNA encoding a novel ATP-sensitive potassium channel subunit expressed in pancreatic beta-cells, brain, heart and skeletal muscle. FEBS Lett 1995; 377(3):338-344.
23. Heinemann C, Chow RH, Neher E et al. Kinetics of the secretory response in bovine chromaffin cells following flash photolysis of caged Ca2+. Biophys J 1994; 67(6):2546-2557.
24. Fujimoto K, Shibasaki T, Yokoi N et al. Piccolo, a Ca2+ sensor in pancreatic beta-cells. Involvement of cAMP-GEFII.Rim2.Piccolo complex in cAMP-dependent exocytosis. J Biol Chem 2002; 277(52):50497-50502.
25. Seino S, Shibasaki T. PKA-dependent and PKA-independent pathways for cAMP-regulated exocytosis. Physiol Rev 2005; 85(4):1303-1342.
26. Wiedenkeller DE, Sharp GW. Effects of forskolin on insulin release and cyclic AMP content in rat pancreatic islets. Endocrinology 1983; 113(6):2311-2313.
27. Ostenson CG, Gaisano H, Sheu L et al. Impaired gene and protein expression of exocytotic soluble N-ethylmaleimide attachment protein receptor complex proteins in pancreatic islets of type 2 diabetic patients. Diabetes 2006; 55(2):435-440.
28. Abdel-Halim SM, Guenifi A, Khan A et al. Impaired coupling of glucose signal to the exocytotic machinery in diabetic GK rats: A defect ameliorated by cAMP. Diabetes 1996; 45(7):934-940.
29. Ostenson CG, Khan A, Efendic S. Impaired glucose-induced insulin secretion: Studies in animal models with spontaneous NIDDM. Adv Exp Med Biol 1993; 334:1-11.
30. Ostenson CG, Khan A, Abdel-Halim SM et al. Abnormal insulin secretion and glucose metabolism in pancreatic islets from the spontaneously diabetic GK rat. Diabetologia 1993; 36(1):3-8.
31. Ostenson CG. The pathophysiology of type 2 diabetes mellitus: An overview. Acta Physiol Scand 2001; 171(3):241-247.
32. Ohara-Imaizumi M, Nishiwaki C, Kikuta T et al. TIRF imaging of docking and fusion of single insulin granule motion in primary rat pancreatic beta-cells: Different behaviour of granule motion between normal and Goto-Kakizaki diabetic rat beta-cells. Biochem J 2004; 381(Pt 1):13-18.
33. Nagamatsu S, Nakamichi Y, Yamamura C et al. Decreased expression of t-SNARE, syntaxin 1, and SNAP-25 in pancreatic beta-cells is involved in impaired insulin secretion from diabetic GK rat islets: Restoration of decreased t-SNARE proteins improves impaired insulin secretion. Diabetes 1999; 48(12):2367-2373.
34. Zhang W, Khan A, Ostenson CG et al. Downregulated expression of exocytotic proteins in pancreatic islets of diabetic GK rats. Biochem Biophys Res Commun 2002; 291(4):1038-1044.
35. Eizirik DL, Korbutt GS, Hellerstrom C. Prolonged exposure of human pancreatic islets to high glucose concentrations in vitro impairs the beta-cell function. J Clin Invest 1992; 90(4):1263-1268.
36. Abderrahmani A, Cheviet S, Ferdaoussi M et al. ICER induced by hyperglycemia represses the expression of genes essential for insulin exocytosis. EMBO J 2006; 25(5):977-986.
37. Boden G. Role of fatty acids in the pathogenesis of insulin resistance and NIDDM. Diabetes 1997; 46(1):3-10.
38. Grill V, Qvigstad E. Fatty acids and insulin secretion. Br J Nutr 2000; 83(Suppl 1):S79-S84.
39. Unger RH. Lipotoxicity in the pathogenesis of obesity-dependent NIDDM. Genetic and clinical implications. Diabetes 1995; 44(8):863-870.
40. Zhou YP, Marlen K, Palma JF et al. Overexpression of repressive cAMP response element modulators in high glucose and fatty acid-treated rat islets. A common mechanism for glucose toxicity and lipotoxicity? J Biol Chem 2003; 278(51):51316-51323.

41. Busch AK, Cordery D, Denyer GS et al. Expression profiling of palmitate- and oleate-regulated genes provides novel insights into the effects of chronic lipid exposure on pancreatic beta-cell function. Diabetes 2002; 51(4):977-987.
42. Zraika S, Dunlop ME, Proietto J et al. Elevated SNAP-25 is associated with fatty acid-induced impairment of mouse islet function. Biochem Biophys Res Commun 2004; 317(2):472-477.
43. Sadoul K, Lang J, Montecucco C et al. SNAP-25 is expressed in islets of Langerhans and is involved in insulin release. J Cell Biol 1995; 128(6):1019-1028.
44. Zhang X, Odom DT, Koo SH et al. Genome-wide analysis of cAMP-response element binding protein occupancy, phosphorylation, and target gene activation in human tissues. Proc Natl Acad Sci USA 2005; 102(12):4459-4464.
45. Briaud I, Lingohr MK, Dickson LM et al. Differential activation mechanisms of Erk-1/2 and p70(S6K) by glucose in pancreatic beta-cells. Diabetes 2003; 52(4):974-983.
46. Pohl SL, Birnbaumer L, Rodbell M. Glucagon-sensitive adenyl cylase in plasma membrane of hepatic parenchymal cells. Science 1969; 164(879):566-567.
47. Rodbell M, Birnbaumer L, Pohl SL et al. Properties of the adenyl cyclase systems in liver and adipose cells: The mode of action of hormones. Acta Diabetol Lat 1970; 7(Suppl 1):9-63.
48. Evans GJ, Morgan A. Regulation of the exocytotic machinery by cAMP-dependent protein kinase: Implications for presynaptic plasticity. Biochem Soc Trans 2003; 31(Pt 4):824-827.
49. Holz GG. Epac: A new cAMP-binding protein in support of glucagon-like peptide-1 receptor-mediated signal transduction in the pancreatic beta-cell. Diabetes 2004; 53(1):5-13.
50. Kashima Y, Miki T, Shibasaki T et al. Critical role of cAMP-GEFII—Rim2 complex in incretin-potentiated insulin secretion. J Biol Chem 2001; 276(49):46046-46053.
51. Renstrom E, Eliasson L, Rorsman P. Protein kinase A-dependent and -independent stimulation of exocytosis by cAMP in mouse pancreatic B-cells. J Physiol 1997; 502(Pt 1):105-118.
52. Allagnat F, Martin D, Condorelli DF et al. Glucose represses connexin36 in insulin-secreting cells. J Cell Sci 2005; 118(Pt 22):5335-5344.
53. Molina CA, Foulkes NS, Lalli E et al. Inducibility and negative autoregulation of CREM: An alternative promoter directs the expression of ICER, an early response repressor. Cell 1993; 75(5):875-886.
54. Stehle JH, Foulkes NS, Molina CA et al. Adrenergic signals direct rhythmic expression of transcriptional repressor CREM in the pineal gland. Nature 1993; 365(6444):314-320.
55. Foulkes NS, Borrelli E, Sassone-Corsi P. CREM gene: Use of alternative DNA-binding domains generates multiple antagonists of cAMP-induced transcription. Cell 1991; 64(4):739-749.
56. Hai TW, Liu F, Coukos WJ et al. Transcription factor ATF cDNA clones: An extensive family of leucine zipper proteins able to selectively form DNA-binding heterodimers. Genes Dev 1989; 3(12B):2083-2090.
57. Gonzalez GA, Montminy MR. Cyclic AMP stimulates somatostatin gene transcription by phosphorylation of CREB at serine 133. Cell 1989; 59(4):675-680.
58. Yamamoto KK, Gonzalez GA, Biggs III WH et al. Phosphorylation-induced binding and transcriptional efficacy of nuclear factor CREB. Nature 1988; 334(6182):494-498.
59. Inada A, Hamamoto Y, Tsuura Y et al. Overexpression of inducible cyclic AMP early repressor inhibits transactivation of genes and cell proliferation in pancreatic beta cells. Mol Cell Biol 2004; 24(7):2831-2841.
60. Inada A, Yamada Y, Someya Y et al. Transcriptional repressors are increased in pancreatic islets of type 2 diabetic rats. Biochem Biophys Res Commun 1998; 253(3):712-718.
61. Miyazaki K, Reisine T, Kebabian JW. Adenosine 3',5'-monophosphate (cAMP)-dependent protein kinase activity in rodent pituitary tissue: Possible role in cAMP-dependent hormone secretion. Endocrinology 1984; 115(5):1933-1945.
62. Reisine T, Rougon G, Barbet J et al. Corticotropin-releasing factor-induced adrenocorticotropin hormone release and synthesis is blocked by incorporation of the inhibitor of cyclic AMP-dependent protein kinase into anterior pituitary tumor cells by liposomes. Proc Natl Acad Sci USA 1985; 82(23):8261-8265.
63. Rougon G, Barbet J, Reisine T. Protein phosphorylation induced by phorbol esters and cyclic AMP in anterior pituitary cells: Possible role in adrenocorticotropin release and synthesis. J Neurochem 1989; 52(4):1270-1278.
64. Berkenbosch F, Tilders FJ, Vermes I. Beta-adrenoceptor activation mediates stress-induced secretion of beta-endorphin-related peptides from intermediate but not anterior pituitary. Nature 1983; 305(5931):237-239.
65. Kjaer A, Knigge U, Bach FW et al. Stress-induced secretion of pro-opiomelanocortin-derived peptides in rats: Relative importance of the anterior and intermediate pituitary lobes. Neuroendocrinology 1995; 61(2):167-172.

66. Kvetnansky R, Tilders FJ, van Zoest ID et al. Sympathoadrenal activity facilitates beta-endorphin and alpha-MSH secretion but does not potentiate ACTH secretion during immobilization stress. Neuroendocrinology 1987; 45(4):318-324.
67. Munck A, Guyre PM, Holbrook NJ. Physiological functions of glucocorticoids in stress and their relation to pharmacological actions. Endocr Rev 1984; 5(1):25-44.
68. McEwen BS. Plasticity of the hippocampus: Adaptation to chronic stress and allostatic load. Ann NY Acad Sci 2001; 933:265-277.
69. Mazzucchelli C, Sassone-Corsi P. The inducible cyclic adenosine monophosphate early repressor (ICER) in the pituitary intermediate lobe: Role in the stress response. Mol Cell Endocrinol 1999; 155(1-2):101-113.
70. Lamas M, Molina C, Foulkes NS et al. Ectopic ICER expression in pituitary corticotroph AtT20 cells: Effects on morphology, cell cycle, and hormonal production. Mol Endocrinol 1997; 11(10):1425-1434.
71. Ngsee JK, Fleming AM, Scheller RH. A rab protein regulates the localization of secretory granules in AtT-20 cells. Mol Biol Cell 1993; 4(7):747-756.
72. Zhao S, Torii S, Yokota-Hashimoto H et al. Involvement of Rab27b in the regulated secretion of pituitary hormones. Endocrinology 2002; 143(5):1817-1824.
73. Aguado F, Majo G, Ruiz-Montasell B et al. Expression of synaptosomal-associated protein SNAP-25 in endocrine anterior pituitary cells. Eur J Cell Biol 1996; 69(4):351-359.
74. Redecker P, Cetin Y, Grube D. Differential distribution of synaptotagmin I and rab3 in the anterior pituitary of four mammalian species. Neuroendocrinology 1995; 62(2):101-110.
75. Oho C, Seino S, Takahashi M. Expression and complex formation of soluble N-ethyl-maleimide-sensitive factor attachment protein (SNAP) receptors in clonal rat endocrine cells. Neurosci Lett 1995; 186(2-3):208-210.
76. Freeman ME, Kanyicska B, Lerant A et al. Structure, function, and regulation of secretion. Physiol Rev 2000; 80(4):1523-1631.
77. Djordjijevic D, Zhang J, Priam M et al. Effect of 17beta-estradiol on somatostatin receptor expression and inhibitory effects on growth hormone and prolactin release in rat pituitary cell cultures. Endocrinology 1998; 139(5):2272-2277.
78. Clifton KH, Meyer RK. Mechanism of anterior pituitary tumor induction by estrogen. Anat Rec 1956; 125(1):65-81.
79. Majo G, Lorenzo MJ, Blasi J et al. Exocytotic protein components in rat pituitary gland after long-term estrogen administration. J Endocrinol 1999; 161(2):323-331.
80. Carlstrom L, Ke ZJ, Unnerstall JR et al. Estrogen modulation of the cyclic AMP response element-binding protein pathway. Effects of long-term and acute treatments. Neuroendocrinology 2001; 74(4):227-243.
81. Zhou J, Zhang H, Cohen RS et al. Effects of estrogen treatment on expression of brain-derived neurotrophic factor and cAMP response element-binding protein expression and phosphorylation in rat amygdaloid and hippocampal structures. Neuroendocrinology 2005; 81(5):294-310.
82. Della Fazia MA, Servillo G, Foulkes NS et al. Stress-induced expression of transcriptional repressor ICER in the adrenal gland. FEBS Lett 1998; 434(1-2):33-36.
83. Axelrod J, Reisine TD. Stress hormones: Their interaction and regulation. Science 1984; 224(4648):452-459.
84. Pfeffer M, Maronde E, Molina CA et al. Inducible cyclic AMP early repressor protein in rat pinealocytes: A highly sensitive natural reporter for regulated gene transcription. Mol Pharmacol 1999; 56(2):279-289.
85. Mioduszewska B, Jaworski J, Kaczmarek L. Inducible cAMP early repressor (ICER) in the nervous system—A transcriptional regulator of neuronal plasticity and programmed cell death. J Neurochem 2003; 87(6):1313-1320.
86. Lewy H, Naor Z, Ashkenazi IE. Rhythmicity of luteinizing hormone secretion expressed in vitro. Eur J Endocrinol 1996; 135(4):455-463.
87. Lewy H, Ashkenazi IE, Naor Z. Gonadotropin releasing hormone (GnRH) and estradiol (E(2)) regulation of cell cycle in gonadotrophs. Mol Cell Endocrinol 2003; 203(1-2):25-32.
88. Castillo PE, Janz R, Sudhof TC et al. Rab3A is essential for mossy fibre long-term potentiation in the hippocampus. Nature 1997; 388(6642):590-593.
89. Greber-Platzer S, Balcz B, Fleischmann C et al. Using the quantitative competitive RT-PCR technique to analyze minute amounts of different mRNAs in small tissue samples. Methods Mol Biol 2002; 193:29-38.
90. Baldwin Jr AS. Series introduction: The transcription factor NF-kappaB and human disease. J Clin Invest 2001; 107(1):3-6.
91. Barnes PJ. Nuclear factor-kappa B. Int J Biochem Cell Biol 1997; 29(6):867-870.

92. Lu M, Seufert J, Habener JF. Pancreatic beta-cell-specific repression of insulin gene transcription by CCAAT/enhancer-binding protein beta. Inhibitory interactions with basic helix-loop-helix transcription factor E47. J Biol Chem 1997; 272(45):28349-28359.
93. Mercurio F, Manning AM. NF-kappaB as a primary regulator of the stress response. Oncogene 1999; 18(45):6163-6171.
94. Mohamed AK, Bierhaus A, Schiekofer S et al. The role of oxidative stress and NF-kappaB activation in late diabetic complications. Biofactors 1999; 10(2-3):157-167.
95. Seufert J, Weir GC, Habener JF. Differential expression of the insulin gene transcriptional repressor CCAAT/enhancer-binding protein beta and transactivator islet duodenum homeobox-1 in rat pancreatic beta cells during the development of diabetes mellitus. J Clin Invest 1998; 101(11):2528-2539.
96. Bernal-Mizrachi E, Wen W, Shornick M et al. Activation of nuclear factor-kappaB by depolarization and Ca(2+) influx in MIN6 insulinoma cells. Diabetes 2002; 51(Suppl 3):S484-S488.
97. Karin M, Lin A. NF-kappaB at the crossroads of life and death. Nat Immunol 2002; 3(3):221-227.
98. Meffert MK, Chang JM, Wiltgen BJ et al. NF-kappa B functions in synaptic signaling and behavior. Nat Neurosci 2003; 6(10):1072-1078.
99. Pahl HL. Activators and target genes of Rel/NF-kappaB transcription factors. Oncogene 1999; 18(49):6853-6866.
100. Li X, Stark GR. NFkappaB-dependent signaling pathways. Exp Hematol 2002; 30(4):285-296.
101. Norlin S, Ahlgren U, Edlund H. Nuclear factor-{kappa}B activity in {beta}-cells is required for glucose-stimulated insulin secretion. Diabetes 2005; 54(1):125-132.
102. Gauthier BR, Wollheim CB. MicroRNAs: «Ribo-regulators» of glucose homeostasis. Nature Medicine 2006; 12:36-38.
103. Baines AJ, Bennett V. Synapsin I is a microtubule-bundling protein. Nature 1986; 319(6049):145-147.
104. Greengard P, Valtorta F, Czernik AJ et al. Synaptic vesicle phosphoproteins and regulation of synaptic function. Science 1993; 259(5096):780-785.
105. Lievens JC, Woodman B, Mahal A et al. Abnormal phosphorylation of synapsin I predicts a neuronal transmission impairment in the R6/2 Huntington's disease transgenic mice. Mol Cell Neurosci 2002; 20(4):638-648.

Index

A

Acrosome 47, 49, 50, 56, 117-128, 131-134, 137, 138, 140-147
Acrosome reaction 47, 49, 56, 117, 119-125, 127, 128, 131-134, 137, 138, 140, 142
Actin 3, 4, 33, 35, 37, 62, 64, 66, 68, 70, 71, 104, 107, 108
Adipocyte 3, 12, 17, 19, 31, 91, 93, 104, 148, 149, 153, 154
Akt 50, 55, 152, 154
AMPA
    (a-amino-3-hydroxy-5-methyl-4-isoxazole propionic acid) 148, 150, 151
AMPA receptor
    (a-amino-3-hydroxy-5-methyl-4-isoxazole propionic acid receptor) 150, 151
Aquaporin 153, 154
ARF6 (ADP ribosylation factor 6) 76, 79-81, 105, 155, 156
Axonal growth 151, 152

B

Brain dependent neutrophic factor (BDNF) 72, 152
Botulinum neurotoxin (BoNT) 92, 133-138

C

CAAT enhancer binding protein (C/EBPb) 168, 169
Ca^{2+} channel 10, 11, 15-17, 20, 34, 62, 91, 94, 163
Calcium 5, 11, 16, 32, 76, 77, 79, 80, 91, 104-108, 110, 117, 119, 123-130, 132-140, 142, 164
Calcium sensor 16, 91, 105, 106, 126, 128, 140
Calmodulin 3, 6, 15-19, 21, 35-37, 63, 79, 106, 127, 128, 150-152
Ca^{2+}-calmodulin-dependent protein kinase II (CamKII) 65, 150-152
cAMP 34, 140, 142, 152-154, 161, 163, 165-168
cAMP signalling pathway 165, 166
Caveolae 84, 87-89, 93, 95

Chromaffin cells 2, 11, 12, 17, 19-21, 32, 33, 35, 45, 76, 77, 79, 80, 92, 94, 106, 128, 132
Complexin 5, 91, 93, 105, 110, 162
CREB (cAMP-responsive element binding protein) 166-168
CREM (cAMP-responsive element modulator) 166, 167
Cytokinesis 148, 149, 155-157
Cytoskeleton 15, 33, 37, 68, 70, 71, 107, 150, 155, 156, 169

D

Dendritic cell 148, 154
Diabetes 96, 81, 142, 148, 149, 164, 166, 168
Diacylglycerol 10, 12, 13, 15, 21, 77, 87, 127, 163
Doc2 18, 19, 21, 43, 44, 56
Domain 5, 6, 13-16, 18-20, 29-37, 42-45, 47-49, 53-56, 62, 63, 66, 69, 71, 84, 87-95, 105-109, 127-130, 132, 135, 138, 139, 166

E

Endocytosis 3, 43, 45, 62, 65, 66, 68, 75, 92-94, 101, 120, 130, 149, 153, 155, 157
Endosomes 29, 30, 75, 84, 95, 148, 150, 155, 156
Enlargeosomes 148, 152-154
Exocytosis 1-3, 5, 6, 10-21, 28, 30-38, 42, 43, 45, 47-50, 53, 55, 56, 62, 65, 67, 68, 70, 71, 75-77, 79-81, 84, 87, 89, 91-95, 100-110, 117, 119-121, 123-136, 140, 142, 148-157, 161-165, 168, 169

F

Fatty acid 13, 14, 20, 89
Fertilization 117-120, 122
Fusion pore 4-6, 45, 78, 80, 84-87, 93, 94, 107, 128
Fusion protein 1, 84, 87, 95, 103, 106, 108, 123, 131, 162